Smart Innovation, Systems and Technologies

Volume 136

The Smart Innovation, Systems and Technologies book series encompasses the topics of knowledge, intelligence, innovation and sustainability. The aim of the series is to make available a platform for the publication of books on all aspects of single and multi-disciplinary research on these themes in order to make the latest results available in a readily-accessible form. Volumes on interdisciplinary research combining two or more of these areas is particularly sought.

The series covers systems and paradigms that employ knowledge and intelligence in a broad sense. Its scope is systems having embedded knowledge and intelligence, which may be applied to the solution of world problems in industry, the environment and the community. It also focusses on the knowledge-transfer methodologies and innovation strategies employed to make this happen effectively. The combination of intelligent systems tools and a broad range of applications introduces a need for a synergy of disciplines from science, technology, business and the humanities. The series will include conference proceedings, edited collections, monographs, handbooks, reference books, and other relevant types of book in areas of science and technology where smart systems and technologies can offer innovative solutions.

High quality content is an essential feature for all book proposals accepted for the series. It is expected that editors of all accepted volumes will ensure that contributions are subjected to an appropriate level of reviewing process and adhere to KES quality principles.

More information about this series at http://www.springer.com/series/8767

Valentina Emilia Balas ·
Sanjiban Sekhar Roy · Dharmendra Sharma ·
Pijush Samui
Editors

Handbook of Deep Learning Applications

Editors
Valentina Emilia Balas
Aurel Vlaicu University of Arad
Arad, Romania

Dharmendra Sharma
University of Canberra
Bruce, ACT, Australia

Sanjiban Sekhar Roy
School of Computer Science
and Engineering
Vellore Institute of Technology
Vellore, Tamil Nadu, India

Pijush Samui
Department of Civil Engineering
National Institute of Technology Patna
Patna, Bihar, India

ISSN 2190-3018 ISSN 2190-3026 (electronic)
Smart Innovation, Systems and Technologies
ISBN 978-3-030-11478-7 ISBN 978-3-030-11479-4 (eBook)
https://doi.org/10.1007/978-3-030-11479-4

Library of Congress Control Number: 2018967433

This Springer imprint is published by the registered company Springer Nature Switzerland AG
The registered company address is: Gewerbestrasse 11, 6330 Cham, Switzerland

Contents

Designing a Neural Network from Scratch for Big Data Powered by Multi-node GPUs

Alcides Fonseca and Bruno Cabral

1 Introduction

Lately, Machine Learning has taken a crucial role in the society in different vertical sectors. For complex problems with high-dimensionality, Deep Learning has become an efficient solution for learning in the context of supervisioned learning. Deep Learning [1] consists in using Artificial Neural Networks (ANN or NN) with several hidden layers, typically also with a large number of nodes in each layer.

ANNs have initially been proposed in 1943 [2], but only recently have been gaining popularity due to decreasing storage costs and the increase of computational power, both in CPU and GPUs. Nowadays, ANNs are used for several tasks, such as image classification [3], character recognition in scanned documents [4], predicting bankruptcy [5] or health complications [6]. More recently, ANNs have been the basis for the software used in self-driving vehicles [7].

In complex problems, both in terms of diversity in instances and in number of features/classes, networks also have a more complex structure and more expensive training process. It is recommended to use a number of instances three orders of magnitude higher than the number of features [8]. Training an ANN consists on applying the NN to several batches of multiple instances as many times as necessary until a good-enough weight distribution is obtained. Thus, training a complex ANN is a computationally intensive operation in terms of processing, memory and disk usage. As the amount of data available for training goes above a terabyte, it becomes Big Data problem [9].

A. Fonseca (✉)
LASIGE, Faculdade de Ciências da Universidade de Lisboa, Lisbon, Portugal
e-mail: amfonseca@ciencias.ulisboa.pt

B. Cabral
CISUC, Departamento de Engenharia Informática, Faculdade de Ciências e Tecnologia da Universidade de Coimbra, Coimbra, Portugal
e-mail: bcabral@dei.uc.pt

V. E. Balas et al. (eds.), *Handbook of Deep Learning Applications*,
Smart Innovation, Systems and Technologies 136,
https://doi.org/10.1007/978-3-030-11479-4_1

The two most common and efficient approaches for performing this computation in useful time is to distribute work across different machines, and to use GPUs to perform ANN training. GPUs are used in ANN training because they are more efficient than CPUs for matricial operations (what they were designed for, in the field of Graphical Computation), and ANN application and training can be described in those operations.

Distributing the computation of the training is also used to train with more instances that fit a single machine (usually in terms of memory, as CPUs and GPUs have limited volatile memories). This approach is possible because training can be subdivided in embarrassingly parallel sub-problems that can be combined at a latter stage.

This chapter will cover the design and implementation of a distributed CPU and GPU-backed Deep Artificial Neural Network for classification problems. This chapter aims to help researchers and practitioners, who are looking to implement alternative ANN configurations or models, in creating efficient GPU-enabled code.

This approach is an alternative to existing tutorials that focus on the usage of readily-available ANN toolkits, such as as Tensorflow [10], Theano [11] or Torch [12], among many others [13]. While these tools make the task of creating ANNs simple without much effort, the space of possible resulting ANN is limited by the framework. In this chapter, the goal is to equip practitioners with the tools to develop new and different approaches to ANNs, so they can explore different research lines in ANN architecture or training systems.

This chapter will briefly introduce ANNs and the back-propagation training process (Sect. 2 briefly introduces ANNs and the back-propagation training process) along with its mathematical background for ANNs (Sect. 3). Then, we will present the problem that will serve as an example for developing our custom Artificial Neural Network. Then, we will cover three different phases of the implementation of the ANN: a single-CPU version (Sect. 5); a parallel, distributed version (Sect. 6); and a GPU version (Sect. 7). Finally, we will discuss the goodness of the implemented design (Sect. 8) and conclude this chapter (Sect. 9) with final remarks and future work.

The runnable source code for this chapter is also made available online at https://github.com/alcides/bigdatagpunn.

2 A Primer on Neural Networks

Neural Networks are used in classification (and also regression) problems, more frequently in supervisioned problems. In these cases, there is an existing dataset with classification labels (*classes*) in each instance. From this data set, a learning algorithm should infer how to classify new *unlabelled* instance from its features. Features are a set of instance properties (when classifying the type of fruit, the color and length of the fruit are candidate features. Orange, apple and banana are some of the possible classes.

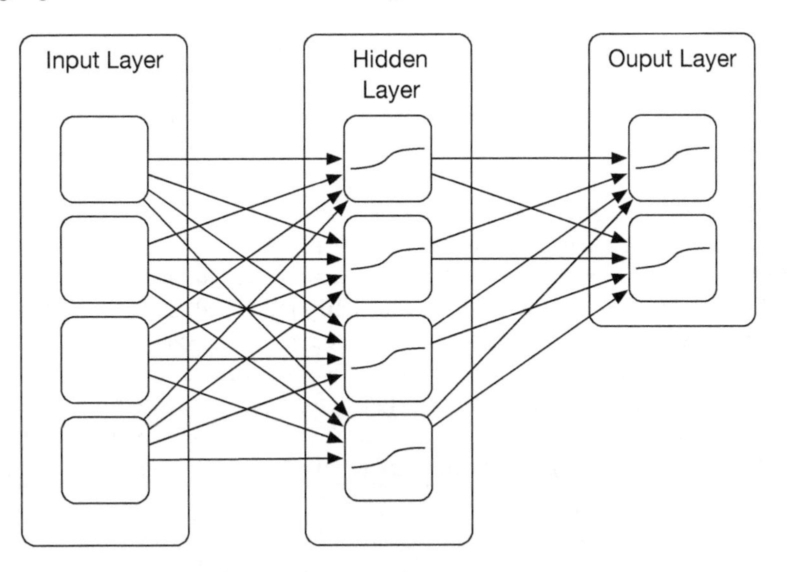

Fig. 1 An ANN with four input neurons, one single hidden layer with four neurons and two output neurons

When using ANNs as the learning algorithms, features are the input of the ANNs, and there can be one output for each class. When feeding the features of an unlabelled new fruit through the ANNs, the highest scoring class output node should be the correct one (Assuming a correct prediction, which unfortunately is not always the case).

ANNs have a directional architecture, in which the processing flows from the input layer to the output layer, passing through several hidden layers. The input layer has one neuron for each considered feature. Hidden layers can have any number of neurons that receive as input the output of the previous layer. Finally, the output layer has as many neurons as desirable for the problem at hand, and receive input from the previous layer outputs. An example of a very simple neural-network is depicted in Fig. 1, consisting of four input neurons, four neurons in the hidden layer and two output neurons.

This is an example of a single-hidden-layer network, but Deep Neural Networks have a large number of hidden layers, representing the abstraction capability of the network. The layout of the network consists on the number of layers and the number of neurons on each layers.

When the number of features increases (high dimensionality), the number of neurons in the hidden layers increases as well, in order to compensate for the possible interactions of input neurons [14]. However, a rule of thumb is to use only one hidden layer [8] with the same number of hidden neurons as there are input neurons. Nonetheless, one might prefer to explore more recent approaches to this problem [15, 16], since nowadays computers are much more powerful than they were in 1995, when this rule was first published.

After defining the layout of a network, it is now necessary to define the weights of each connection. For each non-input neuron, its value is obtained from the averaged sum of all the output values from the previous layer. So each connection between two neurons (of adjacent layers) have a weight. The weights are where the intelligence lays within the neural network. Typically, weights are randomly initialized, which does not result in a good classification performance. Thus, it is necessary to tune the weights to improve the performance of the network.

Supervisioned training with back-propagation is used to obtain better weights. Labelled instances are fed through the randomly initialized network to obtain the predicted outcome. The outcome error compared to the expected label is measured and weights are slightly changed in the direction that reduces error. This process is repeated several times, using different instances to converge on a best weight configuration for the problem. Recently, other alternatives to random weight initialization have been proposed [17–19] that can be more efficient, depending on the context.

3 A Mathematical Formalization of Neural Networks

The goal of Artificial Neural Networks is to approximate some function $f*$ [20]. In our case, a classification problem, $y = f*(x)$ maps x to a class y. Connections between neurons are weighted so the neural network produces good results. As such, we will use f to approximate $f*$, evidencing weights θ with f defined as $y = f(x; \theta)$. This is the definition of a Feed-forward Neural Network, but there are other types of neural networks, such as Recurrent Neural Networks that use feedback connections.

Neural Networks can have several layers, each one represented as $f^{(1)}, f^{(2)}, \ldots, f^{(n)}$, where n is the number of layers, or depth of the network. Deep Neural Networks are those that have a high depth. In this paper we will consider just one hidden layer, but adding more layers is trivial. Layers can be chained, to form $f(x) = f^{(3)}(f^{(2)}(f^{(1)}(x)))$, where x is the input layer and $f^{(3)}$ is the last layer. Layers are vector-based, based on the biologic neuron concept, and $f^{(i)}(x)$ is the result of multiplying x by that layers weights $\theta^{(i)}$.

This approach works for linear models, but in order to support non-linear models, we can introduce $\phi(x)$, a non-linear transformation. Our model is now $y = \psi(x; \theta)$. There are several families of functions that can be used as ψ. In this work we will use one of the simplest, that is also very common: the sigmoid function. Most sophisticated approaches use a rectified linear unit (ReLU) [21], which prevents overfitting.

Because during the training process, one has access only to the input and expected outputs, there is no indication how to distribute weights over the different layers. First, we identify by how much did our $f(x)$ miss the real $f*(x)$. We will consider a simple difference, $J(\theta) = |f*(x) - f(x, \theta)|$, but other metrics, such as MSE loss function, are frequently used.

From the error, one can derive the gradient that needs to be applied to the network weights, so the next prediction is closer to the expected. Because training is done during several iterations (epochs), there is no need to obtain the ideal weights in just one step. This process can be controlled by a learning algorithms, such as Stochastic Gradient Descend.

Back-propagation is the process of correcting the different layers of an ANN from its error [22]. The error is multiplied by the derivative of the last layer $f^{(n)}$, obtaining the gradient $\nabla_x f^{(n)}(x^{(n)}, \theta^{(n)}) = J(\theta^{(n)}) \cdot \frac{\partial \psi(\theta^{(n)})}{\partial x^{(n)}}$, being ϵ the final error. This gradient is then added to the current weights to obtain the new, updated, weights. On the previous layers, back-propagation works via the Chain Rule, that is applied recursively. The Chain Rule considers not only the derivative of the current layer, but of all layers from the output up to the current layer. In this case, the gradient of the layer $n - i$ is define recursively as $\nabla_x f^{(n-i)}(x^{(n-i)}, \theta^{(n-i)}) = \nabla_x f^{(n-i+1)}(x^{(n-i+1)}, \theta^{(n-i+1)}) \cdot \frac{\partial \psi(\theta^{(n-i)})}{\partial x^{(n-i)}}$. In our example, we consider only the final layer and one single hidden layer.

4 Problem and Dataset

An ANN by itself has no purpose or usage, it is trained to perform a particular task with correctly-performed examples of such task. In this chapter, we will use a concrete motivational example to drive our implementation (the methodology is general to apply to other problems). Because of the lengthy process of training networks with large datasets, smaller subsets are used in development.

In this example, we will use the Wine Data Set [23], a dataset for classifying types of wine. This dataset is used to evaluate classifiers in the context of high dimensionality. The dataset contains 178 instances, enough for our purpose, with each instance containing 13 features (real and integer) and three output values (one for each class).

In order to evaluate the performance of our ANN, we will use 20 instances for training the network, and leave the remaining for evaluating the performance of the network, as if they were unseen wine instances. In more realistic scenarios, a larger portion of the dataset (e.g. 70 or 80%) would be used in training.

Given the problem characteristics, the ANN will have 13 input neurons (one for each feature) and three output neurons (one for each class). A single hidden layer will be used, also with 13 neurons, to allow some complexity to be abstracted.

Many other network layouts could have been applied for this problem, or any other. In fact, there are automatic algorithms for designing network layouts, of which HyperNEAT is a recent example [24].

5 A Neural Network in Python

This section presents a very simple implementation in Python of the back-propagation training our example Neural Network. This example relies on just two existing libraries: Numpy and Pandas [25]. These libraries are used to store matrices in more efficient data-structures (similar to C arrays) than Python lists. The code for importing the dataset and defining the global problem constants is in Listing 1.1.

Listing 1.1 Importing the Dataset and defining global constants

```
1 import numpy as np
2 import pandas as pd
3
4 df = pd.read_csv("datasets/wine.txt", sep="\t", header=None)
5
6 instances = df.shape[0]
7 train_instances = 20
8 ndims = 13
9 nclasses = 3
```

The ANN layout for our problem is static (13 input neurons, 13 neurons in a single hidden layer, and 3 output neurons). Neuron-neuron connection weights are the dynamic part of the network that will be trained. Listing 1.2 defines the function that generates the random weights for connections between the input and hidden layers (`weights0`) and between the hidden and output layers (`weights1`). Weights are uniformly distributed in the $[-1, 1[$ interval, recommend for use with sigmoid activation functions.

Listing 1.2 Generation of a random configuration

```
1 def generate_random_config():
2     weights0 = 2 * np.random.random((ndims, ndims)) - 1
3     weights1 = 2 * np.random.random((ndims, nclasses)) - 1
4     return (weights0, weights1)
```

Listing 1.3 defines the training function, based on "A Neural Network in 11 lines" [26]. This function receives a matrix of features for each instance `X` (13 by 20 in our example), an array of known classes for each instance `y` (one of 3 classes for each 20 instances), the initial configuration `conf` (obtained from the function in Listing 1.2) and the number of `iterations`.

Listing 1.3 Training the ANN

```
1  def train(X, y, conf, iterations=6000):
2      weights0, weights1 = conf
3      for j in xrange(iterations):
4          # Feed forward
5          l0 = X
6          l1 = sigmoid(np.dot(l0,weights0))
7          l2 = sigmoid(np.dot(l1,weights1))
8          # Back Propagation
9          l2_error = y - l2
10         l2_delta = l2_error*sigmoid_d(l2)
11         l1_error = l2_delta.dot(weights1.T)
12         l1_delta = l1_error * sigmoid_d(l1)
13         weights1 += l1.T.dot(l2_delta)
14         weights0 += l0.T.dot(l1_delta)
15     return (weights0, weights1)
```

The training process is repeated over `iterations` epochs, improving the performance of the network at each step. In each epoch, the training process is a three-part process. Firstly, the network is applied with the current weights (Feed-forward), then the measured error is used to understand how the weights have to change in each layer (Back Propagation), and, finally, that change is applied to the current weights to obtain new weights for the matrix.

Feed-forward is done through matrix multiplication of the input data `X` and the weights for the hidden layer `l1` (`weights0`). Each neuron uses the sigmoid function as its activation function (other alternative functions exist, but the sigmoid has a derivative easy to computed (Listing 1.4). In Line 7, the same process occurs in the output layer `l2`, using different weights (`weights1`). Being the output layer, each neuron contains a value corresponding to whether that instance belongs to that class or not.

Listing 1.4 The Sigmoid function and its derivative

```
1  def sigmoid(x):
2      return 1/(1+np.exp(-x))
3
4  def sigmoid_d(x):
5      return x*(1-x)
```

Back-propagation occurs after computing the error in the classification (Line 9), and consists in computing the deltas for each weight matrix. The deltas move the weights toward a better classification, hence the use of derivatives of the values from the output layer iteratively to the input layer. Deltas are applied to the weight matrices and the process is repeated the given number of iterations.

The result of this training process is the trained weights that can now be applied in the Feed-forward process to perform new classifications of unseen instances.

An example of driving the training process in the Wine Dataset is shown in Listing 1.5, defining the input matrix `X` and the expected result `y`.

Listing 1.5 An example of a call to the training method

```
1 conf = generate_random_config()
2 X = df.iloc[0:train_instances,0:ndims].as_matrix()
3 y = df.iloc[0:train_instances,ndims:].as_matrix()
4 output_conf = train_fun(X, y, conf_, iterations)
```

6 A Distributed Neural Network Using a Message Queue for Communication

The function presented in the previous section for training an ANN using a given dataset is concise and understandable. Despite relying on numpy arrays for performance, it is limited by the amount of RAM available in the machine. To train datasets larger than the maximum amount of RAM, it is necessary to either use batches, subsets of the dataset that are iteratively used in the training process, or to distribute training over different machines. Distributed training performs the same training process as batches, but does it in parallel, which dramatically reduces the training time.

The most common programming paradigm for Big Data problems is Map-Reduce, popularized by frameworks such as Hadoop and Spark [27]. Map-Reduce is inspired by the homonymous higher order functions that convert elements using a given function and combine values together two at a time respectively. Using this method, a problem is solved in four phases. Firstly, the problem is subdivided in several smaller problems, each one solvable by a single machine at most. Secondly, each machine solves one or more of those smaller problems. Next, these results are aggregated in central machines, which aggregate the partial results. Finally, these partial results are combined together to produce the final result.

Parallelization of ANNs occurs over subsets of the training dataset [28]. Subsets are distributed over different machines, each subset used as input in the training process that occurs on each node. From each of these parallel training processes, different weight matrices are generated. These matrices can be merged on a central master node by averaging the values at the same position in all matrices.

Figure 2 shows a master node that controls the training process, and two worker nodes that are responsible for performing the training process. The master node is responsible for sending the requests with a dataset subset to each worker, waits for the response and merges the results with the current master weight matrices. The most computationally expensive part of the process is the training that occurs in parallel in the worker nodes.

To support asynchronous sending of requests and responses across the network, we propose the usage of a Message Queue, that handles network communication. Our implementation will rely on Redis [29], an in-memory database that also serves as a message queue. Redis has a very straightforward API, but more powerful and complex alternatives are discussed in Sect. 8.

The master node may send different requests to the same node, allowing for training with more data than fits the sum of all nodes memory, or perform load-

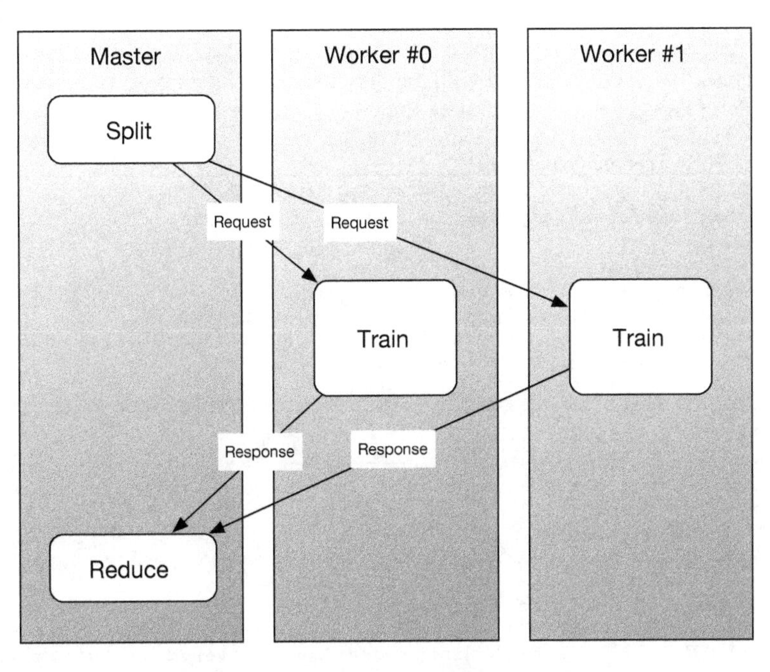

Fig. 2 Example of a Master-Worker model with two workers

balancing across a distributed heterogeneous network. The source code for the master node is shown in Listing 1.6. The master node subdivides the range of the input dataset in as many slices as there are works. Each slice bounds is send to the queue of each worker ("worker_0" for the first worker), along with the configuration matrices and metadata to auxiliary queues.

Listing 1.6 Master Splitting

```
master_conf = generate_random_config()
blocks_per_worker = instances/(workers+2)
for k in range(10):
    for i in range(workers):
        a = blocks_per_worker * i
        b = blocks_per_worker * (i+1)
        print "Scheduling to worker", i, " data from ", a, " to ",
              b
        metadata, data, data2 = encode_req(a, b, 60000,
              master_conf)
        r.rpush("worker_%d" % i, metadata)
        r.rpush("worker_data_%d" % i, data)
        r.rpush("worker_data2_%d" % i, data2)
```

Because requests and responses contain the same kind of information (matrices), the same format is used for both, namely the number of iterations, the bounds of the data subset to use in training, and the configuration matrices.

The matrices sent in requests are the current most up-to-date configurations at the master node, and the ones received in response are the result of applying the deltas locally on the worker node. Matrices are encoded as strings, through an intermediate flat representation concatenated with the dimensions and data type as metadata. An example of encoding and decoding is shown in Listing 1.7.

Listing 1.7 Redis encoding functions

```
1  import redis
2  r = redis.StrictRedis(host='localhost', port=6379, db=0)
3
4  def encode_req(a,b,it,conf):
5      weights0, weights1 = conf
6      metadata = "|".join(map(str,[a,b,it, weights0.shape[0],
7          weights0.shape[1], weights0.dtype, weights1.shape[0],
8          weights1.shape[1], weights1.dtype ]))
9      data = conf[0].ravel().tostring()
10     data2 = conf[1].ravel().tostring()
11     return metadata, data, data2
12
13 def decode_req(metadata, data, data2):
14     a, b, iterations, l, w, array_dtype, l2, w2, array_dtype2 =
15         metadata.split('|')
16     weights0 = np.fromstring(data, dtype=array_dtype).reshape(int
17         (l), int(w))
18     weights1 = np.fromstring(data2, dtype=array_dtype2).reshape(
19         int(l2), int(w2))
20     return int(a), int(b), int(iterations), (weights0, weights1)
```

Listing 1.8 shows the source code for each worker. Workers can optionally loop over this code to continuously process new requests, in the case there are more slices than workers. Each worker decodes the request information and performs the training a given number of iterations. The resulting configuration matrices are send back to the master via symmetrical queues that handle worker-to-master communication (e.g., "master_0").

Listing 1.8 Worker code

```
1  metadata = r.blpop('worker_%d' % wid)[1]
2  data = r.blpop('worker_data_%d' % wid)[1]
3  data2 = r.blpop('worker_data2_%d' % wid)[1]
4  a, b, iterations, conf = decode_req(metadata, data, data2)
5
6  X = df.iloc[a:b,0:ndims].as_matrix()
7  y = df.iloc[a:b,ndims:].as_matrix()
8  output_conf = train(X, y, conf, iterations)
9
10 metadata, data, data2 = encode_req(a, b, iterations, output_conf)
11 r.rpush("master_%d" % wid, metadata)
12 r.rpush("master_data_%d" % wid, data)
13 r.rpush("master_data2_%d" % wid, data2)
```

The master node handles all responses in a similar fashion (Listing 1.9) by averaging the received matrices, and does not need to known which worker the response originated from.

Listing 1.9 Master code to received data from workers

```
1   (...)
2   new_conf = copy.deepcopy(master_conf)
3   for i in range(workers):
4     metadata = r.blpop('master_%d' % i)[1]
5     data = r.blpop('master_data_%d' % i)[1]
6     data2 = r.blpop('master_data2_%d' % i)[1]
7     a, b, iterations, conf = decode_req(metadata, data, data2)
8     diff = op_configs(master_conf, conf, lambda a,b: a-b)
9     new_conf = op_configs(new_conf, diff, lambda a,b: a+b)
10
11    print "Data from worker", i, "had error:", config_error(df,
12       conf)
13    print "Data from worker", i, " merged had error:",
14       config_error(df, new_conf)
15
16  master_conf = copy.deepcopy(new_conf)
```

While this map-reduce approach has been presented in the context of distributed machines, it can be used to distribute work across different cores and processors within the same machine. On a deca-core machine, 10 workers can be executed simultaneously, and training will occur in parallel. The limitation of multicore parallelism is that system memory is shared across all processes.

7 A GPU-Powered Neural Network

Initially, GPUs were introduced to accelerate the generation of 2 and 3D graphics for design, video and game applications. This hardware was design to perform matrix operations as fast as possible. However, this capabilities have been exposed to non-graphics applications through C-like APIs such as Cuda or OpenCL. The idea of using GPUs for non-graphics related computations is called General Purpose GPU Computing (GPGPU). Recent efforts have allowed high-level programming languages to be compiled to the GPU, such as Matlab [30], Haskell [31], Java [32] or Python [33].

Since ANN training consists mostly on matrix multiplication and scalar multiplications, additions and subtractions, GPUs are used as accelerators to speed training compared with just using the CPU for this process. GPUs are also used for their lower power consumption.

In our example, we will use the Numba [34] framework, which supports just-in-time and ahead-of-time compilation of python functions. One of the compilation backends is the Cuda API for NVIDIA GPUs, requiring only the installation of the Numba python module and CUDA SDK. There is also support for HSA AMD GPUs, which has a very similar API to the one presented here for Cuda.

Programming for GPUs follows a different programming model than programming for CPUs. As such the training function defined previously will not run on the GPU. Listing 1.10 shows the GPU version of the training function. One of the major differences is that GPUs have their own memory and are not able to access the host RAM memory. In order for the GPU to operate on the weight and neuronal matrices, it requires explicit memory copying from and to the host. The `to_device` and `to_host` methods take care of memory copies.

Another important aspect of GPGPU is to define the shape of the computation, i. e., defining how many threads (work-items) and how many thread groups (work-groups) will perform this computation. On NVIDIA hardware, a generally good work-group size is the warp size. A warp is the number of hardware threads that share the same program counter and that, ideally, should all execute the same code. In this case, the warp size is 32, leading to a matrix of 32 by 32 work-items, totaling 1024 threads. Not all of these threads will be necessary, but matching the physical layout often improves over having fewer threads.

All of the computation that occurs on the GPU is defined in a special function, called the kernel. This function is called in line 11 using the number of workgroups and work-items as special arguments. This is necessary for the GPU scheduler to start that many threads, each one executing the same kernel function.

Listing 1.10 Host code for driving GPU training of an ANN

```
1  def train_cuda(X, y, conf, iterations=6000):
2      gpu = cuda.get_current_device()
3      weights0, weights1 = conf
4      weights0g = cuda.to_device(weights0)
5      weights1g = cuda.to_device(weights1)
6      Xg = cuda.to_device(X)
7      yg = cuda.to_device(y)
8      rows = X.shape[0]
9      thread_ct = (gpu.WARP_SIZE, gpu.WARP_SIZE)
10     block_ct = [ int(math.ceil(1.0 * rows / gpu.WARP_SIZE)), int(
11        math.ceil(1.0 * ndims / gpu.WARP_SIZE))]
12     train_kernel[block_ct, thread_ct](Xg, yg, weights0g, weights1g
13        , iterations)
14     weights0g.to_host()
15     weights1g.to_host()
16     return (weights0, weights1)
```

Additional auxiliary functions (Listing 1.11) that have to execute on the GPU have to be defined using a special decorator that allows the Numba library to compile them to the Cuda intermediate language (PTX). In this case, function inlining is being enable to reduce the overhead of function calling on the GPU.

Listing 1.11 Sigmoid function and its derivative for the GPU

```
1 @cuda.jit(device=True, inline=True)
2 def sigmoidg(x):
3     return 1/(1+math.exp(-x))
4
5 @cuda.jit(device=True, inline=True)
6 def sigmoidg_d(x):
7     return x*(1-x)
```

These functions are used within the main kernel function (Listing 1.12) that performs the parallel training. Given the restrictive nature of the GPU architecture, not all of the Python language is allowed inside GPU functions. NoPython is the name of the subset allowed in Cuda functions that does not support exception handling or `with` blocks. In the particular case of kernel functions, the resulting type should be void. Inside Cuda functions, it is possible to use GPU-specific features, such as accessing the indices of the current thread within the GPU (via `cuda.grid(2) in line 6`. The 2D coordinates of the read in the work-group grid are returned, so that the current thread can use different inputs than all other threads. In this case, the only interesting threads are those with i between 0 and 24, and j between 0 and 16.

Listing 1.12 Kernel function for GPU training of an ANN

```
1  @cuda.jit()
2  def train_kernel(X, y, weights0, weights1, iterations):
3      l1 = cuda.shared.array(shape=(instances, ndims), dtype=numba.
4          float32)
5      l2_delta = cuda.shared.array(shape=(instances, 3), dtype=
6          numba.float32)
7      l1_delta = cuda.shared.array(shape=(instances, ndims), dtype=
8          numba.float32)
9      i, j = cuda.grid(2)
10     if i < instances and j < ndims:
11         for it in range(iterations):
12             acc = 0
13             for k in range(ndims):
14                 acc += X[i, k] * weights0[k, j]
15             l1[i, j] = sigmoidg(acc)
16             cuda.syncthreads()
17             if j < 3:
18                 acc = 0
19                 for k in range(ndims):
20                     acc += l1[i,k] * weights1[k,j]
21                 l2 = sigmoidg(acc)
22                 l2_error = y[i, j] - l2
23                 l2_delta[i, j] = l2_error * sigmoidg_d(l2)
24             cuda.syncthreads()
25             acc = 0
26             for k in range(3):
27                 acc += l2_delta[i,k] * weights1[j, k]
28             l1_error = acc
29             l1_delta[i, j] = l1_error * sigmoidg_d(l1[i, j])
30             cuda.syncthreads()
31             if j < 3:
32                 acc = 0
33                 for k in range(instances):
34                     acc += l1[k, i] * l2_delta[k, j]
35                 weights1[i, j] += acc
36             acc = 0
37             for k in range(instances):
38                 acc += X[k, i] * l1_delta[k, j]
39             weights0[i, j] += acc
40             cuda.syncthreads()
```

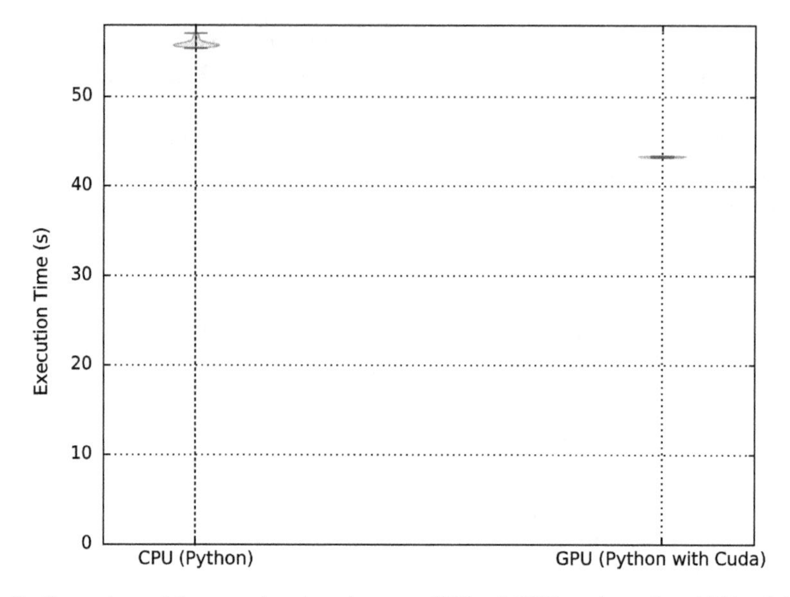

Fig. 3 Comparison of the execution times between CPU and GPU versions of an ANN training

Another difference in GPU programming is the usage of local arrays that have much better performance than shared memory. These local arrays are used in lines 3–5 in intermediate matrices that are not required to be read or written to/from the host.

Because neural network training performs different matrix multiplications, not all threads will be used at all steps. To synchronize operations, the `cuda.syncthreads()` function is used, acting as a barrier. Code between barriers occurs at the same pace within the same warp, but not necessarily across warps.

In each matrix multiplication, it is necessary to select only the threads that are within the same of the output matrix (lines 7, 14 and 28). In order to understand which threads are used at each stage, Fig. 4 presents a visual representation of the threads used at each step.

Executing the code above will result in the kernel and auxiliary functions being compiled to the GPU, while the training function will manage memory copies and scheduling of the necessary threads, all running the same kernel function.

To understand the possible speedup, without any aggressive optimization, the GPU version was compared against the previous CPU version (both using CPython 2.7.6, on a machine with an Intel i7-3520M processor and a NVIDIA GeForce GT 640 LE GPU).

Figure 3 shows violin plots of the distribution and quartiles of execution times of both versions. The GPU version executes faster than the CPU version, showing how this type of programs can be easily parallelized on the GPU with speedups (Fig. 4).

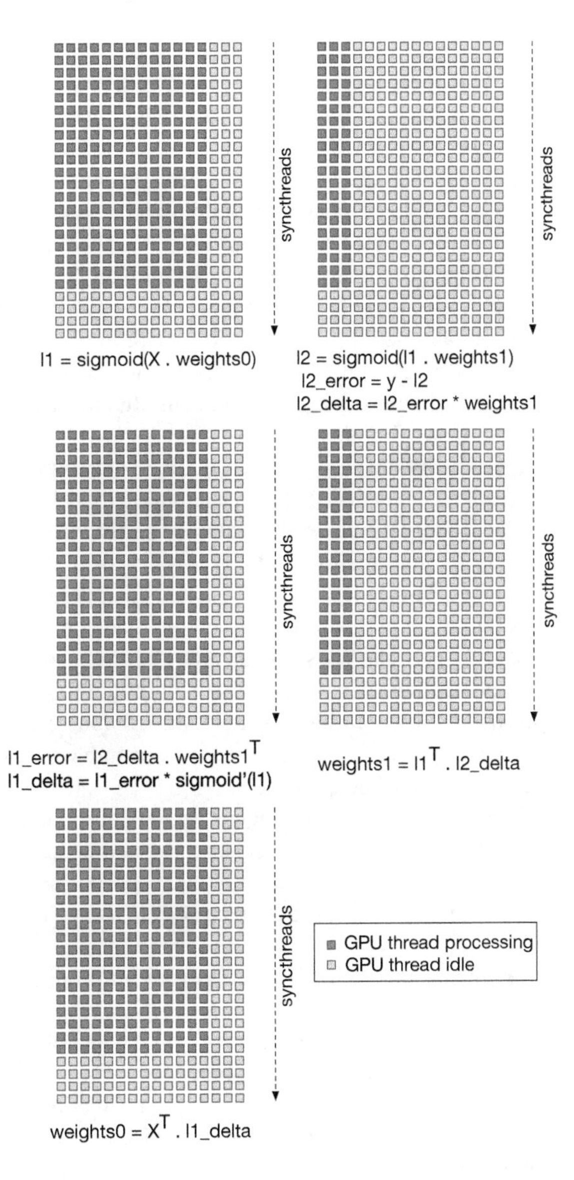

Fig. 4 Visual representation of threads performing neural network training between barriers

8 Discussion and Homework

The presented source code is designed as an educational resource, with its inherent shortcomings and room for further optimization. Different design decisions are discussed in this section, as well as alternatives that the reader can opt to pursue.

Python is frequently used for prototyping, but not for implementing high-performance computing applications. Other languages like C, C++ or Fortran are frequently preferred. Python introduces overhead in interpretation of code, and features dynamic computational expensive data structures, such as lists. Furthermore parallelization has to occur at the process level, and not at the thread level due to the Global Interpreter Lock (GIL). In the presented solution, these drawbacks are not present: matrices are stored in efficient data-structures provided by Numba (programmed in C) and most of the training process occurs on the GPU side, with Python code being compiled to efficient PTX intermediate code. Even the CPU version of training performs matrix operations within the Numpy library, also programmed in C to be efficient. In order to write more efficient Python code, one could use the Cython toolkit to annotate Python functions with the C types to have C-like performance. Additionally, the Numba JIT compiler can also be used to generate CPU versions of functions. Finally, our approach uses multicore parallelism using different processes, thus not being limited by the GIL.

Redis is not the most feature-complete message queue like ActiveMQ [35], but is frequently used in distribution systems due to its low overhead. 0MQ is another low-overhead alternative to Redis that has a similar feature set. None of these approaches is ready to send matrices across the network, all requiring custom encoding and decoding to raw bytes, using the metadata approach presented before.

A limitation of this work is assuming that all nodes will have access to the original dataset. If the dataset is bigger than the local storage of this node, this is not an optimal solution. Alternatives like distributed file systems (NFS) should be used in that case. Workers would download the relevant slice of the dataset before executing the training function. In the developed example, if there are different files for each slice, it is a trivial adaptation.

Regarding the layout of the neural network, our approach had a static layout that depended on the dataset. Deep-learning can be achieved by increasing the number of hidden layers, which has more overhead in training and memory copying times, thus only useful when the problem has that much complexity. Adding more layers and neurons is left as an exercise for the reader. For dynamically selecting layouts there are several alternatives, ranging from pruning of useless neurons [36], using genetic algorithms to evolve ANNs [37] to Monte Carlo methods [38]. In [39] the reader will find an up-to-date survey on ANN processing. These approaches require a more extended study of the subjects.

Another area of improvement is the evaluation of the obtained solution. In our case, we are using 20 instances for training, and the remaining for testing. This is subject to bias in the division of training-testing datasets. A Cross-validation approach is preferable, because instances are used as both training and testing in different iterations. This approach has the advantage of requiring more computation power for the same dataset.

For the sake of simplicity, a naïve GPU version was presented. This code can be further optimized in different ways. First is managing the memory being copied to the GPU. Given the limited amount of GPU memory, it is important to limit the memory copies. Just like Map-Reduce is used to split a larger dataset is smaller tasks, the same can be used to schedule tasks to the GPU in chunks. While the GPU is processing a chunk, it is possible for the CPU to be sending the next chunk to the

GPU. It is advantageous to overlap data transfers and execution, in order to reduce latency in GPU operations. The organization of threads in work-groups and work-items can be improved, since our solution did not utilize all scheduled threads, in order to maximize the usage of cache locality. One way to do this is change the layout of the ANN to match the GPU. This allows for more complex ANNs using the same computational power (excluding memory transfers, assuming they can be overlapped with processing).

The final GPU optimization can be done in scenarios with multiple GPUs on the same machine, directly connected among each other (using SLI, for instance). In those scenarios, it would be necessary to synchronize matrix averaging across GPUs without requiring the computation to synchronize with the GPU. This would reduce the communication overhead, since the CPU would only need the final weights when the training is complete to pass to the master node.

9 Conclusion

In this chapter, we have covered the creation of a distributed GPU-backed Neural Network implementation from scratch. Training was developed for execution on the CPU using back-propagation. A distributed protocol for training in parallel within the same machine, as well as across any number of machines was presented and discussed. Finally, a GPU implementation was also discussed, highlighting the major differences in the programming model. For this tasks, we have relied on a limited set of existing software: Python (with Numpy, Pandas and Numba libraries) and Redis for message communication. All of these tools are open-source, allowing practitioners to even modify the underlying implementations if they feed it is preventing them from exploring more radical new ideas.

Additionally, we have discussed the shortcomings and advantages of this approach, mainly the choice of language, parallelization methods and ANN training and layout methods that could be explored further.

Acknowledgements The first author was supported by the LASIGE Research Unit (UID/CEC/00408/2013).

References

1. J. Schmidhuber, Deep learning in neural networks: an overview. Neural Netw. **61**, 85–117 (2015)
2. W.S. McCulloch, W. Pitts, A logical calculus of the ideas immanent in nervous activity. Bull. Math. Biophys. **5**(4), 115–133 (1943)
3. A. Krizhevsky, I. Sutskever, G.E. Hinton, Imagenet classification with deep convolutional neural networks, in *Advances in Neural Information Processing Systems* (2012), pp. 1097–1105

4. G.L. Martin, J.A. Pittman, Recognizing hand-printed letters and digits using backpropagation learning. Neural Comput. **3**(2), 258–267 (1991)
5. G. Zhang, M.Y. Hu, B.E. Patuwo, D.C. Indro, Artificial neural networks in bankruptcy prediction: general framework and cross-validation analysis. Eur. J. Oper. Res. **116**(1), 16–32 (1999)
6. M.H. Ebell, Artificial neural networks for predicting failure to survive following in-hospital cardiopulmonary resuscitation. J. Family Pract. **36**(3), 297–304 (1993)
7. D.A. Pomerleau, Efficient training of artificial neural networks for autonomous navigation. Neural Comput. **3**(1), 88–97 (1991)
8. W.S. Sarle, *On Computing Number of Neurons in Hidden Layer* (1995)
9. A. Jacobs, The pathologies of big data. Commun. ACM **52**(8), 36–44 (2009)
10. M. Abadi, P. Barham, J. Chen, Z. Chen, A. Davis, J. Dean, M. Devin, S. Ghemawat, G. Irving, M. Isard, M. Kudlur, J. Levenberg, R. Monga, S. Moore, D.G. Murray, B. Steiner, P. Tucker, V. Vasudevan, P. Warden, M. Wicke, Y. Yu, X. Zheng, Tensorflow: a system for large-scale machine learning, in *Proceedings of the 12th USENIX Conference on Operating Systems Design and Implementation. OSDI'16*, Berkeley, CA, USA, USENIX Association (2016), pp. 265–283
11. J. Bergstra, O. Breuleux, P. Lamblin, R. Pascanu, O. Delalleau, G. Desjardins, I. Goodfellow, A. Bergeron, Y. Bengio, P. Kaelbling, *Theano: Deep Learning on GPUs with Python*
12. R. Collobert, S. Bengio, J. Marithoz, *Torch: A Modular Machine Learning Software Library* (2002)
13. B.J. Erickson, P. Korfiatis, Z. Akkus, T. Kline, K. Philbrick, Toolkits and libraries for deep learning. J. Digit. Imaging **30**(4), 400–405 (2017)
14. S. Lawrence, C.L. Giles, A.C. Tsoi, What size neural network gives optimal generalization? convergence properties of backpropagation (1998)
15. F. Cao, T. Xie, The construction and approximation for feedforword neural networks with fixed weights, in *International Conference on Machine Learning and Cybernetics, ICMLC 2010*, Qingdao, China, 11–14 July 2010, Proceedings (2010), pp. 3164–3168
16. N. Guliyev, V. Ismailov, On the approximation by single hidden layer feedforward neural networks with fixed weights (2017)
17. S. Koturwar, S. Merchant, Weight initialization of deep neural networks (DNNs) using data statistics. arXiv:1710.10570 (2017)
18. S.K. Kumar, On weight initialization in deep neural networks. arXiv:1704.08863 (2017)
19. D. Mishkin, J. Matas, All you need is a good init. arXiv:1511.06422 (2015)
20. I. Goodfellow, Y. Bengio, A. Courville, Y. Bengio, *Deep Learning*, vol. 1 (MIT Press, Cambridge, 2016)
21. V. Nair, G.E. Hinton, Rectified linear units improve restricted Boltzmann machines, in *Proceedings of the 27th International Conference on Machine Learning (ICML-10)* (2010), pp. 807–814
22. D.E. Rumelhart, G.E. Hinton, R.J. Williams, Learning representations by back-propagating errors. Nature **323**(6088), 533 (1986)
23. M. Lichman, *UCI Machine Learning Repository* (2013)
24. K.O. Stanley, D.B. D'Ambrosio, J. Gauci, A hypercube-based encoding for evolving large-scale neural networks. Artif. Life **15**(2), 185–212 (2009)
25. W. McKinney et al., Data structures for statistical computing in python, in *Proceedings of the 9th Python in Science Conference*, vol. 445 (2010), pp. 51–56
26. A. Trask, *A Neural Network in 11 Lines of Python* (2013)
27. L. Gu, H. Li, Memory or time: performance evaluation for iterative operation on hadoop and spark, in *2013 IEEE 10th International Conference on High Performance Computing and Communications and 2013 IEEE International Conference on Embedded and Ubiquitous Computing (HPCC_EUC)* (IEEE, 2013), pp. 721–727
28. G. Dahl, A. McAvinney, T. Newhall et al., Parallelizing neural network training for cluster systems, in *Proceedings of the IASTED International Conference on Parallel and Distributed Computing and Networks* (ACTA Press, 2008), pp. 220–225
29. S. Sanfilippo, P. Noordhuis, *Redis* (2009)

30. J. Reese, S. Zaranek, GPU programming in MATLAB, in *MathWorks News & Notes* (The MathWorks Inc., Natick, MA, 2012), pp. 22–25
31. M.M. Chakravarty, G. Keller, S. Lee, T.L. McDonell, V. Grover, Accelerating Haskell array codes with multicore GPUs, in *Proceedings of the Sixth Workshop on Declarative Aspects of Multicore Programming* (ACM, 2011), pp. 3–14
32. A. Fonseca, B. Cabral, Æminiumgpu: an intelligent framework for GPU programming, in *Facing the Multicore-Challenge III* (Springer, 2013), pp. 96–107
33. B. Catanzaro, M. Garland, K. Keutzer, Copperhead: compiling an embedded data parallel language. ACM SIGPLAN Not. **46**(8), 47–56 (2011)
34. S.K. Lam, A. Pitrou, S. Seibert, Numba: a LLVM-based python JIT compiler, in *Proceedings of the Second Workshop on the LLVM Compiler Infrastructure in HPC* (ACM, 2015), p. 7
35. B. Snyder, D. Bosnanac, R. Davies, *ActiveMQ in Action*, vol. 47 (Manning, 2011)
36. E.D. Karnin, A simple procedure for pruning back-propagation trained neural networks. IEEE Trans. Neural Netw. **1**(2), 239–242 (1990)
37. D.B. Fogel, L.J. Fogel, V. Porto, Evolving neural networks. Biol. Cybern. **63**(6), 487–493 (1990)
38. J.F. de Freitas, M. Niranjan, A.H. Gee, A. Doucet, Sequential Monte Carlo methods to train neural network models. Neural Comput. **12**(4), 955–993 (2000)
39. V. Sze, Y. Chen, T. Yang, J.S. Emer, Efficient processing of deep neural networks: a tutorial and survey. arXiv:1703.09039 (2017)

Deep Learning for Scene Understanding

Uzair Nadeem, Syed Afaq Ali Shah, Ferdous Sohel, Roberto Togneri and Mohammed Bennamoun

Abstract With the progress in the field of computer vision, we are moving closer and closer towards the ultimate aim of human like vision for machines. Scene understanding is an essential part of this research. It seeks the goal that any image should be as understandable and decipherable for computers as it is for humans. The stall in the progress of the different components of scene understanding, due to the limitations of the traditional algorithms, has now been broken by the induction of neural networks for computer vision tasks. The advancements in parallel computational hardware has made it possible to train very deep and complex neural network architectures. This has vastly improved the performances of algorithms for all the different components of scene understanding. This chapter analyses these contributions of deep learning and also presents the advancements of high level scene understanding tasks, such as caption generation for images. It also sheds light on the need to combine these individual components into an integrated system.

U. Nadeem · S. A. A. Shah · M. Bennamoun (✉)
Department of Computer Science and Software Engineering,
The University of Western Australia, Crawley, Australia
e-mail: mohammed.bennamoun@uwa.edu.au

U. Nadeem
e-mail: uzair.nadeem@research.uwa.edu.au

S. A. A. Shah
e-mail: afaq.shah@uwa.edu.au

F. Sohel
Discipline of Information Technology, Mathematics & Statistics, Murdoch University, Perth, Australia
e-mail: f.sohel@murdoch.edu.au

R. Togneri
Department of Electrical, Electronics and Computer Engineering,
The University of Western Australia, Crawley, Australia
e-mail: roberto.togneri@uwa.edu.au

V. E. Balas et al. (eds.), *Handbook of Deep Learning Applications*,
Smart Innovation, Systems and Technologies 136,
https://doi.org/10.1007/978-3-030-11479-4_2

Keywords Scene understanding · Deep learning · Object recognition · Face detection and recognition · Text detection · Depth map estimation · Scene classification · Caption generation · Visual question answering (VQA)

1 Introduction

Scene understanding is a major field of computer vision research. The main goal of scene understanding is to equip computers and machines with human like vision i.e., a computer should be able to extract the same amount of information and understanding from an image as a human is able to do. When one looks at an image they can tell whether it is outdoor or indoor, they can infer the location of the scene (e.g., bedroom or dining room), whether there is text in the image and how it relates to the objects in the scene. Humans are able to perceive the precise location and depth (distance from viewpoint) of the objects. We can understand and identify objects and segment or visually separate them from the background.

Deep learning is the latest trend in machine learning. With the availability of ample computational resources and big datasets, it has now become possible to train very deep networks which was never thought possible only a decade ago. Though there is still much room for improvement, deep learning has significantly enhanced the performance of the various components of scene understanding, such as object recognition, text detection in natural scenes, depth map estimation, and face detection and recognition (Fig. 1). These components are required for human like understanding of scenes and also aid in achieving higher level tasks, such as scene classification, caption generation and visual question answering. These sub-tasks are inter-connected and form the essential elements in the framework for the complete understanding of a scene (Fig. 2). Despite the achieved progress in the individual components, there are very few significant works which apply deep learning to develop holistic scene understanding systems. In the following sections, we will analyse the impact that deep learning has made on the various components of scene understanding.

2 Object Recognition

The ability to recognize objects plays a crucial role in scene understanding. When looking at an image of the scene, humans can easily recognize all the objects in a given scene (e.g., a chair, a desk) and interpret such objects as part of a coherent geometrical and semantically meaningful structure (e.g., the office). This is achieved by accomplishing two tasks. **First**, representations of 3D objects are built that allow us to identify objects regardless of their location or viewpoint in the image. This requires that the observer has learnt models that are robust with respect to viewpoint changes to assist the identification of object instances/categories in poses that the observer has not seen before. **Second**, we need to infer the objects' geometric attributes such as pose, scale and shape of objects. These geometric characteristics can provide robust cues for the interpretation of the interactions between objects in the scene, the estimation of object functionalities, and ultimately, to infer the 3D layout of the scenes.

This section outlines the different steps involved in object recognition and review different techniques that are commonly used to recognize objects from images under various conditions.

2.1 *Object Recognition Pipeline*

The major steps of a typical object recognition pipeline are shown in Fig. 3.

2.1.1 Image Acquisition

The image of an object is acquired using 2D (e.g., high resolution cameras) or 3D scanners (e.g., Kinect or Minolta). 2D cameras capture the intensity information of the object such as colour, while 3D scanners provide point clouds, 3D mesh or depth image. Each pixel in a depth image represents the distance of a 3D point of the object from the camera/scanner (as opposed to the intensity information in the case of 2D cameras).

Fig. 1 Deep learning can detect faces, text and objects in an image, but can it describe a scene and determine the various interactions between objects?

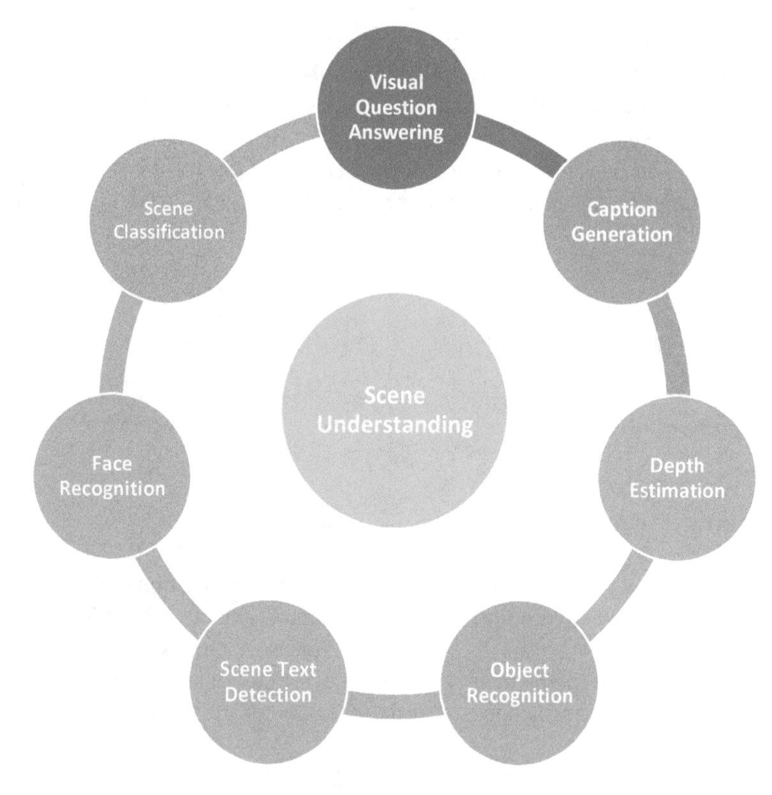

Fig. 2 The main components of scene understanding

2.1.2 Object Scan Pre-processing

The 3D images usually contain significant surface noise (e.g., in the case of the low resolution Kinetic, holes i.e., missing data due to self-occlusion and spikes). Pre-processing steps are used to remove such noise. Then surface interpolation is used to fill small holes. The holes may originally be present in the scan or may have been formed as a result of the removal of data spikes. The final stage of pre-processing is usually surface smoothing (e.g. smoothing surface with Gaussians).

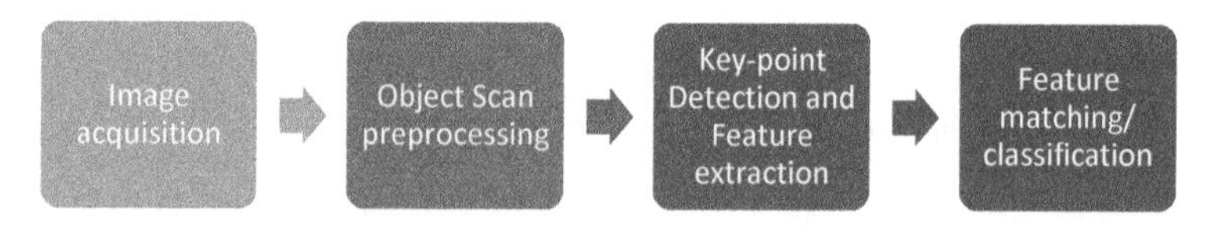

Fig. 3 Different steps in feature based object recognition

2.1.3 Key-Point Detection and Feature Extraction

A set of highly repeatable key-points are first detected and features are then extracted around those key-points. In 2D images, features usually represent the colours, texture and appearances. In 3D, features describe the geometry of the key-point of the object. Features are usually chosen as a trade-off between the descriptiveness of the features that is required for discrimination, and invariance properties of the features (e.g., invariance with respect to rotation or translation). Example of 'features' include the intensity values in the case of 2D, raw depth values for 3D data, surface normals, curvatures, spin images [1] and Scale-Invariant Feature Transform (SIFT) descriptor [2] to name a few.

2.1.4 Feature Matching/Classification

The final step of object recognition is the feature matching/classification phase where machine learning classifiers are used to recognize objects in an image. Some popular examples include Support Vector Machines (SVM), neural nets and k-nearest neighbours (k-NN). K-NN can also be used along with different subspaces (e.g. LDA or PCA).

2.2 *Hand-Crafted Features for Object Recognition*

Hand-crafted features are those which are extracted from an image according to a certain manually predefined algorithm based on expert knowledge. Local Binary Pattern (LBP) [3] and SIFT [2] features are popular examples of hand-crafted features. Here, we shall briefly discuss some of the popular hand-crafted features.

To create LBP feature vector [3], a sliding window is divided into cells. An 8-connected neighbourhood is compared with the centre pixel. The pixels in the neighbourhood are set to '0' if their respective value is smaller than the centre pixel, otherwise they are given a value of '1'. Next, a binary number is constructed by going clockwise as shown in Fig. 4. The centre pixel is then replaced with the decimal value of the binary number (Fig. 4).

LBP is not rotation invariant and the binary number is sensitive to the starting point. In addition, minor changes in illumination can change the decimal value which makes LBP less robust to illumination changes.

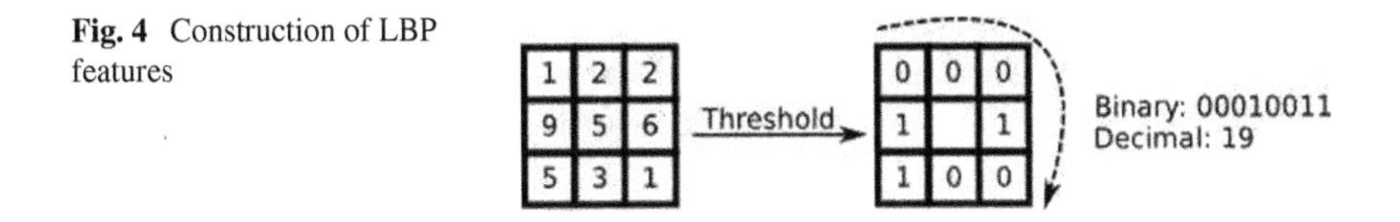

Fig. 4 Construction of LBP features

The Lowe's Scale Invariant Feature Transform (SIFT) [2] provides a set of features of an object that are robust to rotation and scale. The first step of the SIFT computation is "key-point detection". For this purpose, a Difference of Gaussian (DoG) is first used to smooth the images at different scales. Next, these images are searched for local extrema over scale and space. Next, a histogram of the local gradient directions is created at a selected scale. The canonical orientation at the peak of the smoothed histogram is next assigned. Gradients of orientations in an area around the key-point are used to create the orientation histogram which consists of 36 bins with each bin representing 10°. The samples are weighted by their magnitude and by a 'Gaussian-weighted circular window' and then are added to the bins. Next a region is selected around a key-point based on scale. Then the gradient information is aligned with the key-point orientation. 8 orientation bins are calculated at 4×4 bin array, which forms an $8 \times 4 \times 4 = 128$ dimension SIFT feature.

In addition to 2D features, several 3D features have been proposed in the literature. Tombari et al. [4] proposed a descriptor named Signature of Histograms of OrienTations (SHOT). A local reference frame is first constructed for a given key-point p and the neighbourhood space is divided into 3D spherical volumes. A local histogram is then generated for each volume by accumulating the number of points according to the angles between the normal at the key-point and the normals at the neighbouring points. All local histograms are then concatenated to form an overall SHOT descriptor. The SHOT descriptor is highly descriptive, computationally efficient and robust to noise [5]. Guo et al. [6] proposed the ROtational Projection Statistics (RoPS) as a 3D local feature descriptor. A covariance matrix is formed using points lying on a local surface. The Eigen-value decomposition of the covariance matrix is used to define a rotation invariant reference frame. The 3D points are rotationally projected on the neighbouring 2D planes to form the feature descriptor. Depth kernel descriptors (comprising of up to five different cues including, size, 3D shape and depth edges, which are extracted from the depth map and RGB images) were proposed to provide a way of turning any pixel attribute to patch-level features. Depth kernel descriptors were tested on a low resolution Kinect dataset and achieved more than 10% increase in accuracy over the-state-of-the-art techniques at that time [7]. Guo et al. [8] published a comprehensive survey of the feature-based 3D object recognition methods.

2.3 Deep Learning Techniques for Object Recognition

Deep learning has been found very effective and is actively used in several object recognition tasks [9]. Hayat et al. proposed autoencoder based Deep Reconstruction Models (DRM) [10] for image classification. This deep learning framework consists of encoder and decoder layers, which are used for the reconstruction of input images. ADNT has been shown to achieve a superior object recognition performance on the ETH-80 object dataset. Shah et al. proposed Iterative Deep Learning Model (IDLM) and tested it for the task of object recognition [11]. IDLM consists of Pool

Convolutional Layer followed by ANNs applied iteratively in a hierarchical fashion to learn a non-linear feature representation of the input images. The input to IDLM consists of raw images and it has achieved an accuracy of 98.64% on ETH-80 object dataset.

A notable advance in deep learning was achieved by AlexNet [12] in the 2012 ImageNet LSVRC contest. The training set consists of 1.2 million high-resolution images and 1000 different object classes. On the test set, consisting of 150,000 images, AlexNet achieved an error rate considerably lower than the previous state-of-the-art approach. AlexNet is a very deep network, which consists of 60 million weights, and 650,000 neurons, and five convolutional layers together with max-pooling layers [12].

Among the very deep networks, GoogleNet [13] was the first popular model which uses quite a complex architecture with several network branches. This model won the ILSVRC'14 competition with the best top-5 error rate of 6.7% on the classification task. GoogleNet has now several improved variants. He et al. [14] from Microsoft proposed the residual net, which won the ILSVRC 2015 challenge by reducing the top 5 error rate to 3.6% compared to the 6.7% error rate of GoogleNet. The remarkable feature of the Residual architecture is the identity skips connections in the residual blocks, which allow it to easily train very deep CNN architectures.

Qi et al. [15] recently proposed PointNet for 3D object recognition. Unlike other architectures, this new deep learning architecture directly takes point clouds as input, and outputs either class labels for the entire input or per point segment labels for each point of the input. The PointNet architecture well respects the permutation invariance of input 3D points and it has been shown to achieve a comparable performance with the state-of-the-art techniques for 3D object recognition performance.

3 Face Detection and Recognition

Recognizing a person is an important requirement for the full understanding of a scene and obviously the best natural way to identify a person is to recognise his/her face. Face detection and recognition has received an increasing interest from the computer vision community in the past several years. This field has also important applications in biometrics, surveillance and security, crowd analysis and smart user interfaces. The main challenges in the field of face detection and recognition are illumination and pose variations, low resolutions, partial occlusions, inter-class similarities, noises, background similarity, and the availability of sufficient training data. It is to be noted that face recognition can be thought of as a very fine-grained object recognition problem, since even humans many times confuse the faces of different persons.

3.1 Non-deep Learning Techniques for Face Detection and Recognition

The discussion on face detection techniques cannot be complete without discussing the Viola and Jones face detection algorithm [16]. The Viola and Jones face detector consists of three main components: the integral image, a classifier training with AdaBoost and the cascaded classifiers. **First** of all the image is transformed to an integral image. In the integral image, each pixel is the sum of the intensity values of all pixels above and to the right of it in the original image. The integral image is used for an efficient calculation of the Haar like rectangular features. These features are the weighted differences between the sums of intensities of two to four adjacent rectangles. **Then** a variant of AdaBoost (Adaptive Boost) is used with selected rectangular features. In AdaBoost many weak classifiers are combined in a weighted manner to increase the final classification accuracy and form a strong classifier. Decision stumps (single node classification trees) are used as weak classifiers and each node tries to find the optimum threshold for one feature. Then several strong classifiers of increasing complexity are cascaded to form a degenerate decision tree. Only the instances detected as positive (face) are passed to the next subsequent strong classifier in the cascade. In this way each later classifier has less decisions to make than each of its previous classifiers in the cascade. This ensures that the processing time is fast and achieves a rate of 15 frames per second. This technique achieved a comparable accuracy to the state of the art approaches at that time while achieving a significantly faster computational times.

Face recognition also follows the general pipeline of "feature extraction" and "classification". A comprehensive study of the state of the art surface features for the recognition of the human face is presented in [17]. Modern face recognition systems consist of four major steps which are detection, alignment, representation and classification [18]. One of the approaches to the problem of face recognition is by using image-sets i.e., both the training and testing data contain a set of images. This is inspired by the fact that due to the wide availability of mobile devices and CCTV cameras, usually more than one image of the query person is available and the extra information can be used for better decision making. Ortiz et al. [19] use a mean sequence sparse representation for an end-to-end video based face recognition system. Faces are detected and tracked in the input video clip. Then three types of features, LBP, Gabor and HOG, are extracted. Finally, the extracted features are used in a modified version of a sparse representation based classification for dictionary based classification. Shah et al. [20] represent the test images as a linear representation of the images of gallery sets of each class. The most accurate representation is decided as the class of the test image set.

3.2 Deep Learning for Face Detection and Recognition

There are many deep learning based approaches for face detection. However, most of them require high computational resources and time. A cascaded architecture of CNNs provides a good compromise between the two goals of accuracy and efficiency [21]. Li et al. [21] propose a CNN cascade of six CNNs for face detection. Three of the CNNs are binary face/non-face detectors while the other three are used for calibration. The system operates at multiple resolutions, and similar to the Viola and Jones algorithm, verifies the different detections in multiple stages of increasing difficulty. A calibration stage based on CNN is used after every stage to reduce the number of potential face regions in later stages and to improve localization. This technique achieves a classification rate of 14 frames per second, while achieving state of the art results on face detection benchmarks.

Deep learning can effectively be used to learn non-linear transformations for the mapping of an image set into a shared feature space, while maximizing the distance between the different classes [22]. Hayat et al. [10] use deep learning to train separate auto-encoders for each class for image-set based face recognition. The test images are then reconstructed from each auto-encoder. The minimum reconstruction error is used as the measure for classification. The nonlinearity of images in image sets can be modelled by manifolds and deep learning. Depth maps of faces are also used, wherever available, for face recognition which shows that face recognition performance can significantly be improved by using depth information along with RGB face images.

Taigman et al. [18] use a piecewise affine transformation with 3D face modelling. A deep network consisting of nine layers is used to extract face features. Different from the conventional CNNs, the deep network uses locally connected layers without weight sharing. The network involves 120 million parameters which are learned by a data driven approach. A simple inner product of deep features is used to recognize a person. This method produced a big gain in accuracy on datasets of face recognition in unconstrained environments.

Schroff et al. [23] use a deep neural network called FaceNet, to transform face images to a compact Euclidean Space. A triplet loss is used for training the CNN. It is motivated by the idea that the distance between the images of the same class in the transformed Euclidean space should be less than the distance between the images of different classes. A 128 dimensional vector is used in the transformed domain to represent each image. The distances in the transformed domain are taken as a measure of similarity. The features produced by FaceNet can then be used for the tasks of face clustering, recognition or verification, as required. Two different architectures are suggested for the CNN based on the previous works. This system achieves a significantly better performance, compared to the other techniques.

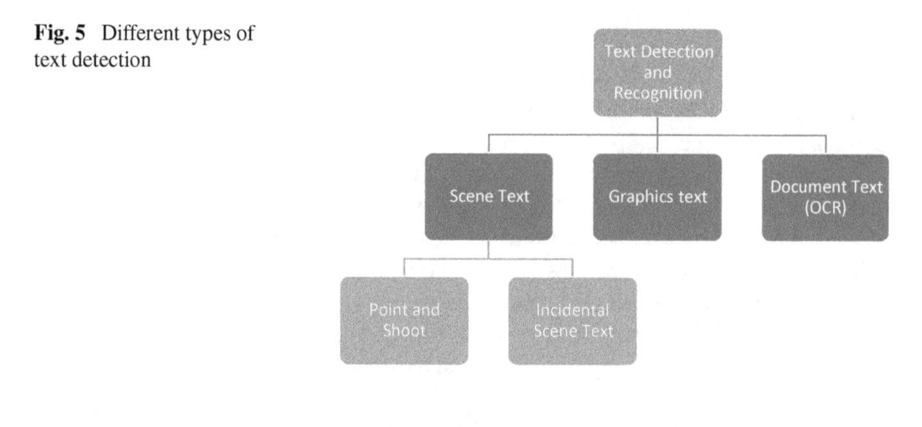

Fig. 5 Different types of text detection

4 Text Detection in Natural Scenes

The computer vision community has been giving lots of attention to the problem of text detection and recognition in images over the past decade. Text Detection problems can be classified into three types: Document Text (OCR), Graphics text such as text in emails, posters, advertisements and scene text (Fig. 5). Although optical character recognition (OCR) for documents is mostly considered as a solved problem, the performance for text detection and recognition in natural scenes is below par [24]. Text detection in natural scenes suffers from all the problems of object detection, in addition to the problems which are inherent to text detection. Challenges in text detection in natural scenes include high variability in size, aspect ratio, shapes of characters, fonts, uneven illumination and interclass variations.

Text detection in natural scenes is significant for personal assistant devices, data retrieval, autonomous driving, driver assistance and for scene understanding. Text recognition can be used to help blind people to read navigation signs, boards and even books. It can also help in the retrieval of images and text from a large database. The importance of text makes its detection an important topic in computer vision. Text detection in images consists of two major steps: text detection (or text localization) and text recognition. A system which accomplishes both of these tasks is called an "end to end text detection system".

Scene text can be classified into "Point and Shoot" and "Incidental Text" [25] (Fig. 5). In Point and Shoot the image, picture or video is captured with the intention to focus on the text. Incidental text refers to text that can occur in any randomly taken pictures or videos and in images where text is usually not prominent.

Stepwise methodologies have four major, distinct steps: localization, verification, segmentation and recognition [25]. In localization, the components are coarsely classified into text regions which are improved with a verification step. Characters are separated using the segmentation approach and fed to the recognition module. In the case of the integrated approach, it is not possible to subdivide the method into distinct steps because the steps reinforce each other and the results are intermediate and incomplete until the final output [26].

Both methods can use a machine learning classifier, a connected component analysis or a combination of both. For a machine learning approach, supervised learning is used to train a classifier to differentiate between text and non-text. The trained classifier is applied on text images using a sliding window to identify the text regions in the image. Another classifier is trained to differentiate between different characters and cases (uppercase or lowercase). The output of the character identifier is used in combination with a predefined lexicon or general English dictionary to form words. The work of Wang et al. [27] is a typical example of this approach.

For connected component analysis, hand crafted features are used to discriminate text from non-text. Feature processing is applied on the whole image and connected component analysis is used to identify the text regions. Features are also used to differentiate among characters which are combined with a lexicon for word formation. Stroke Width Transform [28] is considered as one of the most important features for text detection. Other features that are usually exploited for text detection and recognition are edges, colours, contrast, points of inflections and maximally stable extremal regions (MSER).

4.1 Classical Approaches for Text Detection

Similar to other areas of scene understanding, the classical approaches for Scene text detection extract various types of features from the images and then train a classifier based on those features. Ephstein et al. [29] used the property that most of the text has nearly a constant stroke width to develop the Stroke Width Transform. Many later works used the Stroke Width Transform in its original or modified form and it is still one of the main feature detection techniques for text detection in natural scenes. Another prominent approach is to detect Maximally Stable Extremal Regions (MSER) to identify potential text regions. The detected regions are then filtered using geometric and stroke width information to exclude false-positives [30]. Letters are paired to identify text lines, which are subsequently separated into words. Neumann and Matas [31] detected extremal regions and used morphological features to filter out the non-text regions. Then a system for exhaustive search was used for the final output. Matas and Neumann [32] developed an end to end text detection pipeline by detecting strokes of particular orientations at specific distances. They used bar filters to detect the strokes in an image gradient field. Finally a nearest neighbour classifier was used to achieve the final result of text detection.

Using text for scene understanding is a relatively unexplored topic. Zhu et al. [33] exploit information from text regions in natural scenes to improve object and scene classification accuracy. Their system combined visual features extracted from the full image, with features extracted only from the detected text regions. Karaoglu et al. [34] demonstrated an improvement in the accuracy of object class recognition with the help of text detection.

4.2 Deep Networks for Text Detection

One of the initial applications of convolutional neural networks for the task of text detection is the work by Wang et al. [27]. They extracted characters and background patches from popular text detection datasets to train a CNN architecture. The input images are resized to 32×32. This data was augmented by synthetically generated characters. The network architecture consists of two convolutional layers and one fully connected layer with average pooling layers in-between. They initialized the first layer with unsupervised learning, which was then fixed during the training process. The network was trained by back propagating the L2-SVM classification error. They used two similar but separate CNN for text detection and text recognition. The CNN for text detection was used for a two-way classification (text and non-text) while for text recognition a 62 way classifier was developed. The detection CNN was applied on the test images using a sliding window approach. Non maximum suppression was then used to obtain a set of candidate lines, along with the location of spaces between characters and words. Then a character CNN was used to identify the characters. This information was combined using beam search to obtain end-to-end results.

Jaderberg et al. [26] developed an end-to-end text detection system and text based image retrieval system. It involved a region proposal mechanism for the detection and a deep CNN for recognition. The system uses Edge Boxes and an Aggregate Channel Feature Detector for fast and computationally less expensive generation of region proposals, while maximizing the recall at the cost of precision. This avoids the use of a sliding window for application of the CNN, which is a very computationally expensive step. Edge boxes are inspired by the idea that objects have sharp boundaries (edges), so the number of boundaries, which are wholly contained in a box, can represent the objectiveness of that bounding box. The Aggregate Channel Feature Detector uses eight channels: normalized gradient magnitudes, raw greyscale images and 6 channels of Histogram of Oriented Gradients (HOG) features. The channels are smoothed and divided into blocks. Then the pixels in each block are added together and smoothed again. An ensemble of decision trees were used as weak classifiers and trained using adaboost. Since the channels are not scale invariant, the classifier was applied at different scales along with an approximate calculation of the channels at scales between the two computed scales. However, this process generates thousands of false positives. Therefore, a random forest with decision stump, acting on the HOG features, is used to filter out the generated region proposals. The regions with a confidence level below a certain threshold are rejected to reduce the number of bounding boxes. The bounding boxes are regressed using a CNN to improve the overlap with the ground truth boxes, which completes the text detection part. For text recognition, a CNN trained on synthetic data is used. Nine million synthetic data instances are generated from a dictionary of ninety thousand commonly used English words. The data was generated using 1400 different fonts and involved the steps of border and shadow rendering, base colouring, projective distortion and natural data blending. Noise was also introduced to account for distortions and blur in natural scenes. The convolutional neural network for text recognition has five convolutional layers and three fully connected layers. The CNN is trained using the synthetic data for full word classification. The last layer of the neural network performs 90,000-way classification i.e., there is an output neuron for each word in the dictionary of ninety thousand words. This is a major change from earlier text recognition methods, which recognise characters instead of words, and then use post processing to form those characters into words. This work of Jaderberg [26] is a major breakthrough in terms of accuracy. The technique was evaluated on various datasets and provided 10–20% increase in F-score depending on the dataset.

Zhang et al. [35] applied a fully convolutional network for the task of multi oriented text detection. Their system uses both local and global cues in order to locate text lines in a coarse to fine approach. Two Fully Convolutional Networks (FCN) are used in the system. The first FCN called Text-block FCN generates the holistic saliency map for text regions. It uses the first 5 layers of a VGG-16 Network. These layers have different filter sizes, so each layer captures contextual information at virtually different scales. The output of each convolutional stage is also connected to a 1×1 convolutional layer and an upsampling layer. This creates feature maps of the same size, which are then concatenated. This feature concatenation is followed by a 1×1 convolutional layer and a sigmoid layer, which generates pixel-level predictions. The saliency map is then combined with character components to estimate potential text lines. Scale invariant Maximally Stable Extremal Regions (MSER) are extracted in the text regions detected by the first FCN. Area and aspect ratio of the detected MSER regions are used to filter out the false positives. Then, the component projection is used to estimate the orientation of the text lines within a text block. The method assumes that text occurs in straight or nearly straight lines. Character components which fall across straight lines are counted. A separate count is kept for the various possible orientations of straight lines. Then the straight line's orientation with the maximum number of character components is taken as the orientation of the straight line. The character components are merged into groups, and height and orientation of groups are used to combine them into text line candidates. The second FCN removes the false-positives and predicts the centroid of each character. This second FCN has a similar structure to Text-block FCN but instead of five, it has the first three convolutional layers from the VGG-16 Net. It is also trained with a cross-entropy loss function. Finally confidence levels, number of characters in the text line and a geometric criterion are used to threshold the false-positives. The framework is claimed to be suitable for the detection of text in multiple languages, fonts and orientations. It achieved state of the art F-score on MSRA-TD500, ICDAR 2015 Incidental Scene text and the ICDAR 2013 dataset.

5 Depth Map Estimation

The ability to perceive the distance of objects from viewpoint (depth) is an important sense for humans. It allows them to analyse and recognise the position of objects and their surrounding layout. Recovering depth from RGB cameras has many

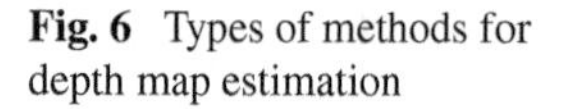

Fig. 6 Types of methods for depth map estimation

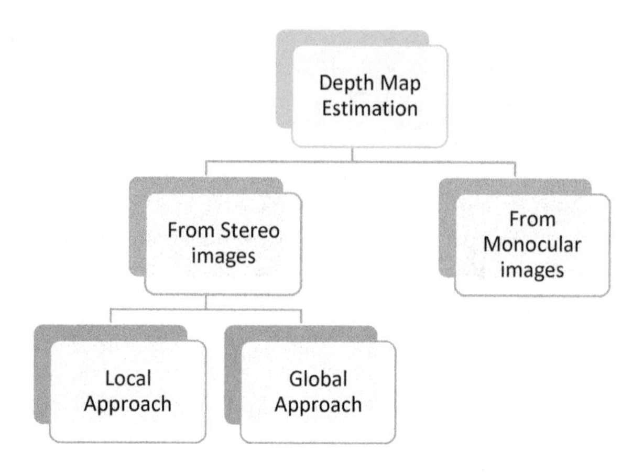

applications, including surveillance, robot navigation, autonomous driving and scene understanding. Depth information can aid in the development of reliable autonomous driving systems and personal assistant devices. It has also applications in gaming, surveillance and robotic navigation. The challenges in depth map estimation include reflections, transparency, occlusions, presence of bland regions, lighting conditions, repetitive textures and indiscriminative background.

There are two main approaches for the recovery of depth from RGB images: (i) depth estimation from stereo (two) or more images (or from videos), and (ii) depth calculation from single (monocular) images (Fig. 6). **(i)** A stereo vision system mostly consists of two horizontally placed cameras. The cameras capture the images, at the same time, which are then processed and compared with each other. Stereo matching is a mathematically ill-defined problem. It is particularly challenging for highly textured or bland scenes. Computer vision and pattern matching techniques are used to determine the disparity and depth map.

(ii) Recovering depth from a single image is an inherently ambiguous task [36]. It requires the use of cues such as object sizes, image position, lighting perspectives, shading, relative sizes of objects, and information about the global view. This is an ill-posed problem since an infinite number of real world scenes can produce the same RGB image. Humans are able to perceive depth even with one eye because humans are very good at dropping out impracticable solutions. But computers need a strategy to do so.

5.1 Methodology of Depth Map Estimation

Depth from stereo depends on the intrinsic and extrinsic parameters of the camera. Computer vision techniques are used to estimate the disparity map from stereo images. Disparity is measured in the number of pixels that an object (or precisely each pixel of an object) is displaced in the left image with respect to its location in the right image. Disparity values can be converted to depth by using the formula:

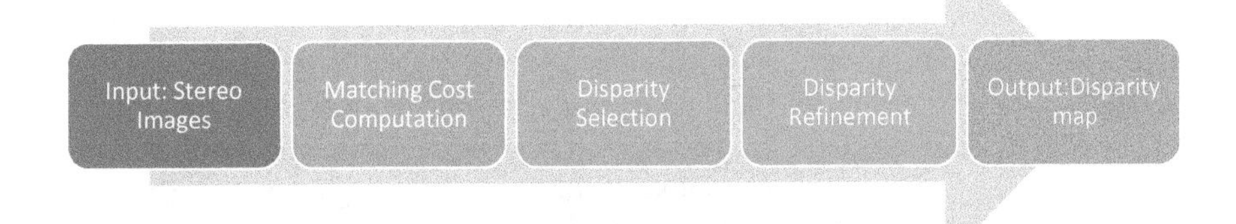

Fig. 7 Pipeline for depth map estimation from stereo images

$$depth = \frac{focal\ length \times baseline}{disparity\ value}$$

The state of the art techniques attempt to calculate disparity at a subpixel level. For stereo matching, epipolar lines are identified to reduce the search space for the matching cost computation. Then pattern matching techniques are used for disparity calculation. There are four major steps in a stereo matching algorithm [37]. These are: Matching cost computation, cost aggregation, disparity computation/optimization and disparity refinement. However there may be modifications in this general pipeline depending on the used algorithm.

Almost all of the algorithms convert RGB images to grey scale. The matching cost is a measure of difference between intensity values at a given disparity such as sum of absolute difference or sum of squared difference. Cost aggregation is done by accumulating matching cost over windows with constant disparity. The calculation of disparities is done by selecting the minimum value of the cost at each pixel. Then the disparity map is refined using several post-processing techniques (Fig. 7).

There are two main types of algorithms: Local and Global (Fig. 6). In a local approach, the disparity calculation at a point or pixel is only based on the intensity values in a predefined local area or local window. They have shorter processing times and complexity. The work of Mattoccia et al. [38] is a typical example of a local approach.

On the other hand, a global method forms a global energy function and attempts to minimize it for all disparity values. This type of methods have two terms in their objective function. One is the usual term which penalizes the output that is inconsistent with the ground truth. The other is a term that smooths the local neighbourhood of the actual pixel to reduce irregularities in the generated depth map. Global methods produce better results than the local methods, but are computationally expensive [39]. They are therefore not suitable for real time systems. These methods usually skip the step of cost aggregation. Most of the global methods use Markov random fields.

5.2 Depth Map Estimation Using Pattern Matching

Before the widespread use of deep learning, depth map estimation techniques used cumbersome feature extraction and pattern matching techniques, which required a lot of parameter tuning. Depth map estimation from stereo images is one of the early ways to recover a depth map. Initially, depth from stereo algorithms did not use ground truth data for training. But the introduction of suitable datasets has opened the possibility of supervised learning. Various methods have been used to calculate the matching cost of the left and right views, such as the sum of squared distances, sum of absolute distances and normalized cross-correlation. Conditional random fields are also popularly used for stereo matching. Li and Huttenlocher [40] use a non-parametric cost function for the conditional random field model and combined it with a support vector machine to learn the parameters of the model. Spyropoulos et al. [41] train a random forest classifier to predict the confidence of the matching cost and used the predictions in a Markov random field to decrease the error of the stereo method.

Depth estimation from a single image is a much more challenging problem. As opposed to stereo correspondence, there are no reliable cues. There are several approaches towards depth estimation from a single image. Saxena et al. [42] developed a system for 3D model generation using super pixels and Markov Random Field (MRF). Their system relied on the assumption that the environment is made up of many small planes. Additional sources of information e.g. repetitive structures, semantic labels or user annotations can help in this task but such information is usually not available. Geometric assumptions can effectively be used e.g. box models are used to estimate a room layout, but these are very simple models which fail with a slight object complexity and are not suitable for detailed 3D reconstructions. Some non-parametric systems [43], search for image patches in a set of known-depth images, which are similar to the input image and combine this information with smoothness constraints to estimate the depth map. This approach has the problem that the smoothness constraints depend on the gradient, which performs poorly for real 3D scene reconstruction. Ladicky et al. [44] use handcrafted features and super-pixels to integrate semantic object labels with monocular depth features to improve performance.

5.3 Deep Learning Networks for Depth Map Estimation

Zbontar and LeCun [45] apply deep learning to estimate depth maps from a rectified image pair. Their work is mainly focussed on the first step of stereo matching algorithms i.e., on the matching cost computation. Ground truth training data is used for training a convolutional neural network by constructing a binary classification dataset. One negative and one positive training example is extracted at the positions where the true disparity is known in order to create a balanced dataset. Then the CNN learns a similarity measure on this dataset of small image patches. This work presents two network architectures, one for fast performance and the other for more precise results. The network architecture consists of two shared-weight sub-networks joined together at the head, called the Siamese network. Each sub-network consists of a number of pairs of a convolutional layer and a layer of rectified linear units. The last convolutional layer of each sub-network is not followed by the rectified linear units layer, and outputs a vector describing the properties of the input image patch. In the fast architecture, the two output vectors are compared using a dot product to produce the final output of the network. However, in the accurate architecture, the two output vectors are concatenated and passed through many fully connected layers. Each fully connected layer is followed by a layer of rectified linear units except the last layer which is followed by a sigmoid layer. The sigmoid layer produces a scalar which is used as the similarity measure between the two patches. At test time, the disparity map is initialized by the output of the convolutional neural network. This initial output is later refined by cross-based cost aggregation, semi-global matching, left-right consistency check, subpixel enhancement, a median filter, and a bilateral filter to achieve the final output. The accurate architecture achieved the least error rate on several benchmark datasets while the fast architecture achieved the least execution time with a reasonable performance.

Information from both global and local cues is required for depth estimation from a single image. Eigen et al. [36] use two deep network stacks. The first network uses the entire image for coarse global prediction. It consists of five convolutional layers and two fully connected layers. The other network is used for the local refinement of the predicted disparity map and consists of three convolutional layers. The raw datasets were used to feed the data hungry deep networks. The method achieved the state-of-the-art results on the NYU Depth and KITTI datasets in 2014, and produced detailed depth boundaries.

Depth map estimation can also aid in the task of surface normal estimation and semantic labelling and the three tasks can complement each other. Eigen and Fergus [46] simultaneously handle three tasks of depth prediction, surface normal estimation, and semantic labelling using a single multiscale convolutional network architecture. In the first step, the deep network uses the complete image to produce a coarse global output. Then local networks refine the initial output. The network involves several convolutional layers and fully connected layers, and works at three different scales. The system produced a better performance on many of the benchmarks for the three tasks. This shows that combining complementary tasks of scene understanding improves the performance.

6 Scene Classification

Scene classification is very different from typical object classification. In scene classification, the images contain numerous objects of different types and in different spatial layout. The variability in size and the different view angles are some of the

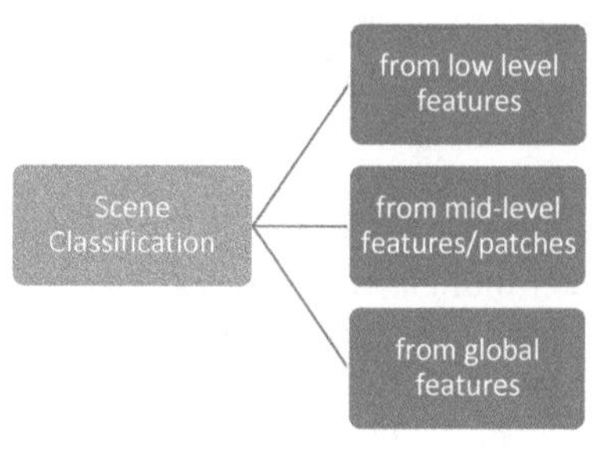

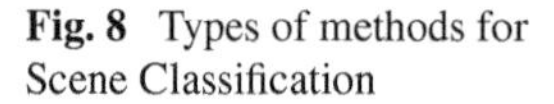
Fig. 8 Types of methods for Scene Classification

other challenges involved in the complex task of scene classification. Scene classification is an active area of research. In recent years, several different approaches have been proposed. Some techniques use local distinctive features, while others work at the global level. However, both types of cues are required for an efficient scene classification. Apart from these, mid-level representations have emerged as potential candidates for scene classification (Fig. 8).

The idea is to discover discriminative mid-level image patches. A feature descriptor then encodes the patches. Since the global layout deformation is a characteristic challenge of scene classification, an important problem is to design layout-invariant feature descriptors. One way is to use locally invariant features such as SIFT followed by bag-of-visual-words models. However, the features learned by deep learning have recently outperformed these local feature representations.

6.1 Scene Classification Using Handcrafted-Features

Parizi et al. [47] use a latent variable model for scene recognition. The idea is to represent the scene in terms of its constituent components e.g., an outdoor scene may contain grass, trees and clouds. These components may be present in any number and at any spatial location in the image, but they will essentially constitute the same type of scene. The image is partitioned into a pre-defined set of regions and a latent variable is used to specify a region model for the image region. The appearance of an image region is described using a bag of words approach. Two training methods are suggested for training: generative and discriminative methods. The Expectation-Maximization (EM) algorithm is used, in the case of generative methods, in order to calculate the model parameters in a supervised setting. A latent structural SVM (LSSVM) is used for the discriminative setting. While the discriminative method produces better results, LSSVMs are not robust against a bad initialization. The generative method can provide the parameter initialization for the LSSVM to overcome this difficulty.

Lin et al. [48] introduced the concept of Important Spatial Pooling Regions (ISPRs). A unified optimization framework is used to learn distinctive region appearances and ISPRs. This method suppresses the false responses in generated feature maps using the statistical information from the training data. Once the false responses are suppressed, simple techniques such as, max pooling can be used to combine feature maps. This mid-level representation is combined with global image features to improve the recognition accuracy.

6.2 Scene Classification Using Deep Features

Previously, it was difficult to use deep learning for scene categorization because sufficient data was not available for training the deep networks. Initial deep learning attempts used the transfer-learning properties of neural networks by using CNNs trained for object recognition for the task of scene classification. However, due to the very different nature of the task, the success of deep learning was not as great as expected [49]. The introduction of Places database [49] has made it possible to train CNNs for the task of scene classification. It contains more than 7 million labelled images of scenes of various categories. This dataset is as dense as other datasets and contains more diversity. The CNN trained on the Scene database achieved 50% Classification accuracy and the deep features extracted from the trained CNN produced state of the art results (using an SVM as classifier) on a number of scene classification datasets.

Hayat et al. [50] use deep learning to extract the spatial layout and scale invariant features. Instead of local or global features, they use an intermediate level of information by extracting mid-level patches from the image. Then scale invariance is achieved by using a pyramidal image representation. This provides multi-level distinctive features for indoor scenes. A dense and uniform patch extraction ensures that most of the information is utilized in making the final decision. To overcome the challenge of spatial layout deformation, the convolutional neural network involves a new "spatially unstructured layer". The CNN consists of five convolutional layers and four fully connected layers. The pyramidal image representation is created by extracting mid-level patches at three different scales. Two CNNs are used for feature extraction. One CNN contains a spatially unstructured layer while the other CNN does not contain that layer. The output vectors of the two CNNs are concatenated to form the final feature vector. The deep features are used to train an SVM which acts as the final classifier. The SVM using the deep learning features achieved state of the art performance on a number of indoor scene classification datasets.

Deep convolutional features have a native ability to retain the global spatial structure. However due to very high variations in the spatial layout of objects, the structure preserving property of deep networks becomes a hindrance in the effective training of the network [51]. A way to overcome this problem is to transform the convolutional features to some other feature space, which is more descriptive for the task of scene classification. The transformed feature space should encode the features as general object categories present in scenes. It should also represent the distinctive aspects of the data. Khan et al. [51] use mid-level convolutional features along with 'Deep Un-structured Convolutional Activations (DUCA)' to overcome the challenge of variability in the spatial layout. First, the dataset is augmented with flipped, cropped and rotated versions of the original training images at three different spatial resolutions, then a sliding window is used to extract dense mid-level patches. A CNN with five convolutional layers and three fully connected layers is used to extract deep features from the images. The resulting feature vectors are highly structured due to the systematic operations in a CNN. To overcome this challenge, feature vectors are represented as multiple code books of the Scene representation patches (SRPs). This increases the effectiveness of deep features. Both supervised and unsupervised learning are used for the formation of code books. The unsupervised SRPs provide information about the distinctive aspects of various scenes while the supervised SRPs are good for providing semantic information. Finally, a one-versus-one SVM is used for classification. This method achieves the best classification accuracy on a number of indoor scenes datasets.

From a holistic point of view, scene classification requires information about both objects and scenes. The category of the scenes (especially indoor scenes), is mainly determined by the objects that are present in those scenes. Therefore, one important challenge is to devise a way to combine the knowledge about objects and the knowledge about scenes in order to improve the decision making process. Since CNNs do not have any inherent ability to cater for significant variations in scale, removing the scale bias is another important step. Experiments show that by combining deep features from Places-CNN and ImageNet-CNN, the overall accuracy of the scene classification increases [52]. However, this boost in accuracy is only achieved when features are extracted at multiple scales and systematically combined while keeping in view the different original scales of the two CNNs which are trained on the Places Dataset and ImageNet dataset [52].

7 Caption Generation

A recently emerging application is to use deep networks for the task of captioning images. It is also known as automatic caption generation. This challenging problem is a big step towards scene understanding, as it is not a typical classification task, and it merges the fields of computer vision and natural language processing. This is closer to the human understanding of surroundings. It requires the analysis of the visual and semantic contents of an image to generate a textual description about the most salient information in the image. A good description should be accurate, concise, comprehensive, and also grammatically correct. The techniques for image caption generation can be classified into three main types [53]: (i) from query image contents, (ii) by retrieving information from images similar to query image and (iii) from videos (Fig. 9).

The methods which generate captions directly from the query image first detect the contents of the image, whose description is generated. This may involve object detection, scene classification, attribute generation, action recognition and semantic segmentation. This information is used in natural language processing systems to

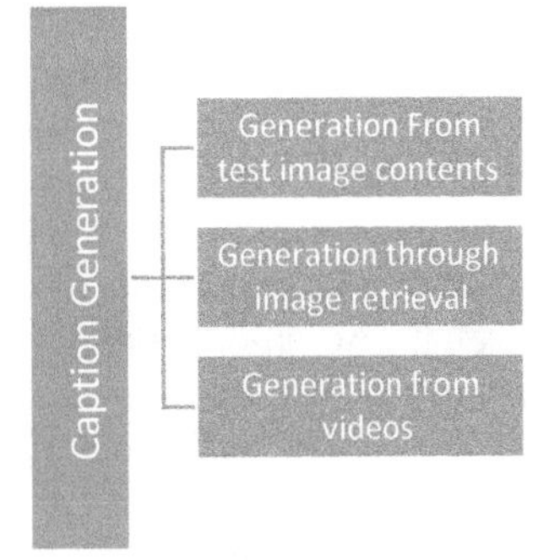

Fig. 9 Types of methods for caption generation

create a description of the image. This approach relies on the accuracy of the used detectors. However, the detection results are not always correct.

The second type of methods are based on similar image retrieval. These methods search for images which are similar to the query image in a large training database to retrieve those images and their respective descriptions (captions). They then create a caption for the query image based on the retrieved description of similar images. This approach requires much more training data to generate good descriptions, compared to the direct caption generation from images.

Caption generation from videos is a much more difficult task as it additionally requires the analysis of the temporal information, and the recognition of the actions, events, and interactions in the videos.

7.1 Deep Networks for Caption Generation

The main idea of a system for caption generation based on deep learning is to use Convolutional Neural Networks to extract visual features from an image, and then use a Recurrent Neural Network (RNN) to 'translate' those features from the visual domain to the textual language. The Long Short Term Memory (LSTM) model is the most commonly used RNN for caption generation tasks. A generative model developed by Vinyals et al. [54] is a representative work of this approach. A deep recurrent architecture is trained to generate the target description sentence from the training image. A deep convolutional neural network is trained for the task of image classification. Then the final classification layer is removed, and RNN is connected to the output to generate a textual description. The resulting network is trained in an end to end manner using stochastic gradient descent and back propagation. The model achieved state of the art performance on many caption generation datasets in 2015.

Word generation using LSTM mainly depends on the current state and last state of the LSTM. Despite the long term memory in the LSTM, if image information is fed at the beginning of sentence generation, its effect gets weaker with the length of the sentence. Global semantic information has proved to be useful to mitigate this effect. Jia et al. [55] use an extension of LSTM called gLSTM. This extension involves the addition of an extra input of semantic features to each block of the LSTM model. In this way, the LSTM produces sentences which are more relevant to the test image. Length normalization is applied to compensate for bias towards short sentences.

The inverse problem of caption generation i.e. visual feature generation from textual descriptions is also an important problem for scene understanding. The idea behind this problem is that the computers should be able to draw or visualize a scene given its textual description. Chen and Zitnick [56] explore the bi-directional transformation between the textual description and visual features of images. A Recurrent Neural Network (RNN) is used to create a visual representation of images while reading or generating captions. The modified RNN with latent variables is capable of retaining long-term visual concepts. Static visual features are fed to the RNN along with a language model for the task of caption generation. For the task of image retrieval, the modified RNN contains a special layer, which generates visual features from the input descriptions. The system can be used for caption generation and to retrieve images based on a textual description.

Most of the caption generation works involving deep learning rely mainly on the availability of databases for caption generation, and cannot describe objects which are not present in the caption generation databases. This limits the type and nature of caption generation to the style and information that is present in the database. Additional information from other sources, e.g., web and Wikipedia, can be effectively used to improve the performance of caption generation systems [57]. Hendricks et al. harvest information from large object recognition datasets and large sources of textual information for a deep learning based system called Deep Compositional Captioner (DCC). It consists of three main components, a lexical classifier, a language model and a caption model. In the lexical classifier a CNN is used to find the relationship and structure of objects in images. Most common adjectives, verbs and nouns are extracted from the caption generation databases. A pre trained CNN, on a large object recognition dataset, is fine-tuned and used as a lexical classifier. Data from the object recognition datasets (in addition to the caption generation databases) is also used for fine tuning. The output features of the CNN correspond to the probability that a particular concept or object is present in the scene. The language model is trained to predict the next word in a sentence, given the previous words of the sentence. Text data from various datasets and sources, such as Wikipedia and the British national Corpus, is used to train the model. It involves a one-hot-vector embedding layer, LSTM and a word prediction layer. The caption model combines the features of the lexical classifier and the language model using a linear affine layer. The simple multimodal combination layer facilitates in comprehending the relationships between the visual and language features. In this way the system becomes capable of generating descriptions for even those objects which are not present in the training set of the caption generation datasets. The system is also used to generate descriptions of video clips.

8 Visual Question Answering (VQA)

A further extension of caption generation for images is the task of Visual Question Answering. VQA also involves the integration of computer vision and natural language processing. In VQA, the system is required to answer any arbitrary question about an image. The answer should directly be inferable from the contents of the image, although more advanced systems are also experimenting with questions whose answers require external information. The scope of the question includes, but is not limited to, the contents of the image such as, the number and types of objects, attributes (e.g., colour) identification, scene classification, and the spatial relationships or interactions between objects. The type of questions include both open ended and multiple choice questions. There has been a recent surge in the efforts of developing various systems for VQA, especially due to the reason that deep learning has improved the performance of various individual tasks, which can now be used to form an integrated system. Generally, a VQA system consists of three main components: extracting features from images, a method of understanding the question, and a methodology to analyse the features for the correct answer [58]. Most of the methods use a CNN, pre-trained on large scale object recognition datasets, for feature extraction from images. There are various approaches for the analysis of the question which include bag-of-words (BOW), recurrent neural networks and long short term memory (LSTM) models. For the answer generation, most of the systems treat VQA as a classification task. The features extracted from the image and the question are used as input features for a classifier, which is trained to output one of the pre-defined answers. However, this approach limits the answers to only those fixed during training. An alternative is to use an RNN to generate multiple word answers.

8.1 Deep Learning Methods for VQA

In order to produce good answers for visual questions, the system should have the ability to focus on the image area which relates to the question asked, rather than the whole image. Therefore VQA requires a deeper understanding of the image, compared to caption generation. Antol et al. [59] developed a large scale VQA database and provided various baseline results using the multilayer perceptron and LSTM models.

Gao et al. [60] developed a system for visual question answering for both the English and Chinese languages. The length of the answers of their system varies from a single word to a complete sentence. It consists of four major components: a CNN, two LSTM models and a fusing component. The CNN is used to extract the visual features from the image. An LSTM model is used to extract the features of the questions. The other LSTM model, which shares the weight matrix with the first LSTM, is used for the answer generation. Finally, the information from the first three components is combined in the fusion component to generate the answer. One-hot encoded vectors [61] are used to represent the words in the questions and answers. A Beginning of Answer <BOA> sign and an End of Answer <EOA> sign are added as two words in the coded dictionary. These signs are added to each answer during training. During testing, the system receives an input image and a question and the <BOA> is passed to the LSTM for the answer generation. Then the model calculates the probability distribution of the next word. This process is repeated until the system outputs an <EOA> sign. Human evaluators ranked 64% of the answers of the system to be at an equal level to answers produced by any human.

The systems which use CNN and LSTM are limited by the nature of the answers produced during testing, which are limited to the words that are available in the training data. Most of these systems have very simple mechanisms to combine the features of images and questions, and do not use high level semantic concepts. Wu et al. [62] incorporate external knowledge to improve the performance of caption generation and VQA systems. Their system also fuses an attribute based representation of high-level concepts into the traditional CNN and RNN based approach. A dictionary of semantic attributes is constructed by using the most common words in the captions of the training data. The learned attributes can be any parts of speech (noun, verb, adjectives, etc.). Then to predict attributes from images, a CNN (pre-trained on large object recognition datasets) is fine-tuned on a multi-label dataset. At test time, the image and a number of sub regions of the test image are fed to the CNN, and the results are aggregated to produce a multi label prediction. This multi label prediction and a LSTM are used for the caption generation. For VQA, the top five attributes that are predicted by the CNN are also used to extract information from external knowledge databases such as DBpedia. The features from the multi-label CNN, the external knowledge, and the generated captions are fed to an LSTM model. The LSTM model also receives the question as an input and generates an answer of the question by using all these inputs. This approach has resulted in considerable improvements in the state of the art performances for caption generation and VQA.

9 Integration of Scene Understanding Components

As discussed in the previous sections, there has been a lot of work on the different individual components of scene understanding. The success of CNNs has resulted in networks which can achieve a very high performance on specialized tasks. Despite these advancements, there have been very few attempts to use deep learning for holistic scene understanding. The different components can be integrated into a unified framework to increase the overall performance.

Depth estimation helps in object recognition, face recognition, scene classification and scene segmentation. Silberman et al. [63] extracted information about major surfaces, objects, and support relations from RGB-D images. They used the depth information in addition to RGB to parse indoor scenes into walls, floor, object regions, and recovered support relationships. Similarly, Hayat et al. [10] used depth information to improve the results for face recognition and object recognition.

Scene text detection can improve the accuracy of object recognition. Karaoglu et al. [34] used text recognition to help in the task of object class recognition. They first performed scene text recognition followed by saliency based object recognition and finally object recognition with the help of the recognized text. There was a clear improvement in the performance of object recognition when text information was used, compared to when only saliency-based features were used.

Networks trained for object recognition also aid in the tasks of text detection, scene classification, caption generation and Visual Question Answering (VQA). In fact it is a common practice to use large networks, trained for object recognition on ImageNet Dataset, as feature extractors for various other tasks of scene understanding [62, 64]. Similarly, scene classification is required for caption generation and VQA. Even caption generation can help to improve the answers of algorithms for VQA [62].

9.1 Non-deep Learning Works for Holistic Scene Understanding

Some of the recent works which combine different components of scene understanding include the following:

- Heitz et al. [65] developed a Cascaded Classification Models (CCM) framework, in which they combined the tasks of 3D reconstruction, scene categorization, multiclass image segmentation and object detection. The cascaded framework learned a set of related models which, in addition to performing their tasks, help each other to improve the output.
- Li et al. [66] extended the work of [65] to Feedback Enabled Cascaded Classification Models (FE-CCM) by maximizing the joint likelihood of the sub-tasks. They introduced a feedback step so that the earlier classifiers could receive feedback from the later classifiers on the types of error modes to focus on. This feedback step improved the performance of the tasks of depth estimation, scene categorization, event categorization, saliency detection, geometric labelling and object detection.
- Yao et al. [67] devised a system for holistic scene understanding. The system provides information about regions, location, the class and spatial extent of objects, the presence of a class in the image, as well as the scene type in an integrated fashion. Segment level learning is used along with auxiliary variables in order to decompose a high order potential into pairwise potentials. The maximum number of states is equal to the number of classes. A convergent message-passing algorithm [67] is used to accomplish the tasks of object detection, scene classification and semantic segmentation. Prior knowledge can be incorporated in the algorithm as it

has neither submodularity restrictions nor requires potential specific moves. This holistic model improved the performance on all the tasks of object detection, scene classification and semantic segmentation.

9.2 Deep Learning Based Works for Holistic Scene Understanding

Compared to non-deep learning techniques, the efforts to integrate scene understanding components using deep learning are rare and modest. Some of these works include:

- Eigen et al. [36] employ two deep network stacks for depth estimation, one for a coarse global prediction based on the entire image and the other to refine the prediction locally. Eigen and Fergus [46] extended [36] to simultaneously handle the three tasks of depth prediction, surface normal estimation, and semantic labelling using a single multiscale convolutional network architecture.
- Machines and robots that interact with the physical environment are not only required to detect and recognize objects in scenes, but they also need to have an understanding of how to use and work with different objects. Ye et al. [64] use a two stage pipeline based on deep learning to localize and recognise the functional areas of various objects in an indoor scene. An attention based selective search algorithm is used to detect the salient regions in an image, which may contain functional regions. Then a convolutional neural network, pre-trained on a large object recognition dataset, is modified by removing the final layer and adding a new classification layer to decide what functions can be performed on the detected regions. The network is then fine-tuned in a supervised manner to produce the final system.
- Khan et al. [68] use multiple convolutional networks to automatically learn features and the dominant boundaries at super-pixel level for shadow detection and removal in natural scenes. Then a conditional random field (CRF) model is used to generate masks for shadows which is followed by a Bayesian formulation to remove shadows.
- Asif et al. [69] propose a system which uses depth information and RGB channels to simultaneously achieve the tasks of object recognition and dense scene reconstruction from videos. First, object proposals are identified which remain spatio-temporally consistent across multiple frames of the video. CNNs are used for global feature extraction, while a Bag of Words (BOW) approach is used to extract mid-level features. These are used for dense scene reconstruction. At the same time, class probabilities of objects are efficiently determined and this information is integrated into a voxel-based prediction hypothesis.
- Object recognition and grasp detection are important for visual perception in robots which interact with their surroundings. Asif et al. [70] propose a depth-based framework of CNN and cascaded forests to integrate the tasks of robotic grasp and object recognition. Pre-trained CNNs are used to extract features from RGBD

object data. The probabilities of each class of object are calculated at different levels of the image hierarchy. A cascaded framework of hierarchical forests is then used to decide on the class of object and grasp it according to its shape.

Most of these works, which use deep learning for integration, are based on the combination of depth information and semantic segmentation. The use of deep learning to combine the tasks of text detection, object recognition, scene classification and caption generation remains an open research field. The need is to integrate these components into a combined framework to aid in the development of a low cost and robust scene understanding system.

10 Conclusion

Deep learning has made its mark in all the components of scene understanding. This chapter has presented a concise survey of deep learning-based techniques for the various components of scene understanding. The evolution of deep learning frameworks for the high level tasks of scene understanding, such as, textual description generation and VQA is also discussed in this chapter. These techniques constitute the state of the art in nearly all the sub-fields of computer vision. Some fields such as Optical Character Recognition in documents is now considered a solved problem, because of the deep networks which can achieve human like accuracy. A commercial application of such systems is the automatic reading of cheques in banks [71] and the automatic reading of postal codes in post offices [72]. For many other fields, deep learning has vastly improved the performance of various tasks, such as object detection and recognition, and face detection and recognition. These advancements have also contributed to the field of autonomous driving, where full autonomy has not been achieved, yet, several companies have already introduced human supervised driving vehicles. Despite of all these advances, there is still much room for improvement. As discussed in the previous sections, most of the state of the art of the deep learning works focus on a single sub-task, rather than on the complete task of scene understanding. Most of the existing works on the various components of scene understanding (e.g., segmentation and object recognition) ignore the physical interactions between objects. A combined neural network architecture which can integrate the various components will greatly help in creating a human-like vision system. The performance of deep learning techniques on high level scene understanding tasks, such as caption generation, visual question answering and even on complex scenarios of scene classification is still far below the human performance.

With more focus on the development of better deep learning systems for scene understanding, many of the current challenges and problems can be solved and many new technologies will become available in the near future e.g., self-driving cars can achieve improved performance by integrating techniques for long range depth map estimation. Such systems also need to read signs and messages on roads to reach human like performance. The "Seeing AI" Microsoft sunglasses (under development) for the blind and the visually impaired will be one of the marvels of advancements in scene understanding. These glasses will be able to detect faces and facial expressions, recognize gender and estimate age, and describe the surrounding environment. It will also read text, answer various types of questions and describe actions being performed by people in the scene. Scene understanding systems will also aid in the field of robotics to develop more 'humanoid' robots. Hence, a combined framework for scene understanding using the state of the art deep networks will aid in the development of low cost and robust scene understanding technologies and will revolutionize many aspects of our daily life.

Acknowledgements This work is partially supported by SIRF Scholarship from the University of Western Australia (UWA) and Australian Research Council (ARC) Grant DP150100294.

References

1. A.E. Johnson, M. Hebert, Using spin images for efficient object recognition in cluttered 3D scenes. IEEE Trans. Pattern Anal. Mach. Intell. **21**(5), 433–449 (1999)
2. D.G. Lowe, Object recognition from local scale-invariant features, in *The proceedings of the IEEE International Conference on Computer Vision* (1999)
3. L. Wang, D.-C. He, Texture classification using texture spectrum. Pattern Recognit. **23**(8), 905–910 (1990)
4. F. Tombari, S. Salti, L. Di Stefano, Unique signatures of histograms for local surface description, in *European Conference on Computer Vision* (Berlin, Heidelberg, 2010)
5. S.A.A. Shah, M. Bennamoun, F. Boussaid, Performance evaluation of 3D local surface descriptors for low and high resolution range image registration, in *International Conference on Digital Image Computing: Techniques and Applications* (2014)
6. Y. Guo, F.A. Sohel, M. Bennamoun, J. Wan, M. Lu, RoPS: a local feature descriptor for 3D rigid objects based on rotational projection statistics, in *International Conference on Communications, Signal Processing, and Their Applications* (2013)
7. L. Bo, X. Ren, D. Fox, Depth kernel descriptors for object recognition, in *IEEE/RSJ International Conference on Intelligent Robots and Systems* (2011)
8. Y. Guo, M. Bennamoun, F. Sohel, M. Lu, J. Wan, 3D object recognition in cluttered scenes with local surface features: a survey. IEEE Trans. Pattern Anal. Mach. Intell. **36**(11), 2270–2287 (2014)
9. L. Deng, A tutorial survey of architectures, algorithms, and applications for deep learning. APSIPA Trans. Signal Inf. Process. **3**, e2 (2014)
10. M. Hayat, M. Bennamoun, S. An, Deep reconstruction models for image set classification. IEEE Trans. Pattern Anal. Mach. Intell. **37**(4), 713–727 (2015)
11. S.A.A. Shah, M. Bennamoun, F. Boussaid, Iterative deep learning for image set based face and object recognition. Neurocomputing **174**, 866–874 (2016)
12. A. Krizhevsky, I. Sutskever, G.E. Hinton, ImageNet classification with deep convolutional neural networks, in *Advances in Neural Information Processing Systems* (2012)
13. C. Szegedy, W. Liu, Y. Jia, P. Sermanet, S. Reed, D. Anguelov et al., Going deeper with convolutions, in *Proceedings of the IEEE Conference on Computer Vision and Pattern Recognition* (2015)
14. K. He, X. Zhang, S. Ren, J. Sun, Deep residual learning for image recognition, in *Proceedings of the IEEE Conference on Computer Vision and Pattern Recognition* (2016)
15. C.R. Qi, H. Su, K. Mo, L.J. Guibas, PointNet: deep learning on point sets for 3D classification and segmentation. arXiv:1612.00593 (2016)

16. P. Viola, M.J. Jones, Robust real-time face detection. Int. J. Comput. Vis. **57**(2), 137–154 (2004)
17. F.R.M. Al-Osaimi, M. Bennamoun, 3D face surface analysis and recognition based on facial surface features, in *3D Face Modeling, Analysis and Recognition* (Wiley, 2013), pp. 39–76
18. Y. Taigman, M. Yang, M. Ranzato, L. Wolf, Deepface: closing the gap to human-level performance in face verification, in *Proceedings of the IEEE Conference on Computer Vision and Pattern Recognition* (2014)
19. E.G. Ortiz, A. Wright, M. Shah, Face recognition in movie trailers via mean sequence sparse representation-based classification, in *Proceedings of the IEEE Conference on Computer Vision and Pattern Recognition* (2013)
20. S.A.A. Shah, U. Nadeem, M. Bennamoun, F. Sohel, R. Togneri, Efficient image set classification using linear regression based image reconstruction, in *Proceedings of the IEEE Conference on Computer Vision and Pattern Recognition Workshops* (2017)
21. H. Li, Z. Lin, X. Shen, J. Brandt, G. Hua, A convolutional neural network cascade for face detection, in *Proceedings of the IEEE Conference on Computer Vision and Pattern Recognition* (2015)
22. J. Lu, G. Wang, W. Deng, P. Moulin, J. Zhou, Multi-manifold deep metric learning for image set classification, in *Proceedings of the IEEE Conference on Computer Vision and Pattern Recognition* (2015)
23. F. Schroff, D. Kalenichenko, J. Philbin, FaceNet: a unified embedding for face recognition and clustering, in *Proceedings of the IEEE Conference on Computer Vision and Pattern Recognition* (2015)
24. D. Karatzas, L. Gomez-Bigorda, A. Nicolaou, S. Ghosh, A. Bagdanov, M. Iwamura et al., ICDAR 2015 competition on robust reading, in *13th International Conference on Document Analysis and Recognition* (2015)
25. Q. Ye, D. Doermann, Text detection and recognition in imagery: a survey. IEEE Trans. Pattern Anal. Mach. Intell. **37**(7), 1480–1500 (2015)
26. M. Jaderberg, K. Simonyan, A. Vedaldi, A. Zisserman, Reading text in the wild with convolutional neural networks. Int. J. Comput. Vis. **116**(1), 1–20 (2016)
27. T. Wang, D.J. Wu, A. Coates, A.Y. Ng, End-to-end text recognition with convolutional neural networks, in *21st International Conference on Pattern Recognition* (2012)
28. B. Epshtein, E. Ofek, Y. Wexler, Detecting text in natural scenes with stroke width transform, in *Proceedings of the IEEE Conference on Computer Vision and Pattern Recognition* (2010)
29. B. Epshtein, E. Ofek, Y. Wexler, Detecting text in natural scenes with stroke width transform, in *Proceedings of the IEEE Conference IEEE Conference on Computer Vision and Pattern Recognition* (2010)
30. H. Chen, S.S. Tsai, G. Schroth, D.M. Chen, R. Grzeszczuk, B. Girod, Robust text detection in natural images with edge-enhanced maximally stable extremal regions, in *18th IEEE International Conference on Image Processing* (2011)
31. L. Neumann, J. Matas, Real-time scene text localization and recognition, in *Proceedings of the IEEE Conference on Computer Vision and Pattern Recognition* (2012)
32. L. Neumann, J. Matas, Scene text localization and recognition with oriented stroke detection, in *Proceedings of the IEEE International Conference on Computer Vision* (2013)
33. Q. Zhu, M.-C. Yeh, K.-T. Cheng, Multimodal fusion using learned text concepts for image categorization, in *Proceedings of the 14th ACM International Conference on Multimedia* (2006)
34. S. Karaoglu, J.C. Van Gemert, T. Gevers, Object reading: text recognition for object recognition, in *European Conference on Computer Vision (ECCV)* (2012)
35. Z. Zhang, C. Zhang, W. Shen, C. Yao, W. Liu, X. Bai, Multi-oriented text detection with fully convolutional networks, in *Proceedings of the IEEE Conference on Computer Vision and Pattern Recognition* (2016)
36. D. Eigen, C. Puhrsch, R. Fergus, Depth map prediction from a single image using a multi-scale deep network, in *Advances in Neural Information Processing Systems* (2014)
37. D. Scharstein, R. Szeliski, A taxonomy and evaluation of dense two-frame stereo correspondence algorithms. Int. J. Comput. Vis. **47**(1–3), 7–42 (2002)

38. S. Mattoccia, S. Giardino, A. Gambini, Accurate and efficient cost aggregation strategy for stereo correspondence based on approximated joint bilateral filtering, in *Asian Conference on Computer Vision* (2010)
39. R.A. Hamzah, H. Ibrahim, Literature survey on stereo vision disparity map algorithms. J. Sens. (2015)
40. Y. Li, D.P. Huttenlocher, Learning for stereo vision using the structured support vector machine, in *Proceedings of the IEEE Conference on Computer Vision and Pattern Recognition* (2008)
41. A. Spyropoulos, N. Komodakis, P. Mordohai, Learning to detect ground control points for improving the accuracy of stereo matching, in *Proceedings of the IEEE Conference on Computer Vision and Pattern Recognition* (2014)
42. A. Saxena, M. Sun, A.Y. Ng, Make3D: learning 3D scene structure from a single still image. IEEE Trans. Pattern Anal. Mach. Intell. **31**(5), 824–840 (2009)
43. K. Karsch, C. Liu, S.B. Kang, Depth extraction from video using non-parametric sampling, in *European Conference on Computer Vision* (2012)
44. L. Ladicky, J. Shi, M. Pollefeys, Pulling things out of perspective, in *Proceedings of the IEEE Conference on Computer Vision and Pattern Recognition* (2014)
45. J. Zbontar, Y. LeCun, Stereo matching by training a convolutional neural network to compare image patches. J. Mach. Learn. Res. **17**(2), 1–32 (2016)
46. D. Eigen, R. Fergus, Predicting depth, surface normals and semantic labels with a common multi-scale convolutional architecture, in *Proceedings of the IEEE International Conference on Computer Vision* (2015)
47. S.N. Parizi, J.G. Oberlin, P.F. Felzenszwalb, Reconfigurable models for scene recognition, in *Proceedings of the IEEE Conference on Computer Vision and Pattern Recognition* (2012)
48. D. Lin, C. Lu, R. Liao, J. Jia, Learning important spatial pooling regions for scene classification, in *Proceedings of the IEEE Conference on Computer Vision and Pattern Recognition* (2014)
49. B. Zhou, A. Lapedriza, J. Xiao, A. Torralba, A. Oliva, Learning deep features for scene recognition using places database, in *Advances in Neural Information Processing Systems* (2014)
50. M. Hayat, S.H. Khan, M. Bennamoun, S. An, A spatial layout and scale invariant feature representation for indoor scene classification. IEEE Trans. Image Process. **25**(10), 4829–4841 (2016)
51. S.H. Khan, M. Hayat, M. Bennamoun, R. Togneri, F.A. Sohel, A discriminative representation of convolutional features for indoor scene recognition. IEEE Trans. Image Process. **25**(7), 3372–3383 (2016)
52. L. Herranz, S. Jiang, X. Li, Scene recognition with CNNs: objects, scales and dataset bias, in *Proceedings of the IEEE Conference on Computer Vision and Pattern Recognition* (2016)
53. R. Bernardi, R. Cakici, D. Elliott, A. Erdem, E.I.-C.N. Erdem, F. Keller, A. Muscat, B. Plank, Automatic description generation from images: a survey of models, datasets, and evaluation measures. J. Artif. Intell. Res. **55**, 409–442 (2016)
54. O. Vinyals, A. Toshev, S. Bengio, D. Erhan, Show and tell: a neural image caption generator, in *Proceedings of the IEEE Conference on Computer Vision and Pattern Recognition* (2015)
55. X. Jia, E. Gavves, B. Fernando, T. Tuytelaars, Guiding the long-short term memory model for image caption generation, in *Proceedings of the IEEE International Conference on Computer Vision* (2015)
56. X. Chen, C. Lawrence Zitnick, Mind's eye: a recurrent visual representation for image caption generation, in *Proceedings of the IEEE Conference on Computer Vision and Pattern Recognition* (2015)
57. L. Anne Hendricks, S. Venugopalan, M. Rohrbach, R. Mooney, K. Saenko, T. Darrell, Deep compositional captioning: describing novel object categories without paired training data, in *Proceedings of the IEEE Conference on Computer Vision and Pattern Recognition* (2016)
58. K. Kafle, C. Kanan, Visual question answering: datasets, algorithms, and future challenges. arXiv:1610.01465 (2016)
59. S. Antol, A. Agrawal, J. Lu, M. Mitchell, D. Batra, C. Lawrence Zitnick, D. Parikh, VQA: visual question answering, in *Proceedings of the IEEE International Conference on Computer Vision* (2015)

60. H. Gao, J. Mao, J. Zhou, Z. Huang, L. Wang, W. Xu, Are you talking to a machine? Dataset and methods for multilingual image question, in *Advances in Neural Information Processing Systems* (2015)
61. D. Harris, S. Harris, *Digital Design and Computer Architecture* (Morgan Kaufmann, 2010), p. 129
62. Q. Wu, C. Shen, P. Wang, A. Dick, A. van den Hengel, Image captioning and visual question answering based on attributes and external knowledge, in *IEEE Transactions on Pattern Analysis and Machine Intelligence* (2017)
63. N. Silberman, D. Hoiem, P. Kohli, R. Fergus, Indoor segmentation and support inference from RGBD images, in *European Conference on Computer Vision (ECCV)* (2012)
64. C. Ye, Y. Yang, C. Fermuller, Y. Aloimonos, What can I do around here? Deep functional scene understanding for cognitive robots. arXiv:1602.00032 (2016)
65. G. Heitz, S. Gould, A. Saxena, D. Koller, Cascaded classification models: combining models for holistic scene understanding, in *Advances in Neural Information Processing Systems* (2009)
66. C. Li, A. Kowdle, A. Saxena, T. Chen, Towards holistic scene understanding: feedback enabled cascaded classification models, in *Advances in Neural Information Processing Systems* (2010)
67. J. Yao, S. Fidler, R. Urtasun, Describing the scene as a whole: Joint object detection, scene classification and semantic segmentation, in *Proceedings of the IEEE Conference on Computer Vision and Pattern Recognition* (2012)
68. S. H. Khan, B Mohammed, F. Sohel, R. Togneri, Automatic shadow detection and removal from a single image. IEEE Trans. Pattern Anal. Mach. Intell. **38**(3), 431–446 (2016)
69. U. Asif, M. Bennamoun, F. Sohel, Simultaneous dense scene reconstruction and object labeling, in *IEEE International Conference on Robotics and Automation (ICRA)* (2016)
70. U. Asif, M. Bennamoun, F.A. Sohel, RGB-D object recognition and grasp detection using hierarchical cascaded forests. IEEE Trans. Robot. (2017)
71. R. Jayadevan, S.R. Kolhe, P.M. Patil, U. Pal, Automatic processing of handwritten bank cheque images: a survey. Int. J. Doc. Anal. Recognit. (IJDAR) **15**(4), 267–296 (2012)
72. G. Dreyfus, *Neural Networks: Methodology and Applications* (Springer Science & Business Media, 2005)

An Application of Deep Learning in Character Recognition: An Overview

Sana Saeed, Saeeda Naz and Muhammad Imran Razzak

Abstract For automated document analysis, OCR (Optical character recognition) is a basic building block. The robust automated document analysis system can have impact over a wider sphere of life. Many of the researchers have been working hard to build OCR systems in various languages with significant degree of accuracy, character recognition rate and minimum error rate. Deep learning is the start of art technique with efficient and accurate result as compared to other techniques. Every language, moreover every script have its own challenges e.g. scripts where characters are well separated are less challenging as compared to cursive scripts where characters are attached with one another. In this chapter, we would take a detailed account of the state of art deep learning techniques for Arabic like script, Latin script and symbolic script.

1 Introduction

Technology has introduced new dimensions for document processing that gave rise to era of automated and paperless office. As automated systems are time saving, require less human efforts, reduce risk of human errors and increase the financial savings for an organization. Processing the textual documents is relatively easier task whereas digital photographic images are complex to deal with. By using the techniques of DIA (Document Image Analysis), paper documents are represented in digital processable format. Paper documents are scanned and stored in the form of digital document images. Document Images are later, converted into their textual

S. Saeed · S. Naz (✉)
Govt. Girls Postgraduate College No. 1, Abbottabad, Pakistan
e-mail: saeedanaz292@gmail.com

S. Saeed
e-mail: saranahid7@gmail.com

M. I. Razzak
University of Technology, Sydney, Australia
e-mail: imranrazak@hotmail.com

V. E. Balas et al. (eds.), *Handbook of Deep Learning Applications*,
Smart Innovation, Systems and Technologies 136,
https://doi.org/10.1007/978-3-030-11479-4_3

equivalent before applying any operation over the content of the document. With the growing trend of paperless offices, there is an urgent need for a robust DIA System. DIA can be divided into two categories; first one is *graphic processing system or layout processing* which aims at recognizing the graphical components including symbols, lines, graphs, diagrams, section of text and all the non-textual material in the document. Other one is *Textual processing*, which includes the locating the textual content in document and converting them into textual format, this phase is also referred as OCR (Optical character Recognition). Our focus of study is "textual processing". DIA applications are also referred as, intelligent character recognition, Dynamic character recognition, On-line character recognition, Real-time character recognition. Gustav Tauschek initially studied OCR in early 1930s. OCR is sub field of Machine learning, intelligent systems, signal possessing and pattern recognition systems [1]. OCR is the automated extraction and conversion of the text form into the digital image. The image may contain printed text, handwritten text, scene text (text painted in the pictures or signboards etc.). Digital camera or scanner usually acquires the digital images after certain processing it is converted into required format i.e. textual form. Input to the OCR application is an image containing text while output would be plain text. Other names of OCR are intelligent character recognition, dynamic character recognition, on-line character recognition or real-time character recognition.

Broadly OCR is classified into two categories off-line character recognition and on-line character recognition [2]. First one deal with the statically stored digital images generally scanned or captured through camera. Whereas, the lateral deals with the recording the handwriting as a time sequence i.e., it is used to record the handwriting when someone is directly writing on digital surface using some input device like stylus of light-pen [2]. Generally, OCR technology refers to offline character recognition only, so in this chapter we will use terminology OCR in the context of offline character recognition only. As per the nature of the text offline OCR can be classified into three categories (1) printed text (2) hand written text (3) scene text.

OCR is robust method for digitizing printed texts, as by using OCR digital information can be handled, managed, processed and stored in a more efficient way [3, 4]. More over by applying Artificial intelligence and Machine Learning techniques to this text, the information contained inside the text can be analysis and processed for example such text can be used for sentiment analyses etc. Almost every language has huge amount of data which is required to build OCR application, so it is feasible to create the OCR application that need massive data to train the model and yield the highest level of accuracy. Over the past few decades, a lot of work has been done but OCR remained among the hottest and challenging research topic for most of the scripts.

The motivation of this chapter is to take an account of deep learning based algorithms for document analysis. The study aims at listing the barriers in way of exploring the full potential of deep learning algorithms and indicating possible future direction in light of present work. This chapter elaborates the basic knowledge and the state of the art deep learning approaches applied for document processing and analysis.

2 Objectives of Document Analysis

Robust Automated Document analyses system can revolutionize many walks of life. As these systems would not only help in extracting the text from images but it also helps to understand the format of the text inside document as well. Few of the objectives of applying document analysis process are following.

2.1 Extraction of Properties (Metadata) for Indexing and for the Provision of Filter Criteria for the Search

Digital images contain text but this text is useless for a computer system until it is properly represented in the computer system in computationally read-able format. After extracting, the text form image the text can be used by indexing and ranking software for indexing the digital images. Additionally this text can be used for keywords based searching by search engines.

2.2 Classification of Documents Based on Specific Categories

In an organization there can be multiple types of documents e.g. in a Bank same person at the same time may be dealing with deposit slips, cash with draw check and utility bills in such situation if the scanned images of all the documents are provided to an automated document analysis system it would automatically classify all the documents in their respective category, Extract the customer credential from the image and perform the required operation. Moreover, in post office OCR based techniques can be integrated with automatic sorting and arranging the letters in different categories as per address specified over envelops.

2.3 Automatic Creation of Company-Specific Dictionaries

A word can be used to convey different meaning in different situations at different places e.g. the word balance normally means to maintain the condition of equilibrium but in a bank the word balance means "The remaining amount of money in a customer account". So deal with such situations after analyzing the text extracted from the document a company specific dictionary can be built for convenience and avoiding miss understandings.

2.4 Statistics on Various Properties of Document Contents

Different statistical operations can be applied in order to manipulate the information provided in the text in order to utilize it in way that is more effective.

2.5 Automatic Translation

The digital documents may exist in different languages; more over with the advent of internet technologies the accessibility of digital documents isn't constrained. it doesn't matter in which language the document was originally created it can be translated automatically without hiring the services of any expert, through automatic translation software, such as, Google translator is the most commonly known translation software. After converting text successfully form the images using OCR techniques then this text can be passed to Automatic translation software to make it accessibility over even more wider sphere. Such a system in [5] is designed for Arabic to English translation but the translation process performed poorly, because of recognition errors which propagated to the translation phase.

3 Application of the Automated Document Analysis

The few of the general application of the automated document analysis system are stated as below.

3.1 Historic Document Analysis

The historic document analysis is a challenging area of OCR, as they are generally hundreds of years old and under gone through number of degradations affecting the legibility of the content, hence more robust and especially designed system are required to deal with such systems. One of the most common degradation which appeared in the historic documents is Bleed through effect [6]. Which is actually caused by oozing of ink from the back side of the paper or the situation may happen when the paper quality is too low. Therefore, text impression from the backside of the paper would appear on the front side. Such type of degradation would make OCR for Historic document analysis more challenging [6, 7].

3.2 Document Layout Analysis

Document layout analysis is another application of OCR, which is used to identify different regions in the image containing different information. The Document layout analysis is not a big problem if all documents are of the same format e.g. official documents in specific format as prescribed by organization. However, if there are irregularities and inconsistency in the document layout, like in the case of dealing with historic documents and literature books there would be obviously great variation in writing style, script and shapes of decorative entities. Previously two approaches for document layout analysis are used (1) Bottom Up method. (2) Top down method [7]. Informer the document analysis starts with the small and basic entities likes pixels and they are slowly grouped into lager region of interests. While in the lateral, one the analysis starts with the whole document image, which is divided into smaller regions of interest. Both the supervised and unsupervised learning method can be used for training the deep learning model here.

3.3 Text Extraction Form Scanned Documents and Digitizing the Information

A scanned document is stored in the form of digital image in a computer system. Digital images need more storage space more over such document are in non-textual format hence they cannot be edited and modified. However, such scanned documents are quite often required to be edited and modified. Without a robust OCR system the editing the an image document in is a cumbersome and odd job, because either the documents are over written using some image editor or they are regenerated in textual format manually. Among both the approaches former is odd one and lateral is tiresome and time consuming. A robust OCR system would surely make it possible to handle the digital document image as textual document after automated conversion, without significant human effort.

3.4 Automated Traffic Monitoring, Surveillance and Security Systems

OCR systems can be used for text extraction for the traffic image and License plate recognition for vehicle recognition. This would help to improving the existing traffic monitoring and surveillance system such as money tolling and traffic rule enforcement. OCR technologies will also help in controlling crimes including abduction, traffic violation etc. by analysis the CCTV footages and automated vehicle recognition based on license plate number.

3.5 Automated Postal-Mail Sorting

Postal mail sorting is pre-sorting of the mail envelops before their delivery base on the address on the envelop mails. The automated sorting of the mail in post offices can drastically reduce the human effort required; hence it reduces the time for overall delivery and increase the organization's profit by reducing the expense and increasing the system efficiency. The OCR systems especially dealing with address recognition [8], address block localization, address line localization [9] and numeric digit recognition [10] with reference to the scripts in which addresses are specified over the envelops, are developed to build for efficient mail sorting in post offices.

4 Significance of Deep Learning over Machine Learning

Deep learning is a representation learning technique, that is in-turn machine learning techniques in which models by using the vary huge data learns the pattern inside the data and later on able to make predictions on the basis of that new data [11]. Concept of deep learning is comparatively newer as compared with its other peer technologies but traditional techniques have limited capacity to process the raw and deep learning techniques can Handel the same data in more robust manner and surpassed traditional techniques, in-terms of accuracy. In 2013 Deep learning was listed among the top 10 breakthrough technologies.

The accurate extraction of the text from the image depends upon the quality of image and nature of patterns to be detected inside image and robustness of the algorithms used. However, the quality of image can be improved with the better scanning and imaging techniques and devices. But the quality of objects inside an image is quite a dynamic and uncontrollable factor, handwriting varies from person to person additionally the typo style of the old books is usually quite peculiar but the book may not be in good condition that would obviously results in the low quality of images. Moreover, the algorithms would return better results for well-separated languages while in case of languages with script cursive in nature become complex to process, hence efficiency of the algorithms drops. Though more complex algorithm are developed to attain the high level of accuracy, but still a robust commercial OCR is on the way.

The development of more sophisticated artificial neural network that adds more and more layers and allows higher level of abstraction for the input image analysis in order to achieve the higher level of accuracy. The development of new mathematical functions used at different layer improved and self-healing nature is like renaissance in the field of big data analysis. Hence, deep learning is widely used application in several fields. Like pattern recognition, weather forecasting application, text analysis and processing, speech recognition, object detection, image analysis.

4.1 Deep Learning Techniques and Architecture

Artificial neural networks are the biologically inspired networks, because the learning process in the human brain is done on the basis of the experience gained from the environmental stimulus. In humans inside the brain and nervous system there are tiny cellular bodies, the neurons which are interconnected with each other. The strength of connection in between the neuron depicts the relationship between the events, this strength is symbolized by a term synaptic weight: higher the weight, more strengthen the connection is and vice versa. Like the abstract architecture of the human brain, ANN consists of neurons and weighted inter connections between the neurons, their weights are known as "synaptic weights". These weights are analogous to the strength of interconnections between the neurons in the humans' nervous system. Learning is achieved by adjusting these synaptic weights, by using values form the previous layer and a mathematical activation function applied on them.

One of the simplest and earliest ANN was Preceptron based upon the mathematical model of the neuron is known as McCulloch and Pitts neurons. McCulloch and Pitts neuron takes the weighted input and passes them through the adder to get the output values whereas the Preceptron is simply combination of McCulloch and Pitts neurons along with an activation function applied to the output of each neuron. The Preceptron one of the earliest artificial neural network consists of only two layers i.e. an input layer which is directly connected to the output layer. This is a simple model but unfortunately, this simple model can work with only simple and linear problems. In order to handle the complex and nonlinear problem more layer known as "hidden layers" are introduced. These layers lies in between the input and output layer. Such ANN is known as deep neural networks. Hidden layers introduce a black box effect as we does not known exactly how by applying the specified parameters these layers mange to predict. In deep neural networks the output of each predecessor layer propagates as input to its successor layer. As the number of hidden layer, increase the level of abstraction in model and complexity of model increases. But the model could handle the more complex problem with higher degree of accuracy as it can diagnose the nonlinear relationships. Though due to massively parallel and complex processing the learning process (adjustment of synaptic weights) is quite slow but able to process data more effectively and yields higher degree of accuracy (Fig. 1).

In case of OCR for different languages DNN outperformed all the other techniques in race of tackling the challenges especially problem of none well separated characters and cursive nature of the text. The deep learning is listed among the popular topic since the groundbreaking paper [12]. The recent advancements and emergence of new models like Convolutional neural networks and recurrent neural networks along with its variant LSTRM (Long short term memory network) and BLSTM network (Bi-directional LSTM) crashed the barriers in the way of growing efficiency for the OCR applications for Arabic script.

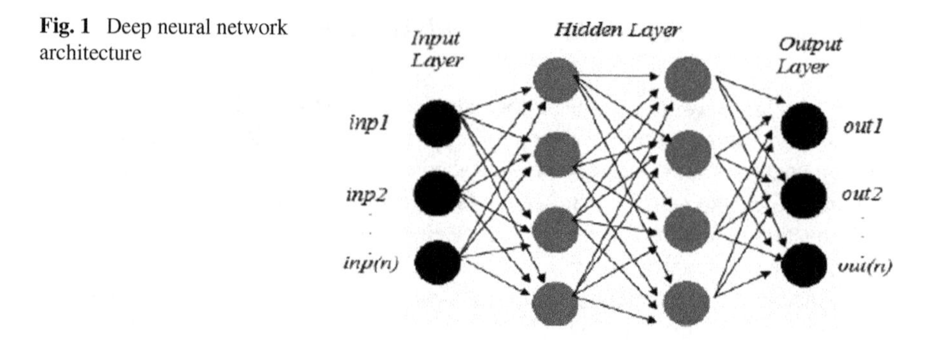

Fig. 1 Deep neural network architecture

5 Peculiarities and Challenges for OCR with Deep Learning

There are number of peculiarities and issues for development of optical character recognition using machine learning and deep learning techniques. In the following, we present each challenge in detail.

5.1 Dataset

The availability of data set is life blood for the any machine learning application; especially for deep learning, as; such application requires huge data sets. The accuracy of results produced by the model is mainly dependent upon the amount of data available; more data, would help the model to set its parameters in more generic sense, hence better would be the learning of the model will be. Besides quantity, the quality of the dataset is a major aspect that largely affects the learning process. Since for many languages like no standard datasets are available. Hence the lack of suitable dataset for a specific language is the greatest barrier in the way of success of Deep learning in the field of OCR. Furthermore, every language and every script has its own peculiarities especially languages with cursive script, where character are not well-separated, have their own challenges, while the languages with Latin script may have challenges like similar characters etc. [13] have their own nature and challenges. Hence document analysis becomes a separate domain for every language to language and script to script. In simple words, the dataset used for English document analysis, can neither be used for Urdu nor for Chinese and vice versa. It should be used for English document analysis. Hence, for developing a multilingual system it become reasonably tedious.

5.2 *Data Interoperability and Data Standards*

Data interoperability and data standards are the one of major barrier. Currently, nature of data differ from language to language thus there exist large variation in images due to device through which data is acquired and format in which data is acquired. Additionally a single language may fallow more than one script e.g. Urdu script can be written in Naskh and Nataleeq script. So to coup with the versatility in this context would mean to be able process the data in any of the available format, in any available script for a given language. Further to make a system suitable for plain hand written document should be used for hand written documents analysis, for computer printed document similar dataset should be used in order to get text form the painted images data set should be used and so on.

5.3 *Build and Integrate Big Image Dataset*

In order to train a deep learning model massively large datasets are required, but unfortunately many languages don't even have any standard data sets. Hence in such situation when model is trained over fewer amounts of data, the overall system would suffer from the efficiency issue. Though building standard dataset for a particular script in a particular situation is not a big issue commonly but for many language the scenario cannot get the due attention and efforts of researcher mainly because it is tiresome and time consuming job additionally it may become expensive for any reason too. In some cases it may become a big issue of privacy e.g. if automated DAS is especially designed to analyses the bank cheques the data for such application would firstly not openly available. I however available in suitable amount. It could not be used openly for building any other application by some other person. Therefore, in such cases, sharing of data is nearly impossible but suitable steps can be taken to overcome such issues.

5.4 *Language and Script Peculiarities*

Every language and scripts has its own character set, format, style more generally script for writing. Every script has its own challenges and peculiarities to deal with e.g. the Latin script where the characters are well separated are easy to segment while for the cursive script where characters are not well separated, combined with certain other issues like, diagonality, bi-directionality and character overlap. Hence it is hardest task, in the present circumstances to build a robust system which can fully deal with all types of languages, so a single OCR system is build to deal with a single language or a single script.

5.5 Black Box and Deep Learning

No doubt, beep learning revolutionized many areas; the automated document analysis applications citations also got the benefited from this amazing technique. Deep learning algorithms opened new avenues for improving the efficiency of the automated document analysis application. However deep learning techniques by itself are mysterious one. However, all the mathematics and rules used for deep learning are quite clear and well defined but how these well-defined rules and laws ultimately make the system intelligent enough to perform AI tasks like document analysis by learning patterns is a unknown phenomena. This problem is known as black box problem. The problem occurs due to the complexities involved during learning and the complicated output for the given input data. Due to black box nature of deep learning it is used by researcher without knowing what actually is done for achieving the desired results.

5.6 Processing Hardware Power

As deep learning system need to process massive amount of data in parallel, this processing not only need huge processing power but also were affected by the floating-point precession of the system. Relatively newer hardware technologies gave birth to faster GPUs and co-processors. In [14] the author states that GPU works 40 times faster than microprocessor while processing the massive amount of data during deep learning process. Author applied inline back propagation algorithm to process the data over the MNIST dataset. Here in this case just the improved hardware with floating point precession of single-precession FP over GPU provided better results over the models with much complex and complicated architectures. If the hardware is more advance and optimized to perform mathematical calculations additionally able to provide more floating point precision, the deep learning application can work better even with relatively small datasets and the techniques for increasing the size of dataset like degradation would no longer be required to apply, hence programming overhead reduces. So the impact of advanced hardware on deep learning is no way less at any stage, for any application either it is handwriting recognition, any other field of pattern recognition, data mining, computer vision and much more.

5.7 Implementation (Available Libraries) Can Be Hardware Dependent

The advent of GPU and co-processors had great impact on the machine learning and deep learning techniques as the new powerful hardware, with more Floating point precision, the growth in the field of ML and deep learning boomed up. But problem is that GPU and co-processors usually are unable to execute the existing set of libraries as they are especially design have their own instruction set, especially a GPU firsts translates the set of mathematical operations and algorithms in the graphical operations. Hence to execute the deep learning algorithms on GPU especial versions of libraries compatible with GPU processing are created e.g. Tensorflow is a python library which have both the versions 'tenserflow with CPU support' and 'tenserflow with GPU support'. The code for GPU application is written in a C like programming language called compute unified Device Architecture (CUDA).

6 Machine Learning, Deep Learning and Optical Character Recognition

In the following subsections, different studies are presenting for OCR for Arabic script, Symbolic Script and Latin script using machine learning and deep learning.

6.1 OCR for Arabic like Script

Urdu, Arabic, Persian and Pashto languages spoken over a large area of the largest continent of Asia including Middle East and south Asia. All the languages though not same but have common Arabic like scripts [15]. Urdu is the language of the over 70 million peoples. Besides it is the national language of Pakistan and official language of 5 states of India as well. All the above stated languages are enrich with the literature more over the religious holy book of Muslims and the other related material like Ahadith are in Arabic language [16].

Such popularity of languages invoked the digitization of the literature in order to keep pace with the current advancement of the technology world. The digitizes literature would preserve the new the books as well as old books for the longer period of time and make the searching easy using algorithms later. In this context, the document image analysis emerged by which the scanned image of a document is analysis for patterns drawn over it. These patterns may include text, simple figures, pictures etc. This versatile combination makes the analysis even complex. So in order to interpret the objects in the image a number of techniques are used. For getting the text a well know technique OCR (optical character recognition is used) the technique of OCR is used to read the document image and convert them into plain text. Hence, it can be said that OCR systems, can recognize all the primary and secondary blocks of write-ups, including characters, words and sentences. OCR has been an active field of research for a few decades. Different languages based on Latin script (characters are well separated) have successfully developed the commercial OCR systems with high accuracies due to the huge progress and evaluation of better methods in the field of image processing and computational intelligence. The languages with Arabic script, which is cursive in nature lags behind, mainly due to the complexity, involved due to their cursive nature. The major challenges and peculiarities of the Arabic script can be listed as Bidirectionality, Diagonality, Graphism multiplication/Context sensitivity, Cursive style, Characters' overlap, Upper/lowercases, The number and position of dots, Complex placement of dots, Stretching, Positioning, Spacing, Filled Loops, False Loops, Ambiguous Segmentation Cue Point, Direction of segmentation, Text line Segmentation and Interline Spacing [17]. Many of the complexities arise at the joining points of the two words (ligatures). Various techniques has been used for the pattern recognition but deep learning methods out passed all of them in terms of performance and accuracy. Popular deep learning models used for pattern recognition include), convolution neural network (CNN), extreme learning model (ELM), recurrent neural network (RNN) and the variants of RNN like BLSTM and MDLSTM. Among the stated models many of them suffer from certain problems including single instance learning, lack of memory and non-cyclic connections of hidden nodes e.g. ELM, traditional Neural Network and CNN lacks the memory hence don't have any ability to store the sequence. Hence HMM and LSTM are the only models that can handle the sequence learning problems. Different studies including [18–22] illustrates that LSTM outperformed other techniques.

In [23] combination of line and ligature segmentation techniques are used in order to extract the connected components. The algorithm relies on baseline detection for line segmentation and showed significant boost up in locating dots and diacritic with accuracy of 99.17%.

6.1.1 Feed Forward Neural Network

Feed Forward neural network (FFNN) is simplest form of ANN. FFNNs are just Multi-layer preceptors (MLP) except the fact that FFNN has additional hidden layer that are lacking in MLP. The study in [24] elaborates the complex nature of character and resulting challenges concluded that most of the errors occurred due to the end characters because there is no proper concept of Capital and small character in Arabic like script except the end characters. To tackle this issue sample of 100 Characters segmented on the basis of pixel strengths, is feed to a simple Feed forward neural network with a single hidden layer with 2000 neurons, the which resulted in average accuracy of 70%.

In another study Seuret et al. [25] adopted PCA (Principal component analysis) initialized Deep neural network approach over the annotated manuscripts form two datasets including DIVA-HisDB and pages from the manuscripts SG8573, SG5624, and GW105. The experiments showed an improved accuracy over the Xavier-initialized networks.

6.1.2 Convolution Neural Network

Convolutional neural networks is a especial class of deep feed forward neural network in which hidden layer consist of convolutional layers, pooling layers and fully connected layers. CNN mainly focuses on instance learning; hence CNN can work better for isolated character recognitions. In [26], convolutional neural network is used for feature extraction and classification over an Arabic dataset named EAST (English-Arabic Scene Text) dataset. Images contained in the data set are complex and captured under various illumination conditions; hence robust segmentation of image is required for the correct classification. The empirical methods for segmentation failed to deal with such complex cases so the images are for s manually segmented with respect to the text line or words. The best resulted obtained using the proposed methodology with the lowest error rate of 0.15%.

Convolutional Neural Network Committees are used for handwritten MIST classification in [27], where among the 7-CNN committees each committee member is a CNN, all the Network are provided with the same image but distorted in some form. The committee averages the results of all the CNN. The lowest error rate reported by the experiments over MINST is 0.35%.

6.1.3 Auto-encoder

An Auto-encoder is a simple MLP with a restriction that it would have the same number of inputs and outputs form the network. The hidden layers may have different number of neurons then the number of neurons Input and output layer has, if the number of neurons are less than the previous layer the networks also impart here a sort of convolutional effect because in this case the next layer would receive the compressed input. An Auto encoder is actually a hybrid network consisting of feed-forward and non- recurrent neural network. In [28] Auto encoder is used for Urdu language character recognition. The network is first trained by using a feed forward pass after which for adjusting the weights the error propagation is used, where the difference of values between the input value a1 and a1 is used to measure the error as there is one correspondence between the input and output value. In order to make the system more robust the model is passed partially corrupted input. For de-noising the input the input is passed to MLP after back propagation is applied for further refinement. The experimental results are comparing with the SVM with RBF kernel. The three layered ULR-SD achieved the ligature recognition accuracy of 96% where the SVM got 95% accuracy for 80 * 80 input dimensions. Further three layered architecture show higher accuracy as compared to the 2 layered architecture.

6.1.4 Hidden Markov Model

Segmentation of a cursive text is very challenging task and it is found that error in segmentation process reduces over all accuracy of the system. In order to avoid segmentation error the holistic approaches are adopted. Holistic approaches are segmentation free approaches. In [4, 16, 29] author adopted holistic approach and ligatures i.e. partial words served as units of recognition instead of characters. The author applied the HMM to extract the statistical features of the ligatures. The system resulted in ligature recognition rate of 92%.

6.1.5 Recurrent Neural Network

In [30], Bi-dimensional Long Short Term Memory (BLSTM) is trained on a Urdu-Nastaliq handwritten dataset (UNHD) collected from 500 individuals. The system reported error rate of 6.04–7.93%. In another study [5], BLSTM is trained on 11% of Arabic handwritten NIST dataset and achieved the recognition rate of 52%. The results of BLSTM–RNN are compared with HMM, the former with accuracy of 0.5% surpassed the lateral one with the accuracy rate of 0.3%. Two deep learning algorithm including BLSTM and MDLSTM are applied to the a Pushto dataset "Katib's Pashto Text Imagebase" which contains 17,015 images of text in [31]. The results of the experiments showed that MDLSTM outperformed BLSTM for Pushto script.

In [32], zoning features are used with variant of RNN called 2-Dimensional Long Short Term Memory (2DLSTM) learning classifier. Zoning features are statistical features that provide significant information with low complexity and high speed. Zoning features are extracted by exploiting the sequences learning property of the LSTM network by applying sliding window of size 'n' over a vertical strip of size 'n'. The character recognition rate of 93.38% is reported on Urdu Nasta'liq UPTI dataset.

In [33], multi-dimensional recurrent neural network with LSTM is connected with output layer of CTC. The model exploited the 15 sets of statistical features by using a sliding window moving from right to the left direction (MDRNN and LSTM and CTC output layer) and statistical features. By using UPTI dataset model is trained on various sets of features and the highest recognition rated reported on the test set is 94.97% with the training error laying in between 3.72 and 5.03% on various set of features. In another similar study [34], multi-dimensional recurrent neural network (MD-RNN) along with LSTM is used to extract 12 statistical features by using sliding windows and which are finally feed to an CTC output layer. The model is feed by the standard UPTI database and reported the highest recognition rate of 96.4%.

Multidimensional deep learning approach (MDLSTM) is used for automated extraction of raw pixels by Naz et al. [3, 35, 36] for Urdu Nasta'liq text line recognition. The purposed technique outperformed the manually extracted statistical features by reducing the error rate up to 50% over UPIT dataset. In [37], gated LSTM used pixel values for recognition over degraded UPTI and achieved the recognition accuracy of 96.71%.

The system in [34] worked better as compared to other systems including HMM-based OCR, OCRopus and Tesseract (Figs. 2 and 3).

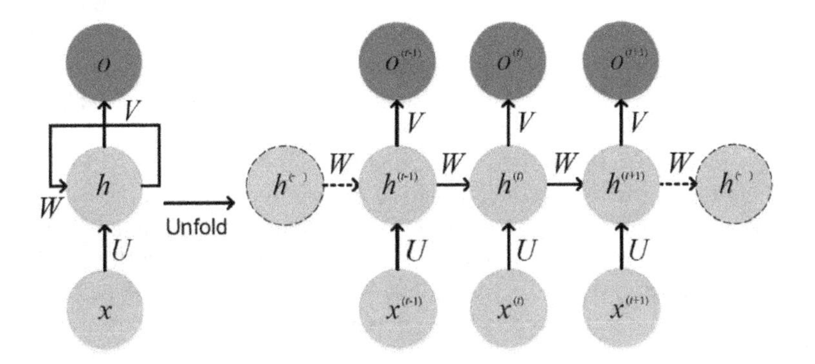

Fig. 2 Recurrent neural network architecture [37]

Fig. 3 a EAST-dataset image [65]. b Segmented image [65]

6.1.6 Hybrid Approach

In order to exploit the properties of multiple deep learning models are used in combination, which can increase the accuracy of the system over all. A hybrid approach consisting of multidimensional LSTM combine with connectionist temporal classification (CTC) and a hierarchical layer structure is used in [19]. The model gets raw pixel data as input. Model designed in such a flexible way that it can work with different languages, and tested for English and Arabic dataset. The system showed 91.4% accuracy over IFN/ENIT database of handwritten Arabic words. In another study [12] DBN (Dynamic Bayesian Network) are coupled with input from HMM. The hybrid model exploited, statistical features based on pixel distributions and local pixel configurations and structured features based on presence of loops, diacritic points, ascenders and descenders. The model achieved accuracy of 90.42% over Arabic dataset IFN/ENIT database.

Low-resolution images are adversely affecting the accuracy of the system. The study in [15] adopted a sophisticated approach for the recognition of low-resolution Arabic handwritten word recognition. The model for feature extraction and classification exploited the MDRNN with LSTM and CTC respectively. The model was feed with the data form Arabic printed text image database (APTI). The APTI database is specially designed with greater variation in data, it contains different types of data including single-font/single-size, single-font/multi-size, and multi-font/multi-size [38]. The experimental results depict the best result with multi font/multi size data as with such data system can generalize its parameters in better way. The model achieved over all mean word recognition rate of 98.32% and mean character recognition rate of 99.69.

Unlike traditional statistical and holistic features and pixel in [39] CNN is used to extract low level translational invariant features, that are provided to MDLSTM classifier which extracts the higher order features. The model is fed with Printed Text-line Image (UPTI) dataset. This hierarchical combination of CNN and MDLSTM achieved the recognition rate of 98.12% for 44-classes defined in dataset.

Systematic review is presented in Table 1 for Arabic like script based character recognition systems.

6.2 *OCR for Symbolic Script*

In the light of fact that languages like Chinese and Japanese have thousands of character classes. Hence building accurate and fast OCR is extremely challenging for such scripts. To take with this issue dimensionality reduction could not work efficiently, so Odate et al. [46] devised candidate reduction techniques for Nearest neighbor and an improved tree based clustering. Reduced candidate region has high dimensionally feature vector which is reduced by multiclass Linear Discriminant Analysis (LDA).

Chinese character recognition for E-commerce plate form is studied in [47]. On ecommerce platforms more information is presented in the form of images. In order to extract the information the images are processed by using canny operator, genetic algorithms and normalization process and them stroke characteristic features are passed to the OCR font library in MODI attached to office 2003 for identification. The overall accuracy of the system over a self-generated database of 25 samples is 74.9%.

6.2.1 Convolutional Neural Network

CNN with multi pooling layer with flexible strides and kernel size is used for 3755 classes of Chinese characters in 280 fonts and 120 selected font in [48]. This modified architecture recognition rates of 94.38% for 280 fonts and 99.74% for 120 selected fonts. In [49, 50] CNN based transfer learning model is applied over

Table 1 Review of OCR based on Arabic like script

Study	Method	Text type	Model	Dataset	Results
Ul-Hassan et al. [40]	Pixels	Printed Urdu	BLSTM	UPTI (10,000 Urdu Nastaliq text lines)	86.4% (shaped) 94.8% (unshaped)
Graves et al. [19, 41]	Pixels	Handwritten Arabic	MDLSTM	IFN/ENIT (32,492 city names = 30,000 train-set and 2492 validation set)	93.37%
Rashid et al. [15]	Pixels	Printed Arabic	MDLSTM	APTI	99%
Pam et al. [42]	Pixels	Handwritten Arabic	MDLSTM	OpenHaRT	90.1%
Morillot et al. [5]	Features	Handwritten Arabic	BLSTM	NIST (training set = 16,000 and testing set = 12,644)	52% (word)
Chherawala et al. [43]	Features	Handwritten Arabic	MDLSTM	IFN/ENIT (32,492 city names)	89%
Rez et al. [44]	Features	Handwritten Arabic	BLSTM	IFN/ENIT (32,492 city names)	87.4%
Ahmed et al. [30]	Features	Handwritten Urdu	BLSTM	UNDH	Error rate of 6.04–7.93%
Naz et al. [19]	Statistical Features	Printed Urdu	MD RNN with LSTM	UPTI	Recognition rate of 96.4%
Naz et al. [32]	Zooning features	Printed Urdu	2D LSTM	UPTI	93.38% of character recognition rate
Naz et al. [45]	Low level translational invariant features	Printed Urdu	CNN and MDSTM	UPTI	Recognition rate of 98.12%

(continued)

Table 1 (continued)

Study	Method	Text type	Model	Dataset	Results
Kh et al. [12]	Statistical and structured features	Handwritten Arabic	DBN with HMM	IFN/ENIT database	Accuracy of 90.42
Rasid et al. [15]	Not specified	Low-resolution Arabic handwritten	MDRNN with LSTM and CTC	APTI	Mean word recognition rate of 98.32% and mean character recognition rate of 99.69%

historical Chinese character recognition, where output of one CNN is provided to another CNN [50]. Semi supervised learning approach is test over the Chinese Printed Dataset. MCDNN is feed with ICDAR 2013 competition on recognizing offline handwritten Chinese characters in [51], the system robustly classified 3755 classes of Chinese character with low error rate of 4.2%.

Math formula recognition is a challenging task in degraded document images. Liu et al. [52] in order to achieve the higher accuracy with the CNN the formula elements are over segmented iteratively into recognizable unit in order to analyze the mathematical formal structure. The model showed precision of 82.46, recall of 87.85 and F-measure of 85.37 over MFR100 [53].

Multi-spatial-context fully convolutional recurrent network (MC-FCRN) based implicit model is used in [54] for online hand written Chinese character recognition on the basis of analytical and statistical properties of pen stroke. The model is tested over CASIA and ICDAR Dataset, with correction rates of 97.10% and 97.15%, respectively. In [55], Relaxation CNN model is used which varies from the traditional CNN model in the sense that in traditional model for a single feature map on a convolutional layer there would be only one convolutional kernel on the other hand in the case of RCNN to generate a sing feature map for a convolutional layer the convolutional kernel varies from each stride to stride. The process of relaxation has improved the overall learning ability of the model as this approach can deal with the deformation caused by during writing e.g. uncontrolled and slightly irregular movement of hand muscles.

A special variant of ATR-CNN or Alternate training CNN is used in this method for each weight matrix an arbitrary learning i rate is selected and during tanning learning rate varies from i to 0 once the learning has been completed i.e. after the back-propagation phase the learning rate is back adjusted to i. The alternative training helps to deal with the sudden increase in number of parameters due to the relaxation process besides it also improved the accuracy of the system and better parameter tuning. The model has been designed, implemented and evaluated for handwritten digit and Chinese character. For hand written digits MNIST dataset is used and the Architecture of the model used [55] is In-32Conv5-32MaxP2-64Conv3-64MaxP2-64RX3-64RX3-Out. The error rate was error rate is 0.24%.

The architecture of the model used for Chinese character is In-64Conv5-64MaxP2-128Conv3-128MaxP2-128RX3-128MaxP2-256RX3-256Full1-Out. CASIAHWDB and ICDAR'13 Competition Dataset are used for training Testing and validating the model. The lowest error rated reported 3.94%. The error rate is nearly as reported by the human natural senses 3.87% which is very close the error rated of the model.

6.2.2 Recurrent Neural Network

Multi-Dimensional Long-Short Term Memory Recurrent Neural Networks (MDLSTM-RNN) is used for handwritten text recognition system for Chinese characters in [56]. As the Chinese characters are already in the segmented form hence the segmentation would be a useless overhead and leaves the remaining process more prone to the errors. The model is trained and evaluated using CASIA Off-line Chinese Handwriting Databases. The dataset contains images of isolated characters and lines of text. The experiments shows character error rate of 16.5% but after applying character n-gram Language Models (LM) on textual data, then using this language model at the character level reduces the error rate to 10.6%. However, the proposed approach cannot perform better than the previous approaches but it is more start of the art technique. Moreover the errors were mainly due to the characters isolations, if somehow some characters in isolation could be avoided the overall performance would certainly boost up.

Handwritten Chinese text without explicit segmentation of the characters is done with Multi-Dimensional Long-Short Term Memory Recurrent [57]. The model is feed with data from 4 sources including CASIA database, PH corpus [58] with newswire texts, the Chinese part of the multi- lingual UN corpus [59] and the Lancaster Corpus of Mandarin Chinese [60]. There are total 50,000 characters, hence the task character recognition for such scripts is extremely challenging. However these characters usually have similar appearance and share common graphical components inside the characters. Graphical components of the characters are called Radicals. In [61] radical extraction network based on CNN is feed with data from CASIA [62] and the best result reported on the dataset are 93.5%.

Self-Organizing Map (SOM) and Multiple Timescales Recurrent Neural Network (MTRNN) are used one after another by author in [63] for online Japanese handwritten character recognition. SOM is used for feature extraction while MTRNN is used for dynamics learning. Input data is acquired by pen tablet Intuos4 ptk-640 (WACOM) and character is written 10 times. The experiments depicted the ability of model to predict the handwriting sequences, affirmatively. The OCR based on symbolic scripts are reviewing in Table 2.

Table 2 Review of OCR based on symbolic script

Study	Method	Text type	Model	Dataset	Results
Messina et al. [56]	Features	Chinese handwritten	RNN-LSTM	CASIA	Error rate to 10.6%
Nishide [63]	Features	Japanese handwritten text	Multiple timescales recurrent neural network (MTRNN)	Online recognition	–
Wu et al. [55]	Features	Handwritten Chinese characters	CNN	CASIAHWDB and ICDAR'13 competition dataset	Error rate 3.94%

6.3 OCR for Latin Script

The hand written digits classification was performed with only 0.35% error over MINST digits data set. In order to increase the available amount of data for the deep learning deformation techniques of affine and elastic distortion are applied only. The author executed the algorithms over Single precision Floating Point GPU in order to get the better accuracy, which was about 0.4–0.5% better as compared to the previously published techniques. Hence, it was observed that for implementing a deep learning model more advanced hardware play significant role in terms of accuracy of the results.

6.3.1 Convolutional Neural Network

Variants of CNN can tackle challenges of latent script along with Cusie and symbolic scripts. Fuzzy Convolution Neural Network a variant of CNN is used in [64] for Czech Language. The results from FCNN are compared with number of other approaches, including KNN, Multiple logistic regression (MLP) and ANN. The test set error rate for each of the approach is stated in Table 3. The results depicts that FHCNN performed the best with minimum error rate of 0.60%.

In order to evaluate the performance of the CNN system based on a novel voting method Chen et al. [65] generated a dataset by gathering articles in 5 different languages from the Internet and translating them into different languages using Google translate which are then printed and scanned for getting document images. CNN combined with novel voting mechanism performed language and orientation analysis with up to 99% accuracy.

Table 3 Error rate for different methods

Method	Error rate (Test set) (%)
KNN	34.65
MLP	31.06
SVM	27.86
ANN (one hidden layer)	27.39
ANN (two hidden layer)	14
FCNN	1.50
HFCNN	0.60

The detailed used architecture used in [66] is 2x48x48-100C5-MP2-100C5-MP2-100C4-MP2-300N-100N-6N. Where the MCDNN is provided with two input images and of size 48 $\times$ 48, and terms like 100C5 and 100C4 depicts the convolutional layer with 100 maps and 5 $\times$ 5 and 4 $\times$ 4 filter respectively. MP2 refers to Max Pooling layer with non over lapping region of size 2 $\times$ 2. Activation functions used at different layers are; hyperbolic tangent activation function at fully connected layers and convolutional layer, linear activation function at max-pooling layers and finally soft-max activation function at output layer. The architecture is evaluated over different datasets. Datasets alone with their relative improvement in whole learning process are; MNIST digits 41%, NIST SD 19 (30–80%), NORB (46%), GTSRB traffic sign dataset (72%), CIFAR10 natural color images dataset (39%).

A convolutional model is used with activation function Relu at both convolutional and fully connected layers is used by Kang et al. [67]. Relu activation improved efficiency and reduced the overall training time required as compared to the Sigmoid and Tanh activations. Where ReLU is simply $f(x) = \max(0, x)$ where $x = input\ value$. The model is employed to learn hand crafted features from the normalized images. The author evaluated the proposed architecture on two datasets including Tobacco litigation and NIST tax-form dataset. Few of the images are shown in Fig. 4 dataset consisting of 10 classes including form, report, resume, memo, scientific, news, advertisements, email letter and different sort of notes note. Table 4 contains the Class-confusion matrix for genre classification on Tobacco dataset, the overall performance resulted in accuracy of 65.35% on the Tobacco dataset.

Document images usually contain both graphical and textual information. Image content analysis can recognize and distinguish both the regions in the document images. In [68] combination of visual and textual features are used for document's image content analysis on two datasets including Loan dataset, (provided by an Italian loan comparison website company) and Tobacco dataset. The dataset are tested with the stated features are tested on the CNN and achieved an accuracy rate of 70% by using Loan Dataset. "AlexNet" is employed by the author in [69] to specially address two questions regarding document image analysis

(1) To which book photographed document images belongs to?
(2) What is the type of the book to which the document image belong to?

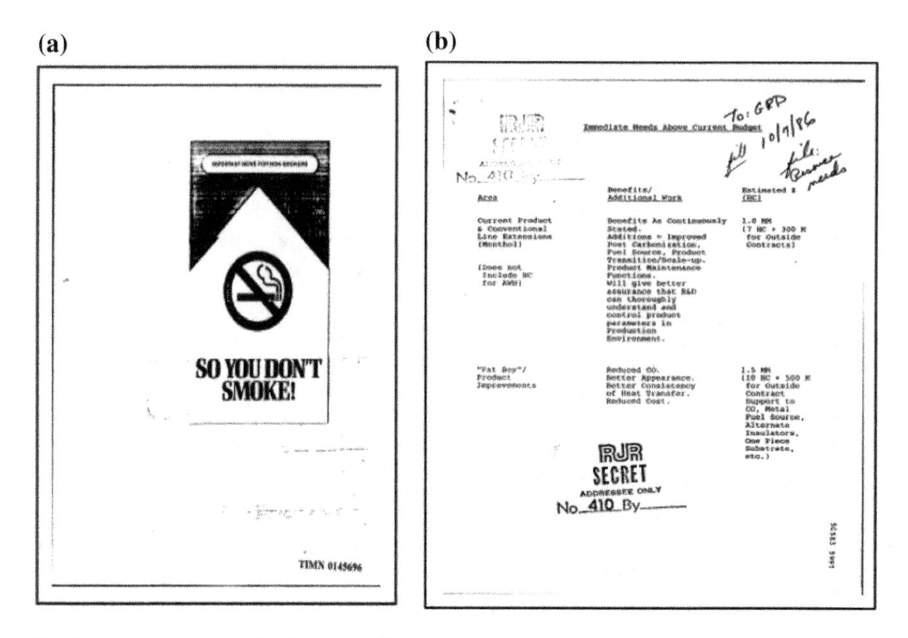

Fig. 4 **a** Advertisement [67]. **b** Report [67]

The model successfully answered the first question with the accuracy of 95.54% and the second question with accuracy of 95.42%. The author gathered the dataset consisting of images of 6 categories including picture book, musical book, Chinese book, English book, mathematical book and newspaper.

CNN with two convolutional layers and two pooling layer connected to one another alternatively are employed by author in [70]. The results from pooling and

Table 4 Class-confusion matrix for genre classification on Tobacco dataset [67]

	Ad	Email	Form	Letter	Memo	News	Note	Report	Resume	Scientific
Ad	**104**	0	1	1	0	9	2	2	0	3
Email	1	**435**	7	3	13	0	4	3	1	0
Form	2	0	**145**	5	37	7	8	7	0	14
Letter	0	8	6	**297**	43	0	1	14	0	10
Memo	1	7	33	51	**294**	6	3	9	0	18
News	19	1	21	13	6	**45**	8	2	0	16
Note	2	10	24	8	31	5	**63**	0	0	11
Report	1	15	34	65	32	11	5	**103**	5	38
Resume	0	7	24	13	12	1	1	13	**13**	6
Scientific	0	16	36	11	52	4	6	12	1	**45**
Accuracy (%)	80.0	87.2	43.8	63.6	56.6	51.1	62.4	62.4	65.0	28.0

convolutional layers are provided to classifier. The model is evaluated using ICDAR 2003 dataset and Street View Text (SVT) datasets. Certain deformation techniques including affine and elastic deformations are used. The method resulted in precision of 0.54, recall of this 0.30 and F score of 0.38, with an overall accuracy of 83.9%.

6.3.2 Recurrent Neural Network

A variant of multi-dimensional variant of RNN with multidimensional hidden layer used in [21] empowered the network to handle the multi-dimensional contextual information performance of a MDRRN is compared with convolutional neural network over the MINST Dataset the results of the experiment over the wrapped MISTS showed that MDRNN with error rate 6.8% surpassed the convolutional method with error rate 11.3%. In another study Dropout layer is carefully employed with the RNN in [42] such that it does not affect the recurrent connections in the layer, but effectively prevented the over fitting. The model is evaluated over Rimes and OpenHaRT databases. Addition of dropout at top most layers in RNN showed decrease of error rate about 10–20% which can reach up to 30–40% with drop out at multiple LSTM layer. In Table 5, OCR based on Latin Script are summarized.

6.4 OCR for Nagari Script

In [73], Deep Bidirectional LSTM are applied to Oriya script. The author adopted a script independent and segmentation free approach for Connectionist Temporal Classification (CTC) for the learning of the un-segmented sequences. With this model lowest error rate reported is 0.1% after 150 epochs of training. Table 6 summarizes the OCR based on Nagari script.

6.5 OCR for Multiple Scripts

Efforts are being made to build a robust system able to recognize the text of every script from the image for this different researchers applied different techniques. Though still no universal system exist able to recognize text of every type but a single system able to deal with two or more language work with reasonable accuracy.

Deep CNN and named it PHOCNet (Pyramidal Histogram of Characters) for extracting words from the document images by Sudholt et al. [75]. The architecture of the purposed model is shown in Fig. 5. The model is evaluated on different datasets including IFN/ENIT database, Esposalles database, IAM Handwritten Database, George Washington dataset. The stated techniques out preformed traditional techniques and showed its robustness for Latin as well as Arabic script.

Table 5 Review of OCR based on Latin script

Study	Method	Text type	Model	Dataset	Results
Morillot et al. [71]	Features	Handwritten French	BLSTM	RIMES (12,107 text lines, Training set 10,329, validation set = 1000 and Testing set = 778)	43.2% (word)
Lickwi et al. [18]	Features	Online English	BLSTM	IAM-OnDB	74%
Graves et al. [22]	Pixels features	Online English	BLSTM	IAM-OnDB	69.9% (word)
			BLSTM+LM		77.2% (char)
			BLSTM		74% (word)
			BLSTM+LM		79.6% (char)
Graves et al. [72]	Features	Online English	BLSTM	Online IAM-OnDB (13,040 text lines)	86.1 (char)
					79.7% (word)
	Features	Offline handwritten English		Offline handwritten IAM-DB	81.8% (char)
					74.1% (word)
Pam et al. [42]	Pixels	Handwritten French	MDLSTM	RIMES	91.1%
		Handwritten English		IAM	85.6%
Chaudhuri and Ghosh [64]	Features	Chez handwritten text	FHCNN	–	Error rate of 0.60%
Cen et al. [65]		5 languages	CNN	–	99% accuracy
Cires and Meier [66]	Features	English	MDCNN	MNIST digits	41%
				NIST SD 19	30–80%
				NORB	46%
				GTSRB traffic sign dataset	72%
				CIFAR10 natural color images dataset	39%

Table 6 Review of OCR based on Nagari script

Study	Method	Text type	Model	Dataset	Results
Anupama et al. [73]	Pixels	Printed Oriya	BLSTM	–	95.85%
Roy et al. [74]	Features	Indian language	HMM with DBN	–	66.48% (recognition rate)

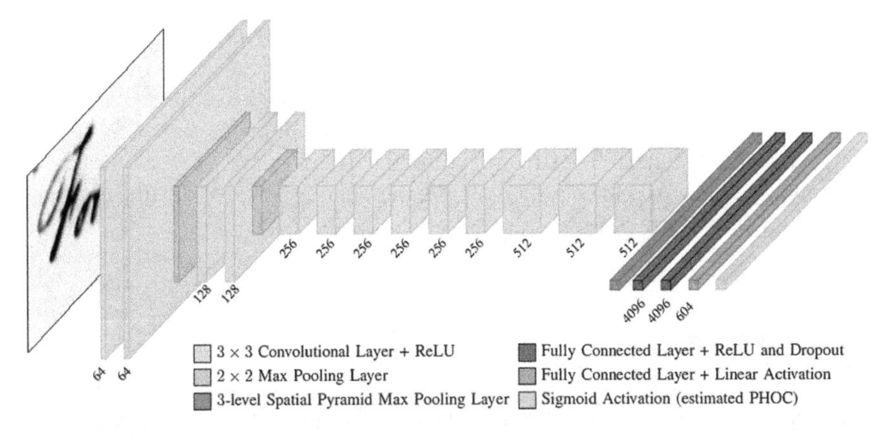

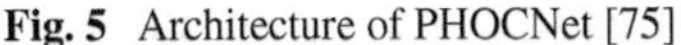

Fig. 5 Architecture of PHOCNet [75]

A Tandem approach of HMM with DBN is employed by the author in [74]. The model is trained and evaluated on three different datasets for three different languages including RIMES dataset for Latin script, IFN/ENIT for Arabic like script, and a dataset for Indian Devanagari script. The model showed the highest recognition rate of 76.98% over RIMES, over 89.41% IFN/ENIT and 66.48% on Devanagari script.

BLSTM networks connected with CTC (connectionist temporal classification) output layers are evaluated on non-Cursive Latina script (English script) and cursive Urdu script in [45]. The model for Latin script resulted with the character recognition accuracy of 99.17% on UNLV-ISRI database. For Urdu cursive script do the complexity involved the character recognition accuracy was relatively low on the same model i.e. 88.94% without position information and 88.79% with position information.

7 Open Challenges and Future Directions

OCR for most of the languages in one or another way is still a open area of research her we would list few of them

- Certain languages including Arabic script languages is an open research are as no perfect method exist. Building a robust method even for simple text extraction would be great job [34, 76].
- Challenges associated with camera capture images e.g. detecting cursive text lines, dealing with painted text in side pictures [13].
- Text extraction from the images taken in outdoor illumination conditions.
- Besides printed text images hand written text recognition is an open challenge. Writer variation in handwriting recognition makes it more challenging and inters class variation needed to deal with.

- For the enhancement of the recognition and accuracy post processing techniques including dictionary validation and grammar verification can be applied [34].
- A method applied to one language or script can be tested with the dataset from other script of language. Additionally the same problem with same data set and model can be evaluated by extracting different set and number of features.
- Applying different language modeling techniques for writer identification and dealing with other high-level problems.

8 Conclusion

The chapter provided a comprehensive insight to the application of deep learning in automatic document analysis. Building a robust and perfect OCR system is still on the way. Deep learning algorithms open avenues for improving the character recognitions rate, accuracy and efficiency of the OCR system. The automated OCR would make great progress and efficient management in different walks of life including business intelligence, efficient surveillance and monitoring systems, preserving the literatures and digitizing the information. Number of deep learning algorithms including CNN, LSTM network, gated RNN coupled with certain techniques like CTC and PCA have shown the higher level of accuracy and character recognition but perfect OCR of certain languages like Urdu, Persian and Arabic is a far cry. Deep learning algorithms are showing betterment in results by leaps and bounds but they still have their own issues one of the basic and major issue that deep learning application need huge datasets, this problem may not be eliminated but can be overcome by applying certain techniques to increase the available number of instances. As currently computational complexity is remained no more a problem more and more computationally intensive deep learning techniques would possibly result in a perfect OCR application.

References

1. S. Naz, A.I. Umar, S.H. Shirazi, S.B. Ahmed, M.I. Razzak, I. Siddiqi, Segmentation techniques for recognition of Arabic-like scripts: a comprehensive survey. Educ. Inf. Technol. **21**(5), 1225–1241 (2016)
2. M.I. Razzak, S.A. Husain, A.A. Mirza, A. Belaid, Fuzzy based preprocessing using fusion of online and offline trait for online Urdu script based languages character recognition. Int. J. Innov. Comput. Inf. Control **8**(5), 21 (2012)
3. S. Naz, S.B. Ahmed, R. Ahmad, M.I. Razzak, Arabic script based digit recognition systems, in *International Conference on Recent Advances in Computer Systems (RACS)* (2016), pp. 67–73
4. M.I. Razzak, M. Sher, S.A. Hussain, Locally baseline detection for online Arabic script based languages character recognition. Int. J. Phys. Sci. **5**(7), 955–959 (2010)
5. O. Morillot et al., The UOB-Telecom ParisTech Arabic Handwriting Recognition and Translation Systems for the OpenHart 2013 Competition To cite this version (2014)
6. N. Stamatopoulos, G. Sfikas, *Historical Document Processing*, vol. 2014 (2016)
7. S. Naz, A.I. Umar, M.I. Razzak, Lexicon reduction for Urdu/Arabic script based character recognition: a multilingual OCR. Mehran Univ. Res. J. Eng. Technol. **35**(2), 209 (2016)

8. S. Chen, Using multiple sequence alignment and statistical language model to integrate multiple chinese address recognition outputs (2015), pp. 151–155
9. M.I. Razzak, S.A. Hussain, M. Sher, Z.S. Khan, Combining offline and online preprocessing for online Urdu character recognition, in *Proceedings of the International Multiconference of Engineers and Computer Scientists*, vol. 1 (2009), pp. 18–20
10. M.I. Razzak, S.A. Hussain, A. Belaïd, M. Sher, Multi-font numerals recognition for urdu script based languages. Int. J. Recent Trends Eng. (IJRTE) (2009)
11. S.B. Ahmed, S. Naz, S. Swati, M.I. Razzak, A.I. Umar, A.A. Khan, UCOM offline dataset—an Urdu handwritten dataset generation. Int. Arab J. Inf. Technol. **14**(2), 239–245 (2017)
12. A. Kh, A. Kacem, A. Bela, M. Elloumi, A. Kh, Arabic handwritten words off-line recognition based on HMMs and DBNs Arabic handwritten words off-line recognition based on HMMs and DBNs, in *13th International Conference on Document Analysis and Recognition* (2015), pp. 51–55
13. S. Naz, M.I. Razzak, K. Hayat, M.W. Anwar, S.Z. Khan, Challenges in baseline detection of Arabic script based languages, in *Intelligent Systems for Science and Information* (Springer, Cham, 2014), pp. 181–196
14. D.C. Cires, Deep, Big, Simple Neural Nets for Handwritten, vol. 3220 (2010), pp. 3207–3220
15. S.F. Rashid, M.-P. Schambach, J. Rottland, S.V.D. Nüll, Low resolution Arabic recognition with multidimensional recurrent neural networks, pp. 1–5
16. S. Naz, K. Hayat, M.I. Razzak, M.W. Anwar, H. Akbar, Arabic script based language character recognition: Nasta'liq vs Naskh analysis, in *2013 World Congress on Computer and Information Technology (WCCIT)* (IEEE, 2013), pp. 1–7
17. S. Naz, K. Hayat, M.I. Razzak, M.W. Anwar, S.A. Madani, S.U. Khan, The optical character recognition of Urdu like cursive script Post Ph.D. View project. Pattern Recognit. 1–20 (2014)
18. M. Liwicki, A novel approach to on-line handwriting recognition based on bidirectional long short-term memory networks
19. A. Graves, Offline handwriting recognition with multidimensional recurrent neural networks, pp. 1–8
20. A. Graves, Offline Arabic handwriting recognition with multidimensional recurrent neural networks
21. A. Graves, S. Fern, Multi-dimensional recurrent neural networks (2013), pp. 1–10
22. A. Graves, S. Fern, M. Liwicki, H. Bunke, Unconstrained online handwriting recognition with recurrent neural networks, pp. 1–8
23. I. Ahmad, X. Wang, R. Li, M. Ahmed, R. Ullah, Line and ligature segmentation of Urdu Nastaleeq Text, vol. 5 (2017)
24. Z. Ahmad, J. Khan, I. Shamsher, Urdu compound character recognition using feed forward neural networks (2009)
25. M. Seuret, M. Alberti, R. Ingold, M. Liwicki, PCA-initialized deep neural networks applied to document image analysis
26. S.B. Ahmed, S. Naz, M.I. Razzak, R. Yousaf, Deep learning based isolated Arabic scene character recognition, in *2017 1st International Workshop on Arabic Script Analysis and Recognition (ASAR)* (2017), pp. 46–51
27. D.C. Cires, U. Meier, L.M. Gambardella, Convolutional neural network committees for handwritten character classification, vol. 10 (2011), pp. 1135–1139
28. I. Ahmad, X. Wang, R. Li, S. Rasheed, Offline Urdu Nastaleeq optical character recognition based on stacked denoising autoencoder (2016), pp. 146–157
29. M.I. Razzak, F. Anwar, S.A. Husain, A. Belaid, M. Sher, HMM and fuzzy logic: a hybrid approach for online Urdu script-based languages' character recognition. Knowl.-Based Syst. **23**(8), 914–923 (2010)
30. S.B. Ahmed, S. Naz, S. Swati, M.I. Razzak, Handwritten Urdu character recognition using one-dimensional BLSTM classifier. Neural Computing and Applications, pp. 1–9 (2017)
31. R. Ahmad, M.Z. Afzal, S.F. Rashid, M. Liwicki, T. Breuel, A. Dengel, KPTI: Katib's Pashto text imagebase and deep learning benchmark (2016)

32. S. Naz, S. Bin, R. Ahmad, M. Imran, Zoning features and 2DLSTM for Urdu text-line recognition. Procedia—Procedia Comput. Sci. **96**(September), 16–22 (2016)
33. S. Naz, A.I. Umar, R. Ahmad, S.B. Ahmed, S.H. Shirazi, M.I. Razzak, Urdu Nasta'liq text recognition system based on multi-dimensional recurrent neural network and statistical features. Neural Comput. Appl. 2015
34. S. Naz et al., Offline Cursive Urdu-Nastaliq script recognition using multidimensional recurrent neural networks author's accepted manuscript. Neurocomputing (2016)
35. S. Naz, A.I. Umar, R. Ahmed, M.I. Razzak, S.F. Rashid, Urdu Nasta'liq text recognition using implicit segmentation based on multi-dimensional long short term memory neural networks. SpringerPlus (2016)
36. S.B. Ahmed, S. Naz, S. Swati, M.I. Razzak, Handwritten Urdu character recognition using one-dimensional BLSTM classifier. Neural Comput. Appl. 1–9 (2017)
37. I. Ahmad, X. Wang, M. Guang, H. Ahmad, R. Ullah, Ligature based Urdu Nastaleeq sentence recognition using gated bidirectional long short term memory. Cluster Comput. (2017)
38. F. Slimane, R. Ingold, S. Kanoun, A.M. Alimi, J. Hennebert, A new Arabic printed text image database and evaluation protocols (2009), pp. 0–4
39. S. Naz, A.I. Umar, R. Ahmad, I. Siddiqi, Urdu Nastaliq recognition using convolutional-recursive deep learning. Neurocomputing (2017)
40. S.B. Ahmed, S. Naz, M.I. Razzak, R. Yusof, T.M. Breuel, Balinese character recognition using bidirectional LSTM classifier, in *Advances in Machine Learning and Signal Processing* (Springer, Cham, 2016), pp. 201–211
41. H. El, A. Volker, ICDAR 2009-Arabic handwriting recognition competition (2011), pp. 3–13
42. V. Pham, Dropout improves recurrent neural networks for handwriting recognition
43. Y. Chherawala, P.P. Roy, M. Cheriet, Feature design for offline Arabic handwriting recognition: handcrafted vs automated? (2013)
44. T.I. Society, O. Engineering, A comparison of 1D and 2D LSTM architectures for the recognition of handwritten Arabic (2015)
45. S.B. Ahmed, S. Naz, M.I. Razzak, S.F. Rashid, M.Z. Afzal, T.M. Breuel, Evaluation of cursive and non-cursive scripts using recurrent neural networks. Neural Comput. Appl. **27**(3), 603–613 (2016)
46. R. Odate, H. Goto, Fast and accurate candidate reduction using the multiclass LDA for Japanese/Chinese character recognition, in *2015 IEEE International Conference on Image Processing (ICIP)* (IEEE, 2015), pp. 951–955
47. Y. Lu, J. Li, H. Zhang, S. Lin, Chinese character recognition of e-commerce platform pictures (2017), pp. 28–31
48. Z. Zhong, L. Jin, Z. Feng, I. Engineering, Multi-font printed Chinese character recognition using multi-pooling convolutional neural network (2015), pp. 96–100
49. Y. Tang, L. Peng, Q. Xu, Y. Wang, A. Furuhata, CNN based transfer learning for historical Chinese character recognition
50. X. Yu, W. Fan, J. Sun, S. Naoi, Semi-supervised learning feature representation for historical Chinese character recognition (2017), pp. 73–77
51. D. Cires, T.R.N. Idsia, D. Cires, Multi-column deep neural networks for offline handwritten Chinese character classification (2013)
52. N. Liu et al., Robust math formula recognition in degraded Chinese document images (2017)
53. D.K. Ning Liu, D. Zhang, X. Xu, L. Guo, L. Chen, W. Liu, MFR100 dataset
54. Z. Xie, Z. Sun, L. Jin, H. Ni, T. Lyons, Learning spatial-semantic context with fully convolutional recurrent network for online handwritten Chinese text recognition
55. C. Wu, W. Fan, Y. He, J. Sun, S. Naoi, Handwritten character recognition by alternately trained relaxation convolutional neural network (2014)
56. R. Messina, Segmentation-free handwritten Chinese text recognition with LSTM-RNN
57. R. Messina, J. Louradour, Segmentation-free handwritten Chinese text recognition with LSTM-RNN (2015), pp. 171–175
58. G. Jin, The PH corpus

59. A. Eisele, Y. Chen, Multi UN: a multilingual corpus from United Nation documents, in *Proceedings of the Seventh International Conference on Language Resources and Evaluation (LREC'IO)*, ed. by N.C.C. Chair, K. Choukri, B. Maegaard, J. Mariani, J. Odijk, S. Piperidis, M. Rosner, D. Tapias (European Language Resource, Valletta, Malta)
60. L.A. McEnery, Z. Xiao, The Lancaster Corpus of Mandarin Chinese: a corpus for monolingual and contrastive language study. Eur. Lang. Resour. Assoc. (2004)
61. Z. Yan, C. Yan, C. Zhang, Rare Chinese character recognition by radical extraction network (2017), pp. 924–929
62. C.-L. Liu, F. Yin, D.-H. Wang, Q.-F. Wang, Casia online and offline Chinese handwriting databases, in *2011 International Conference on Document Analysis and Recognition (ICDAR)* (IEEE, 2011), pp. 37–41
63. S. Nishide, H.G. Okuno, T. Ogata, J. Tani, Handwriting prediction based character recognition using recurrent neural network (2011), pp. 2549–2554
64. A. Chaudhuri, S.K. Ghosh, Optical character recognition system for Czech language using hierarchical deep learning networks, vol. 1
65. L. Chen, S. Wang, W. Fan, J. Sun, N. Satoshi, Deep learning based language and orientation recognition in document analysis (2015), pp. 436–440
66. D. Cires, U. Meier, Multi-column deep neural networks for image classification (2011)
67. L. Kang, J. Kumar, P. Ye, Y. Li, D. Doermann, Convolutional neural networks for document image classification (2014), pp. 3168–3172
68. L. Noce, I. Gallo, A. Zamberletti, A. Calefati, Embedded textual content for document image classification with convolutional neural networks (2016)
69. G. Zhong, H. Yao, Y. Liu, C. Hong, T. Pham, Classification of photographed document images based on deep-learning features, in *ICGIP 2016*, vol. 10225 (2017), pp. 1–6
70. T. Wang, D.J. Wu, A.Y. Ng, End-to-end text recognition with convolutional neural networks, in *ICPR* (2012), pp. 3304–3308
71. O. Morillot, L. Likforman-sulem, O. Morillot, L. Likforman-sulem, New baseline correction algorithm for text-line recognition with bidirectional recurrent neural networks neural networks
72. A. Graves, M. Liwicki, S. Ferna, R. Bertolami, H. Bunke, A novel connectionist system for unconstrained handwriting recognition. IEEE Trans. Pattern Anal. Mach. Intell. **31**(5), 855–868 (2009)
73. A. Ray, Text recognition using deep BLSTM networks
74. P.P. Roy, G. Zhong, M. Cheriet, Tandem hidden Markov models using deep belief networks for offline handwriting recognition. Front. Inf. Technol. Electron. Eng. **18**(61403353), 978–988 (2017)
75. S. Sudholt, G.A. Fink, A.W. Spotting, PHOCNet: a deep convolutional neural network for word spotting in handwritten documents (2016), pp. 277–282
76. A. Rehman, S. Naz, M.I. Razzak, Writer identification using machine learning approaches: a comprehensive review, in *Springer Multimedia Tools and Applications* (2018)
77. W. Feng, N. Guan, Y. Li, X. Zhang, Z. Luo, Audio visual speech recognition with multimodal recurrent neural networks (2017), pp. 681–688

Deep Learning for Driverless Vehicles

Cameron Hodges, Senjian An, Hossein Rahmani and Mohammed Bennamoun

Abstract Automation is becoming a large component of many industries in the 21st century, in areas ranging from manufacturing, communications and transportation. Automation has offered promised returns of improvements in safety, productivity and reduced costs. Many industry leaders are specifically working on the application of autonomous technology in transportation to produce "driverless" or fully autonomous vehicles. A key technology that has the potential to drive the future development of these vehicles is deep learning. Deep learning has been an area of interest in machine learning for decades now but has only come into widespread application in recent years. While traditional analytical control systems and computer vision techniques have in the past been adequate for the fundamental proof of concept of autonomous vehicles, this review of current and emerging technologies demonstrates these short comings and the road map for overcoming them with deep learning.

Keywords Deep learning · Autonomous · Driverless · Convolutional · Neural networks · Machine learning · Machine vision

---

C. Hodges (✉) · M. Bennamoun
School of Computer Science and Software Engineering, University of Western Australia, Crawley, Australia
e-mail: camhodges@gmail.com

M. Bennamoun
e-mail: mohammed.bennamoun@uwa.edu.au

S. An
School of Electrical Engineering, Computing and Mathematical Sciences, Curtin University, Bentley, Australia
e-mail: s.an@curtin.edu.au

H. Rahmani
School of Computing and Communications, Lancaster University, Bailrigg, UK
e-mail: h.rahmani@lancaster.ac.uk

V. E. Balas et al. (eds.), *Handbook of Deep Learning Applications*,
Smart Innovation, Systems and Technologies 136,
https://doi.org/10.1007/978-3-030-11479-4_4

1 Introduction

Automation is becoming a large component of many industries in the 21st century, in areas ranging from manufacturing, communications and transportation. Automation has offered promised returns of improvements in safety, productivity and reduced costs. Many industry leaders are specifically working on the application of autonomous technology in transportation to produce "driverless" or fully autonomous vehicles. Motivations for the introduction of automation vary broadly between different industries and applications although the immediate motivations for driverless vehicles can be seen in improvements in safety and productivity. The aim of this chapter is to outline the practical need for this technology and summarise some of the state of the art approaches to overcome the current technical challenges.

The world's current obsession with road transport, both freight and personal vehicles, comes at a cost of approximately 3300 deaths per day (2013 data) [1]. A report by the World Health Organisation (WHO) listed major risk factors towards road fatalities as driver fatigue, poor adherence to road traffic laws and drug and alcohol use [2], potentially all factors that could be addressed by automation technologies. Amongst wider society the potential to drastically reduce this figure is one of the most significant drivers of the development of driverless vehicles. Safety based motivations for the development of driverless vehicles are also shared with the Mining industry. Whilst safety standards in both underground and surface mining have increased significantly during the 20th century, it is still considered significantly more dangerous than other areas of general industry. Data from the United States shows that in 2015 there were approximately 3,500 mining related injuries and 17 deaths [3]. The mining industry thus has become one of the early adopters of fully driverless technology (Level 4 autonomy) in order to improve safety. Major mining companies such as BHP Billiton, Rio Tinto and Fortescue Metals Group are all utilising autonomous mining solutions in their Iron ore operations.

In addition to significant safety improvements the use of automation has also resulted in some significantly beneficial improvements in productivity. Recently Fortescue Metals Group has demonstrated that its fleet of 54 autonomous haul trucks operating at its Solomon Hub in Western Australia have provided a 20% productivity improvement. This is Compared to when those same trucks were operated by human operators [4]. This drive for autonomation in mining is not unique to Australia or the iron ore industry specifically, recently Suncor Energy Inc has also committed to the introduction of driverless trucks at it's oil sands operations in Canada [5]. Productivity improvements of driverless vehicles are also predicted to have a significant effect on the wider transportation industry. It has been reported that automation to facilitate slip streaming of road trucks in addition to other technological improvements has the potential to reduce energy consumption by up to 20% [6].

Structure of This Book Chapter

Whichever industry or environment driverless vehicles are intended to operate in the system architecture must address three major operational pillars. These can be summarised in Positioning, Path planning and control, and Obstacle detection. Prior to the recent acceptance of deep learning solutions these three elements were consistently addressed with traditional hard coded approaches. In Sect. 2 of this chapter we will summarise the current solutions used in both industry and academia and the challenges which are faced inpractice. In Sect. 3 we will outline the current state of research into deep learning based alternatives that could address many of the challenges faced with current technology.

2 Summary of Current Technology

2.1 Positioning

Among all the challenges faced in designing a driverless vehicle system perhaps the most fundamental is that of positioning. Whilst for the human brain perceiving our own position relatively to the surrounding environment appears to be second nature, for an autonomous system doing so to a high degree of accuracy is an incredibly complex task. Current solutions to this can be summarised into three categories; Global Navigation Satellite System (GNSS) based localisation, inertial navigation and perception based localisation.

With the phase out of selective availability on the US provided GPS system from May 2000 onwards positioning accuracy of less than 9 m horizontally is possible [7]. This when used solely on it's own can provide sufficient accuracy for many of today's applications although cannot be relied upon for driverless vehicle guidance. As a result of this there has been the development of GNSS systems with improved accuracy via the use of locally generate corrections [8, 9] known as differential GPS (DGPS). This approach has allowed for the initial development of autonomous driving systems in closed environments. Although it is susceptible to inaccuracy in environments where there is limited visibility of GNSS satellites or deflection of GNSS signals cause a multi path effect.

To further improve the robustness of GNSS based navigation there have been many examples of the use of inertial navigation. Systems are already commercially available from venders such as Canadian based Applanix [8]. They produce a system to argument the previously mentioned DGPS navigation with a combination of inertial measurement units and distance measurement units (based on wheel rotations) in order to provide high precision navigation and dead reckoning in the event of loss of GNSS navigation, a precision of 0.035 m in operation and 0.69 m in GNSS challenged conditions [8]. This technique for initial aided GNSS has also been the subject of research in academic groups. Jo et al. [10] demonstrated a proposed model in which GNSS information is combined with inertial sensors to provide high accuracy localisations. The main contribution of this work is to demonstrate the use of both dynamic and kinematic models to create localisations at different speeds.

2.2 Path Planning and Control Systems

Once an accurate position is determined then the next step is to undertake path planning and movement control. With the initial development of autonomous vehicle systems, the path planning approaches that were developed for traditional robotics were adapted for use. Algorithms such as [11] as well as variations of A* and D* path planning models [12] provided an important starting point for the control systems for autonomous vehicles but had some key limitations. Approaches such as A* [13] and D* have intrinsic ability to plan around obstacles or area of undrivable ground although in general do not account for the curvature limits of automotive steering. In contrast the Reeds-Shepp curves [11] do account for curvature limitations although they assume the drivable environment is a continuous plane without obstacles.

Broadly planning techniques can be categorized into 2 differing approaches. Local coordinate systems which determine paths relative to local features detected with a perception system [14] or in contrast global based coordinates that can have sensor imports from a positioning system such as GPS or inertial aided GPS [12]. Work by [14] demonstrates the advantages of the locally based path planning approach as it allowed for the dynamic planning of paths around detected obstacles. The proposed algorithm worked in stages to initially propose a base frame path before creating a selection of alternative paths offset from this base frame. The key following step that distinguishes this approach from the Reeds-Shepp curve [11] is to assess each alternative path with a cost function that can determine the probability of collision with detected obstacles. In addition to this improved functionality by planning dynamically as the vehicle moves it limits the memory consumption that becomes a significant challenge when utilising a global planning algorithm over a large environment.

A practical application of these concepts can be seen in [15] who demonstrated an approach called hybrid A* which was based on a multistage search algorithm. The fundamental initial step is based around traditional A* [13] although with a confinement that forced the algorithm to develop paths which are driveable with in the steering curvature limits of the vehicle. Following this there is an alternating step of applying a Reed-Shepp curve to determine at which point when approaching the goal is there an obstacle free path which conforms with the final vehicle heading required. The author then applies a gradient decent based optimisation step to ‘smooth’ the final path to limit oversteering of the vehicle. This optimisation step which produces a more continuous and smooth result introduces some additional complexities. Due to the nature of the ‘smoothing’ process some precision in the original path can be lost which can increase the probability of collision with an obstacle in regions where the optimum path passes close to an obstacle.

2.3 *Obstacle Detection*

The third pillar of an autonomous vehicle system is the obstacle detection system. This is crucial as driverless vehicles are operating in complex environments and they require a system that will allow them to interact with and respond to a multitude of hazards, including other vehicles, people, and the elements of the physical environment. Different forms of these object detection systems have been proposed to utilise a range of computer vision systems, including laser scanners, visible light/Infrared cameras and radar. The use of laser scanners (also known as lidar) has become synonymous with driverless vehicle technology due to their use on the Waymo (Formally Google Car) project and with the current development of autonomous mining trucks.

There has been a significant amount of literature demonstrating the use of traditional computer vision based obstacle detection systems, dating as far back as 2000, with work done by the CSIRO [16]. In these early examples lidar point clouds were analysed for basic geometry and gradient to assess the driveability of the surface and detect basic obstacles. This type of approach has been extensively covered in the years since [17, 18]. This literature review is based on its input data source.

2.3.1 Lidar

Point Cloud Clustering—[19] demonstrated an attempt to detect roadside obstacles. In this example housing bricks were the target object, by clustering analysis of lidar point clouds. Raw lidar data was processed through several filters to normalise for elevation along the contours of the road surface before the application of some heuristics to pair point clusters with what the author refers to as 'holes' or regions that were shadowed from the lidar. Subsequently from the cluster/hole pairs 16 handcrafted specific features were extracted for classification by a random forest tree machine learning algorithm. While this was successful under test conditions, the reliance on handcrafted features can be very labour intensive to code and cannot guarantee to be effective on a wide range of ambiguous conditions.

Line Segment Extraction—[18] like the previous work by [19] based their object detection algorithm around lidar data. In contrast to [19], instead of working around point clusters [18] attempted to extract line segment features from the lidar data. The observed lidar data is analysed to extract lines based on hard coded thresholds. These lines are then classified as obstacles or drivable road segments. This functionality is important when compared with other obstacle detection strategies for 2 key reasons:

1. The ability to positively identify drivable road segments, as opposed to only seeing them, as an absence of obstacles allows this system to locate itself on the road and provide left to right vehicle guidance.
2. The utilisation of lidar data relies on the road geometry and therefore is functional on surfaces without painted road markings and lines.

The major disadvantage of this type of approach is the reliance on hardcoded thresholds. For example, the system requires 12 specific thresholds in order to classify the line segments into only 2 classes (as either obstacles or road).

2.3.2 Single and Stereo RBG Cameras

Selective Search—[20] documented a process referred to as selective search for object classification. This method takes advantage of the hierarchal nature of images and proposed a method of segmenting increasingly larger portions of the images based on several grouping strategies. Initially image regions are grouped based on a measure of the similarity of pixel colours. While this allows for the segmentation of homogenous objects it can struggle with more complex classifications. The example given by the author is a car of which the wheels are of a dramatically different colour to the car body but are part of the same object for the purposes of classification. To improve on these shortcomings a subsequent step of grouping by a strategy known as bag-of-words is used. This is done by classifying objects based on the rate of occurrence of predefined "vocabulary" of image patches that are matched to known object classes.

Active Contour Modelling—Work by [21] demonstrated the application of active contour modelling for object detection and classification. Using depth maps extracted from stereo cameras the author then, by means of the active contour models, a series of edges were detected from each object in the frame. This allowed the classification of objects into classes (vehicles, pedestrians or other) based on manually specified features such as height and aspect ratio to the observer. The study achieved a true positive rate of 92.2% and a false positive rate of 2.7% in the test cases. This type of detection and classification can be effective. However, it can be difficult to scale to more classes as the number of manually coded features becomes unmanageably large.

While these types of detection strategies have been demonstrated to be effective, both academically and in practical examples, such as the DARPA Urban Challenge [22] it is still characterised by the reliance on hard coded thresholds by which data is analysed. It therefore can be limited in complex environments and produces binary results, by that obstacles are only determined to be either obstacles or not obstacles. There is very little context provided to allow for an autonomous system to react with any nuance. As a result of this a more data driven approach similar to the current machine learning techniques needs to be proposed in order to address these shortcomings.

3 Deep Learning for Driverless Vehicles

A key technology that has the potential to overcome the challenges previously discussed is deep learning. Deep learning has been an area of interest in machine learning for decades now [23, 24] but has only come into widespread application in recent years. One supervised deep learning approach known as Convolutional Neural Networks (CNN) that allows for the recognition of complex patterns based on training data has become the state of the art approach for many elements of driver vehicle technologies.

Deep learning based approaches have become the preferred solution for many problems in driverless vehicles [25]. The basic application of deep learning in driverless vehicles can be characterised in two separate ways, a compartmentalised architecture and an end to end architecture. A compartmentalised architecture can be described as an approach in which the major functions of the driving system (obstacle detection, positioning and path planning) are designed as discrete processes. While each process may consist of one or more neural networks the interactions between each major system is hard coded in a traditional sense. In contrast an end to end architecture allows all three major functions to be handled by a single neural network [26]. Next we describe current state of research in the compartmentalised approach before discussing the end to end architecture in detail.

3.1 Deep Learning Based Approaches for Positioning

As driverless technologies are deployed into an increasingly wide range of environments the need for robust positioning is becoming ever more important. As previously detailed the majority of positioning requirements are addressed using GNSS solution. These are often supplement with inertial based aids although these approaches are often still challenged in many environments. While the focus of deep learning for driverless vehicle applications is on areas of control and vision, positioning is still an area of keen interest.

Work by [27] demonstrated the use of a neural network to define a model for position based on WIFI signal within an interior space. The author focuses on a two-phased process, initially during the training phase data samples are taken of the wifi signal including the environmental induced noise at known locations in the space. Based on this in the testing phase the receiver's location is probabilistically determined based on the known signal characteristics through the space. Zhang et al. [28] also applied a deep net model and demonstrated some key improvements over traditional measurement based and machine learning approaches. The proposed method allows the training of positioning models based on very large data sets which can be evaluated by means of a feed forward process by the receiver with very limited resources. In contrast machine learning processes such as k-nearest neighbour requires the receiver device to store and evaluated the entire training data set for each positioning attempt.

These deep learning based approaches have demonstrated positive results in applications such as localisation of consumer smart devices when compared with traditional time of flight or direction methods. While these approaches may have some suitability to the challenges of vehicle positioning there is no significant body of work demonstrating it. In contrast vision based localisation, or sometimes called perception based localisation, has shown greater promise. Work by [29] demonstrated a methodology where single monocular images were used as a input for a deep neural network to estimate pose of the camera. This approach constructs a model of pose within a 3D environment based on a large volume of training images from which a 3D point model is determined. This 3D point model, with some subsequent compression for efficiency, provides high accuracy localisations in both indoor, outdoor and freeway driving settings. In the authors experimental evaluation, it is shown that an accuracy of 37.2 cm is achieved in subsequent trials. The work also illustrated the approach working based on images from an automotive dash cam as we would expect for a driverless vehicle application, although there is not specific data on accuracy in this use.

3.2 *Deep Learning Based Approaches Obstacle Detection*

The requirement to assess and react to complex surrounding environments can be very difficult to hard code in a control function. Therefore, obstacle detection is an element of driverless vehicles that is ideally suited to a deep learning solution. The fundamental process of extracting features on a pixel by pixel basis that was demonstrated by [23]. It is still core to object detection systems for driverless vehicles and modern adaptations have significantly improved the performance.

Prior to the current widespread adoption of Convolutional neural networks there were some attempts at implementations of these for driverless vehicle systems, although these attempts were considered exceptions to the more standard practice of hand crafted feature extraction. Work by [30] demonstrated the use of a CNN for detection of obstacles in a YUV encoded image. The proposed application in this example was an unmanned rover in off-road terrain including grass and wooded areas. 3 channel input images were passed through the series of convolutional layers outputting 20 feature maps followed by max pooling which inputs into an additional convolutional layer outputting 100 feature maps.

The resurgence of Convolutional Neural Network techniques in the recent time can be traced to a publication by [31]. In this work the authors proposed an 8 layer deep network which consisted of 5 convolutional layers and 3 fully connected layers. With this architecture the authors based their learning process on 1.2 million images from the Imagenet data base. These training examples allowed the models weights to be optimised using a gradient decent type optimisation based on manually selected learning rates. This literature is particularly pivotal in the development of deep learning techniques, as it demonstrated the value of CNN architecture for image recognition based on very large learning data sets. Newer image databases such as ImageNet [32] have allowed for the training of large deep nets that previously were not possible due to lack of suitable data and computing power.

The work by [31] was later further improved by [33] with small twists on the network architecture, these included increase the stride in the first 2 layers, removing the overlapping pooling layers and no including contrast normalisation. A particularly important aspect that is first demonstrated by the author is the concept of a sliding frame to apply the CNN across the entire image. This is in contrast to the [31] which applied the CNN to the entire image simultaneously. This significantly improved the efficiency of the classification of the scene as convolutions could be retained for multiple overlapping bounding boxes. The performance of these networks were tested in the ILSVRC challenge (Fig. 1).

The work by [34] is a particularly important milestone in the development of deep learning techniques for obstacle detection because the author was able to demonstrate that the architecture shown by [31], and later developed by [33], could be applied to object detection within a scene and not only to image classification as demonstrated in the earlier work. This is particularly important as it allows for the identification of individual objects within a scene, while earlier work was limited to labelling the entire image (although with some elements of localisation within the image). While the CNN architecture was largely taken from the work of [33] it was applied in a slightly different manner. In order to identify objects within a scene the input image was broken into approximately 2000 proposed regions using a selective search method. From this the CNN extracts features for classification by a Support Vector Machine (SVM). The performance of this approach was demonstrated with the PASCAL VOC challenge for object recognition with a mean average precision of 53.7%. The author coined the term R-CNN to describe this new technique. While this was an important step in the development of a CNN architecture for detection, it did have some key limiting factors. Selective search can be described as a traditional computer

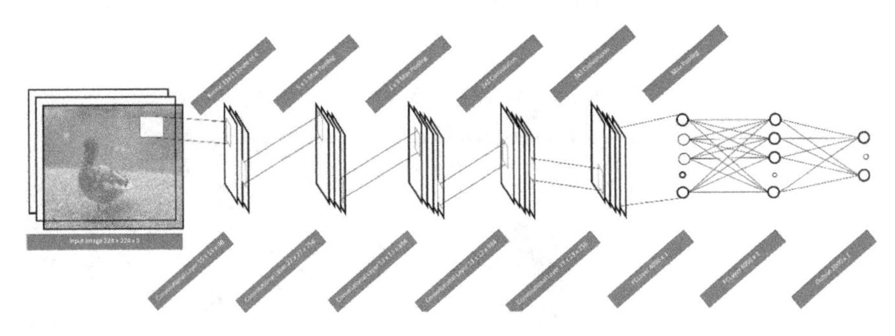

Fig. 1 Alexnet architecture demonstrating convolution, pooling and fully connected layers [31]

vision approach and is slow to generate proposal regions. Additionally, it was a very computationally intensive process by approaching each of the 2000 proposed regions individually for feature extraction.

While most studies in this area have focused on network architectures of up to ten layers [31, 33] the work by [35] demonstrated the value of very deep networks (up to 19 layers) for image recognition. Based on what is now referred to as the VGG16 architecture, the author outlined a network of multiple Conv-pooling-ReLU layers based around a fixed kernel size of 3×3 pixels and stride of 1 pixel. This arrangement is both much deeper than previous works and works with a much smaller kernel and stride than typically seen. Upon evaluation of the networking using the ILSVRC 2012 dataset the authors came to two significant conclusions

1. When comparing multiple network architectures, ranging from 11 layers to 19 layers, it was found that based on this dataset the deep networks performance was consistently better.
2. The 3×3 kernel size is important as it provides spatial context of up, down, left and right, in comparison to a 1×1 kernel in 3 layers the 3×3 kernel performs better.

With respect to the first conclusion the authors made a specific qualification that the network became saturated at 19 layers based on the training dataset. It could be hypothesised that with a larger training set additional layers could be valuable.

Whilst the body of work already conducted primarily focuses on general image and scene understanding in contrast work by [36] is an excellent example of the application of deep learning in driverless vehicles for two critical reasons;

- The work identified key speed improvements compared to the work of [34].
- There is a clear confirmation that the fundamental concepts of object recognition and segmentation that allow for scene understanding and labelling can provide a driverless vehicle with perceptive situational awareness (Fig. 2).

In the area of processing speed [36] outlines the shortcomings of the work by [34]. In this previous work the CNN was used for feature extraction on each regions of interest individually without any attempt to reuse convolution layers for overlapping regions of interest. By improving this, [36] was able to demonstrate a functional system at 10 Hz which is suitable for a highway driving environment. In addition,

Fig. 2 Illustration of lane and vehicle detection demonstrated by [36]

this sliding frame approach could classify multiple instances of the same object within the scene. Moreover, the work also demonstrated the use of actual driving environment data as opposed to image databases. The system was able to identify and localise multiple other cars as well as road markings.

Inspired by the work of [34], work by [37] applies a variation of the R-CNN approach to traffic sign detection in an automotive application. The approach shown is again a regional proposals type technique in which first regions of interest (ROI) are proposed, in this case using a RGB space thresholding technique followed by edge detection and connected component analysis in order to isolate traffic signs for classification, as seen in Fig. 3. This author acknowledges that this method is more efficient then the selective search used by [34] although is only acceptable in the application due to the predictable and consistent design of traffic signs, it is unclear if this would be suitable for more diverse object recognition. These regional proposals are then classified using a CNN framework largely taken from earlier work by [38], this architecture can be seen in Fig. 4: 3 Layer CNN architecture.

Training of the network has two distinct phases. Firstly, initialization of the network is based on randomly generated weightings following by training based on a dataset of 283,304 images from 96 classes, which is constructed from 3 publicly available image databases. Following this there is a fine tuning stage based on a specific Chinese traffic sign image dataset, this dataset contained again 96 classes across 100,000 samples with a 80:20 split for training and testing. This concluded with a detection and classification accuracy of 97.6% and 98.6% respectively.

Follow previous work [39] proposed an improved approach to the authors previous work in order to better address the slow speed of the R-CNN algorithm, this is known as Fast R-CNN. While not a complete architecture it is instead an additional layer to the VGG16 architecture.

More recent work has demonstrated the expanded capabilities of the CNN architecture. Zhu et al. [40] used RGB camera data in order to identify multiple classes of traffic signs (45 different classes in total). The testing dataset included 10,000 images taken in real world street environments. These 10,000 images included a total of 30,000 individual traffic signs. This training data set was further expanded

Fig. 3 Visual representation ROI proposals for classification [37]

synthetically through data augmentation (introducing random elements of skew, rotation and scale). Using this dataset, the authors propose an 8 layer fully convolutional network, this architecture does not include any fully connected layers, to detect and classify street signs and compared two methods for classification, Fast CNN-R and their new method. The new method showed consistently higher recall at all accuracy levels.

The work in [40] demonstrated the CNN architectures ability to classify many different classes of objects which was a crucial step compared to previous work such as [36, 41] which only focused on approximately 4 classes or less. The authors in [42] proposed a parallel algorithm based on the VGG16 deepnet. In which a 2 stage CNN architecture allows for obstacle detection, classification and segmentation based on a common convolution network. This work was particularly important as it illustrated that using a single convolutional network multiple, perception tasks that are required for driverless vehicles can be accomplished at a rate required for real work applications (10 Hz). All training and testing for this network was done with the KITTI dataset [43]. A comparison work to this can be seen in [44]. The author documents a multi stage approach to detection of rear automotive brake lights that is applicable to both driver assistance or driverless applications. This multistage approach follows a similar framework as seen in earlier object recognition examples such as [34] in which conventional computer vision methods. In this case a Gaussian model which takes an initial sample of road surface from immediately in front of the vehicle which is assumed to be driveable. It then uses the characteristics of this to segment the surrounding area into either road or no road sections, this initial segmentation is primarily intended to improve the rate of false positives. Once this

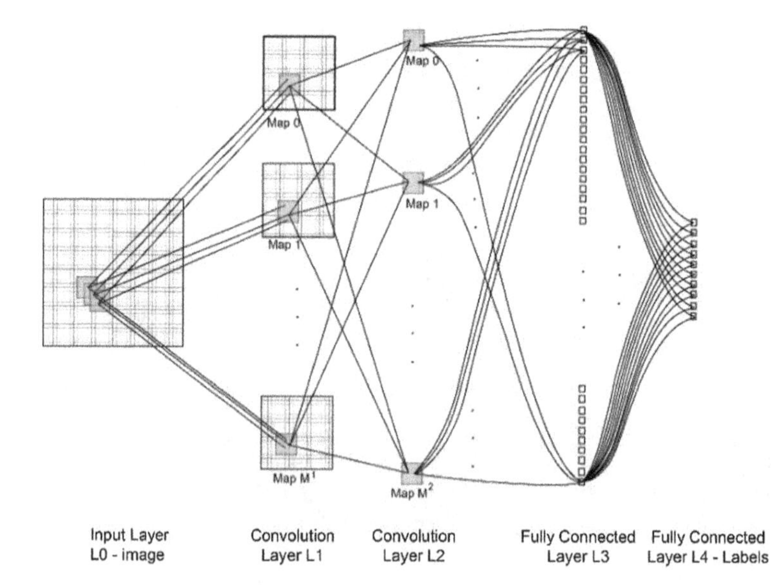

Fig. 4 3 Layer CNN architecture [38]

step is completed the authors then propose a vanishing point detection method to propose regions of interest for classification, in contrast [34] used selective search for this process. By using an assumed vehicle size and aspect ratio the method was able to propose only regions of interest that had a high probability of containing a vehicle. This in turn reduced the computation load on the subsequent CNN classifier. The final stage of the detection process is to apply an 8 layer Alexnet CNN framework as initially demonstrated by [31] to classify the selected ROI into classes of either brake lights on or brake lights off. This work demonstrated two important characteristics;

1. Successfully demonstrated detection accuracies of 89%.
2. Operated at a rate of 26–33 FPS which is sufficient for real world driving environments.

In addition to the many academic examples, industrial examples of deep learning for object detection are also beginning to appear. Nvidia Corporation [45] in their 2016 CES keynote demonstrated their Drivenet architecture. This deepnet was trained in a two stages process, initially the basic feature extractor was trained using a modified version of the Imagenet data set. Subsequently it was trained specifically for a driving environment with the KITTI dataset. This presentation is particularly important for the development of deep learning in driverless vehicles because it includes an example of challenging real world data. While most academic examples, and NVidia's earlier work, include real world street environments for testing, they are generally taken in clear weather with very little environmental effects. In contrast NVidia's key note demonstration culminates with an example of a highway environment with heavy fog, snow and road spray which makes the detection task very difficult. In this environment, the deepnet proves capable of detecting cars are a distance that is even challenging for the human eye.

3.3 Control Systems/End to End Deep Learning Architecture

Whilst the majority of research thus far into the application of deep learning in autonomous vehicles has been for segmented systems (i.e. object detection) there is a new field of research into utilising deep networks for complete vehicle control, this is known as End to End Architecture. This approach is different in that the specific functions of the autonomous system are not individually considered, instead the input to the deep net consists of sensors such as cameras and the weight parameters are trained based a training signal input from a person driving the vehicle [46]. Some early examples of this include [24] demonstration of vehicle steering control by deep learning and the DARPA-IPTO project. In this later example, a small remote control car was training using a 5 layer, as seen in Fig. 5, CNN network in order to navigate passed obstacles in a controlled real world environment [47].

This work was limited due to the type of computer hardware that was available at the time. A more modern demonstration of this concept can be seen in [48] who worked on a project by the Nvidia Corporation.

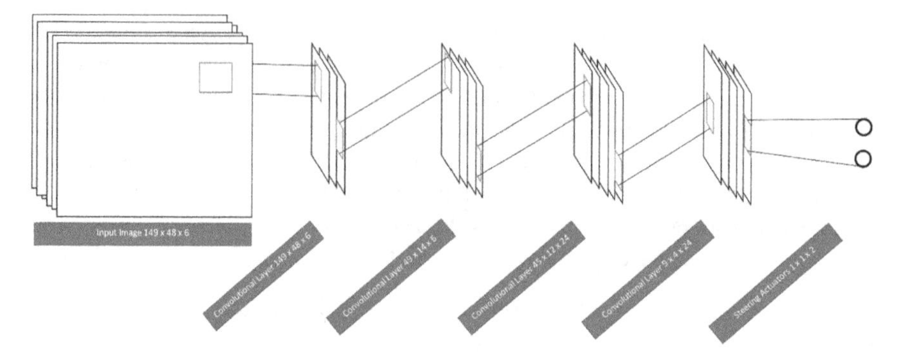

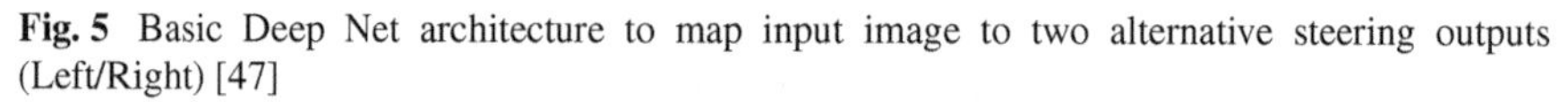

Fig. 5 Basic Deep Net architecture to map input image to two alternative steering outputs (Left/Right) [47]

This project demonstrated two key aspects;

1. The CNN architecture can be trained with limited training data to control steering of the vehicle.
2. Without specific programming, the algorithm can detect important features (such as road edges) within a scene with only the steering input signal.

This work used an architecture based on 7 convolutional layers and 4 fully connected layers in order to map the steering angle training signal to the raw pixel input from a front facing RGB camera.

Nvidia's demonstration system was trialled on highway conditions and was able to control the vehicle autonomously 98% of the time. Similar work is also underway by California based company Drive AI, the trial vehicles have been operated on roads around Silicon Valley. Although this application of deep learning is still in its initial stages with very limited amounts of publicly available literature.

El Sallab et al. [49] provides a more detailed view at a possible End-to-End reinforcement learning solution for vehicle control. The author is able to provide two alternative algorithms for mapping a training signal of steering angle to raw pixel inputs. The two proposed algorithms referred to as Deep Q-network Algorithm (DQN) and Deep Deterministic Actor critic (DDAC) are based around a series of discrete inputs and continuous inputs respectively. Based on the works scenario of controlling the steering input to a car operating in a simulated environment the DQN algorithm demonstrates the basic function of creating a steering model based on the raw pixel input which is function on straight sections of road. In contrast as the model must work with discrete inputs in which the real continuous input must be approximated to the best fit for the available training examples it's performance is degraded in corners as it cannot respond to continuous curves. In contrast to this the DDAC algorithm allows for a continuously interpretation of the raw pixel input data and therefore a more continuous response.

Despite these rapid advancements in end-to-end architecture there are still some challenges for this approach. In an article [25] on the progress of Silicon Valley start-up Drive AI who are developing driverless vehicle technologies based around deep networks they acknowledges some of challenges presented by the black box nature of deep learning architectures. As the functionality of these networks are based on an accumulation of huge amounts of training data it is difficult to conceptualise and challenge its decision-making process. As a result of this the Drive AI system does not implement an entirely end-to-end architecture but instead compartmentalises to some degree the core vehicle functions to allow them to be interrogated independently when testing input data.

4 Conclusion

The development of driverless vehicle technologies has become a key motivation for the improvement and innovation in deep learning techniques. While traditional analytical control systems and computer vision techniques have in the past been adequate for the fundamental proof of concept of autonomous vehicles, the exponential increase in complexity as systems attempt to capture every nuance of the surrounding environment is approaching their fundamental limitations. At the current state of research deep learning techniques are becoming comprehensively part of the state of the art technology for all elements of autonomous driving. Although development efforts must still address some challenges in deep learning including the construction of large application specific training datasets and more robust and transparent techniques for validating system architectures and functions.

References

1. World Health Organisation, Road traffic deaths—2013. http://apps.who.int/gho/data/node.main.A997. Accessed 26 Mar 2017
2. World Health Organisation, *World Report on Road Traffic Injury Prevention* (World Health Organisation, Geneva, Switzerland, 2004)
3. Mine Safety and Health Administration, Mining Industry Accident, Injuries, Employment, and Production Statistics and Reports, 3 Mar 2017. https://arlweb.msha.gov/Stats/Part50/WQ/MasterFiles/MIWQ-Master-2015-final.pdf
4. Fortescue Metals Group Limited, Investor Briefing 2016, 25 Oct 2016. http://fmgl.com.au/media/2899/161025_investor_briefing.pdf. Accessed 15 Nov 2016
5. A. Topf, Canadian oil sands giant testing autonomous haul trucks (2016). http://www.mining.com/canadian-oil-sands-giant-testing-autonomous-haul-trucks/. Accessed 5 May 2017
6. Business Wire, Peloton Announces U.S. Department of Energy-funded Team to Cut Heavy Truck Fuel Use by 20 Percent with Smart Powertrains and Platooning, 14 Nov 2016. http://www.businesswire.com/news/home/20161114006693/en/Peloton-Announces-U.S.-Department-Energy-funded-Team-Cut
7. U.S Airforce, Global Positioning System Standard Positioning Service Performance Standard (2008). http://www.gps.gov/technical/ps/2008-SPS-performance-standard.pdf. Accessed 20 May 2017

8. Applanix, Position and Orientation System for Land Vehicles—Datasheet, 26 Mar 2017. https://www.applanix.com/downloads/products/specs/POSLV_DS_feb_2017_yw.pdf
9. S. Ericson, Vision-Based Perception for Localisation of Autonomous Agricultural Robots (2017)
10. K. Jo, K. Chu, K. Lee, M. Sunwoo, Integration of multiple vehicle models with an IMM filter for vehicle localization, San Diego (2010)
11. J.A. Reeds, L.A. Shepp, Optimal paths for a car that goes both forwards and backwards. Pac. J. Math **145**(2), 367–393 (1990)
12. D. Ferguson, A. Stentz, The field D* algorithm for improved path planning and replanning in uniform and non-uniform cost environments (2005)
13. P.E. Hart, N.J. Nilsson, B. Raphael, A formal basis for the heuristic determination of minimum cost paths. IEEE Trans. Syst. Sci. Cybern. **4**(2), 100–107 (1968)
14. K. Chu, M. Lee, M. Sunwoo, Local path planning for off-road autonomous driving with avoidance of static obstacles. Intell. Trans. Syst. **13**(4), 1599–1616 (2012)
15. D. Dmitri, S. Thrun, M. Montemerlo, J. Diebel, Path planning for autonomous vehicles in unknown semi-structured environments. Int. J. Robot. Res. **29**(5), 485–501 (2010)
16. J. Roberts, P. Corke, Obstacle detection for a mining vehicle using a 2D laser, in *Proceedings of the Australian Conference on Robotics* (2000)
17. A. Stentz, A. Kelly, P. Rander, H. Herman, O. Amidi, Real-time, multi-perspective perception for unmanned ground vehicles (2003)
18. J. Han, D. Kim, M. Lee, M. Sunwoo, Enhanced road boundary and obstacle detection using a downward-looking LIDAR sensor. Trans. Veh. Technol. **61**(3), 971–985 (2012)
19. J. Tang, A. Zakhor, 3D object detection from roadside data using laser scanners, in *Three-Dimensional Imaging, Interaction, and Measurement* (2011)
20. J.R.R. Uijings, K.A. van de Sande, T. Gevers, A.W.M. Smeulders, Selective search for object recognition. Int. J. Comput. Vis. **104**(2), 154–171 (2013)
21. Y. Huang, S. Liu, Multi-class obstacle detection and classification using stereovision and improved active contour models. Inst. Eng. Technol. **10**(3), 197–205 (2016)
22. U. Ozguner, C. Stiller, K. Redmill, Systems for safety and autonomous behavior in cars: the DARPA grand challenge experience. Proc. IEEE **95**(2), 397–412 (2007)
23. L.D. Jackel, D. Sharman, C.E. Stenard, I.B. Strom, D. Zucket, Optical character recognition for self-service banking (1995)
24. D. Pomerleau, T. Jochem, Rapidly adapting machine vision for automated vehicle steering. IEEE Expert 19–27 (1996)
25. E. Ackerman, How Drive.ai is mastering autonomous driving with deep learning, 10 Mar 2017. http://spectrum.ieee.org/cars-that-think/transportation/self-driving/how-driveai-is-mastering-autonomous-driving-with-deep-learning. Accessed 26 Apr 2017
26. Mobileye, The Three Pillars of Autonomous Driving, 20 June 2016. https://www.youtube.com/watch?v=GZa9SlMHhQc. Accessed 20 Apr 2017
27. X. Wang, L. Gao, S. Mao, S. Pandey, CSI-based fingerprinting for indoor localisation: a deep learning approach. IEEE Trans. Veh. Technol. **66**(1) (2016)
28. W. Zhang, K. Liu, W. Zhang, Y. Zhang, J. Gu, Deep neural networks for wireless localization in indoor and outdoor environments. Neurocomputing **194**, 279–287 (2016)
29. K.-W. Chen, C.-H. Wang, X. Wei, Q. Liang, C.-S. Chen, Vision-based positioning for internet-of-vehicles. IEEE Trans. Intell. Transp. Syst. **18**(2), 364–376 (2017)
30. R. Hadsell, A. Erkan, P. Sermanet, M. Scoffier, U. Muller, Y. LeCun, Deep belief net learning in a long-range vision system for autonomous off-road driving, in *International Conference on Intelligent Robotic Systems* (2008), pp. 628–633
31. A. Krizhevsky, I. Sutskever, G. Hinton, ImageNet classification with deep convolutional neural networks. Adv. Neural. Inf. Process. Syst. **2**, 1097–1105 (2012)
32. Stanford Vision Lab, ImageNet (2017)
33. P. Sermanet, D. Eigen, X. Zhang, M. Mathieu, R. Fergus, Y. LeCun, OverFeat: integrated recognition, localization and detection using convolutional networks (2013)

34. R. Girshick, J. Donahue, T. Darrell, J. Malik, Rich feature hierarchies for accurate object detection and semantic segmentation, in *Computer Vision and Pattern Recognition* (2014)
35. K. Simonyan, A. Zisserman, Very deep convolutional networks for large-scale image recognition, in *International Conference on Learning Representations* (2015)
36. B. Huval, T. Wang, S. Tandon, J. Kiske, W. Song, J. Pazhayampallil, M. Andriluka, P. Rajpurkar, T. Migimatsu, R. Cheng-Yue, F. Mujica, A. Coates, A. Ng, An empirical evaluation of deep learning on highway driving (2015)
37. R. Qian, B. Zhang, Y. Yue, Z. Wang, F. Coenen, Robust Chinese traffic sign detection and recognition with deep convolutional neural network, in *International Conference on Natural Computation* (2015), pp. 791–796
38. D. Ciresan, U. Meier, J. Masci, J. Schmidhuber, A committee of neural networks for traffic sign classification, in *Proceedings of International Joint Conference on Neural Networks* (2011), pp. 1918–1921
39. R. Girshick, Fast R-CNN, in *International Conference on Computer Vision* (2015)
40. Z. Zhu, D. Liang, S. Zhang, X. Huang, B. Li, S. Hu, Traffic-sign detection and classification in the wild, in *Computer Vision and Pattern Recognition* (2016)
41. D.V. Prokhorov, Road obstacle classification with attention windows, in *Intelligent Vehicles Symposium,* June 2010
42. M. Teichmann, M. Weber, M. Zollner, R. Cipolla, R. Urtasun, Real-time joint semantic reasoning for autonomous driving (2016)
43. A. Geiger, P. Lenz, R. Urtasun, Are we ready for autonomous driving? The KITTI vision benchmark suite, in *IEEE CVPR* (2012)
44. J.-G. Wang, L. Zhou, Y. Pan, S. Lee, Z. Song, B.S. Han, V.B. Saputra, Appearance-based Brake-Lights recognition using deep learning and vehicle detection, in *IEEE Intelligent Vehicle Symposium (IV)*, 19–22 June 2016
45. Nvidia Corporation, CES 2016: NVIDIA DRIVENet Demo—Visualizing a Self-Driving Future (Part 5), 1 Jan 2016. https://www.youtube.com/watch?v=HJ58dbd5g8g. Accessed 9 Apr 2017
46. Mobileye, Autonomous Car—What Goes into Sensing for Autonomous Driving? (2016). https://www.youtube.com/watch?v=GCMXXXmxG-I. Accessed 28 Apr 2017
47. Net-Scale Technologies Inc, *Autonomous Off-Road Vehicle Control Using End-to-End Learning* (DARPA/CMO, Arlington, VA, 2004)
48. M. Bojarski, D. Del Testa, D. Dworakowski, B. Firner, End to end learning for self-driving cars (2016)
49. A. El Sallab, M. Abdou, E. Perot, S. Yogamani, End-to-end deep reinforcement learning for lane keeping assist, *Presented at the Machine Learning for Intelligent Transportation Systems Workshop* (2016)

Deep Learning for Document Representation

Mehran Kamkarhaghighi, Eren Gultepe and Masoud Makrehchi

Abstract While machines can discover semantic relationships in natural written language, they depend on human intervention for the provision of the necessary parameters. Precise and satisfactory document representation is the key to supporting computer models in accessing the underlying meaning in written language. Automated text classification, where the objective is to assign a set of categories to documents, is a classic problem. The range of studies in text classification is varied, ranging from studying a sophisticated approach for document representation to developing the best possible classifiers. A common representation approach in text classification is bag-of-words, where documents are represented by a vector of the words that appear in each document. Although bag-of-words is very simple to generate, the main challenge in such a presentation is that the resulting vector is very large and sparse. This sparsity and the need to ensure semantic understanding of text documents are the major challenges in text categorization. Deep learning-based approaches provide a fixed length vector in a continuous space to represent words and documents. This chapter reviews the available document representation methods that include five deep learning-based approaches: Word2Vec, Doc2Vec, GloVe, LSTM, and CNN.

Keywords Deep learning · Document representation · Word2Vec · Doc2Vec · GloVe · LSTM · CNN

1 Introduction

The current wide ranging application of text mining includes the domains of social media, information retrieval, legal document processing, and marketing. Document representation establishes the computational cost and performance of tasks such as machine-based translation, content analysis, clustering, and classification [1]. In natural language processing, the well-known bag-of-words approach, in which every

M. Kamkarhaghighi (✉) · E. Gultepe · M. Makrehchi
University of Ontario Institute of Technology, Oshawa, ON, Canada
e-mail: mehran.kamkarhaghighi@uoit.ca

V. E. Balas et al. (eds.), *Handbook of Deep Learning Applications*,
Smart Innovation, Systems and Technologies 136,
https://doi.org/10.1007/978-3-030-11479-4_5

document is represented as a bag of words, is heralded as a principal document representation model [2]. Although offering both speed and a low cost, this approach fails to focus on either grammar or word order. As a consequence, it is deemed to carry the "Curse of Dimensionality", in that even short sized sentence needs a high dimensional sparse feature vector for representation. In this situation, machine learning algorithms may lose their power of discrimination because of the curse of dimensionality. An alternative approach to the bag-of-words is word embedding, where words are mapped to fixed length dimensional vectors in a continuous space. Two popular deep learning-based word embedding models are Word2Vec [3] and Global Vector (GloVe) [4], which present a fixed-length vector for each word of the training data. In order to work correctly, these models need to be trained with large-sized corpuses. Moreover, neither the Word2Vec nor GloVe models disregard the relationship between non-co-occurred terms.

Calculating the summation or the average of a document's word vectors are common approaches used to represent a document based on its word vectors. Nevertheless, these approaches cannot reflect the context of the document. For instance, the word vector for "blackberry" as a company is equivalent to its word vector as a fruit. For each document or paragraph, the Doc2Vec model [5] offers a vector that is trained according to the local data. It can comprise the context of the document without the use of background knowledge. However, the high computational cost of having to produce a model each time is a major weakness of this approach. In contrast, Word2Vec and GloVe have the capacity to create one-time only models, based on the training corpus.

2 Traditional Document Representation Methods

The bag-of-words approach, in which each document is represented as a bag of words, is the most well-known representation models in natural language processing (NLP). The n-gram document representation model uses a continuous sequence of words, which are chunks of words that are used as features to represent a document. Each element of the document representation vector is a set of two or more neighboring words in a document repository. Another similar approach, known as n-gram, involves using a fixed sequence of letters where a unique sequence represents elements in the feature vector for each document. This approach has applications in spelling error detection and language identification. The bag-of-words approach is low in computational cost, rapid and naive, although it does not consider word order and grammar. The bag-of-words approach also suffers from the "Curse of Dimensionality", causing classifiers to lose their discriminatory power. Such a context highlights the importance of feature transformation and selection tools, for example Latent Dirichlet Allocation (LDA) [6], principal component analysis [7], latent semantic indexing [8], independent component analysis [9], and document frequency [10]. For documents with references, such as Internet web pages with hyperlinks as well as scholarly documents, another method can be used to represent a document, as references are related to the document contents. The weight of the references can be renewed according to frequency of use and location in the document. In this approach, each reference is one dimension of the feature vector, which has a much lower dimensionality in examination to the bag-of-words and the n-gram approaches [11].

Explicit Semantic Analysis (ESA) is a mixed approach based on references and the contents of a document. In ESA, the similarity of documents is calculated according to their reference sets. In the feature vector, each element is weighted in relation to specific documents within the reference sets [12]. The similarity measure technique, which is based on compression [13], is another approach for computing the representation of documents. This approach is on the basis of the hypothesis that the similarity of two files can be estimated by comparing: (1) the compressed size of the concatenated version of the two files and (2) the summation of the compressed size of each file. The elements of the representation vector consist of the similarity between the documents in the repository.

In an attempt to capture the sentiment meaning and semantic of words, Maas et al. used a word representation model [14], in which vectors are formed by a probabilistic-based and unsupervised approach. A similar representation of words that congregate can be found in most documents. The subsequent stage uses a supervised-based learning method. Here, the sentiment of words is included and the model is trained.

3 Deep Learning-Based Document Representation Models

An alternative approach to bag-of-words is word embedding, where phrases and words are mapped to fixed-length vectors in a continuous space. Two of the most successful deep learning-based word embedding models are Word2Vec and GloVe. These models present a continuous vector for each word in a training data by training with vast corpuses. In order to represent a document based on its word vectors, two commonly adopted approaches involve calculating the summation or the average of word vectors. However, these approaches do not take into account either word order or the context of the document. As an illustration of the latter issue, the word vector for "bass" as a fish is equivalent to its word vector as an instrument.

In this chapter, five document representation models that are based on deep learning, namely Word2Vec, Doc2Vec, GloVe, CNN, and LSTM, are introduced and compared.

3.1 Word2Vec Model

Word2Vec is a two layer neural network-based approach that learns embedding for words. Negative sampling has been used in the softmax step of the output layer. The objective function maximizes the log probability of a context word ($\mathbf{w_O}$), given its

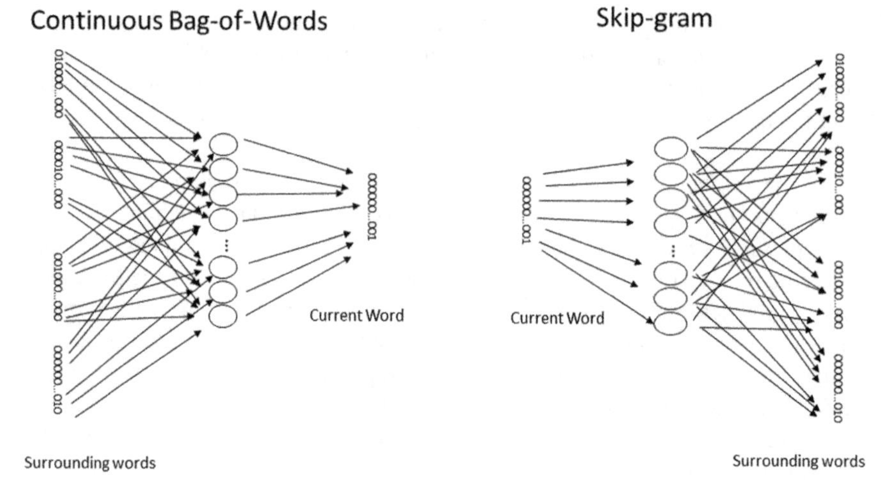

Fig. 1 Word2Vec architectures [15]

input words ($\mathbf{w}_\mathbf{I}$). By using negative sampling, the objective function is to maximize the dot product of $\mathbf{w}_\mathbf{I}$ and randomly selected negative words, while minimizing the dot product of $\mathbf{w}_\mathbf{I}$ and $\mathbf{w}_\mathbf{O}$. The output is a vocabulary of words from the original document and an n-dimensional fixed-size vector representation. Co-occur words in the training corpus are located adjacent to each other in a vector space. It can be observed in Fig. 1, how Word2Vec creates word vector representation by use of two architectures: Continuous Bag of Words (CBOW) and Skip-gram. The CBOW architecture model, which according to the surrounding context words, can predicts a word and also works faster than Skip-gram, which predicts the surrounding words by a center word, in a fixed-length window. For infrequent words, the Skip-gram architecture works better.

Word2Vec generates vector representation only for words, while for document representation, a representation for the entire document is needed. Averaging or summation of all the word vectors of a given document can be a naive solution for creating document representation.

3.2 Doc2Vec Model

Le and Mikolov [5] presented Doc2Vec, also known as Paragraph2Vec. An extension of Word2Vec, the Doc2Vec model represents document, sentence, and paragraph by a fixed-length vector. The authors described two approaches, both of which are developed from vector representation of words. The first approach, commonly known as the Paragraph Vector with Distributed Memory (PV-DM) and the second approach, termed the Distributed Bag of Words (PV-DBOW).

PV-DBOW works in the same way as Skip-gram. The difference is that the input is a unique vector that represents the document and the order of words is ignored. PV-DM works in the same way as CBOW. The additional vectors used by PV-DM are a concatenation of document vectors and several target words. The objective is to use the concatenated document and word vectors to predict a context word.

Doc2Vec or Paragraph2Vec modifies Word2Vec into an unsupervised learning approach to create a continuous representation for an entire document. Different to averaging Word2Vec vectors, the Doc2Vec approach can preserve the order of words and gain more accurate results [5].

By concatenating of both vectors that were generated from this approach and employing them as the features, the most effective outcomes is expected. The objective of a subsequent study by Hong [16] was to better the performance of Paragraph2Vec by use of two approaches: the addition of both a hidden layer and a tensor layer to the paragraph vector, such that it can interact with word vectors, both complexly and non-linearly.

The Doc2Vec model can acknowledge the context of the document, but cannot benefit the background knowledge. However, the high computational cost of creating a model each time is a distinct weakness of this model. In comparison, Word2Vec and GloVe allow pre-trained models to be used multiple times by fine-tuning the input feature vectors to a specific task.

3.3 Glove Model

Pennington et al. [4] introduced an unsupervised word embedding model, known as GloVe, for word representation. GloVe captures global corpus statistics (the frequency of word co-occurrences within a specific window in a large text corpus) to generate linear dimensions of meaning and uses local context window methods and global matrix factorization. This model, which offers a local cost function, includes a weighting function that is used to balance the rare co-occurrences. Optimization methods are used to minimize the cost function.

Based on the hypothesis that similar words have similar distributions, it is expected that general word vectors can be used to measure the semantic similarity. As in the case of Word2Vec, averaging the vectors of words in a document is an option for generating a fixed length vector for document representation.

3.4 Long-Short Term Memory

Nonetheless, the dependency between words in terms of the syntactic and semantic meanings is not that clear. There may be long and short term dependencies in a sentence. As a result, the common approach in neural network methods is to consider sentences as a sequence of tokens, then process them with a recurrent neural network

(RNN). The tokens are processed in a sequential order for which the RNN will learn the compact feature embedding of the word sequences. A long short-term memory (LSTM) [17] network is a common type of RNN, commonly used in sentiment analysis [18, 19]. LSTMs are able to model long-range dependencies for NLP [20, 21]. However, LSTMs are known to be difficult to train since they are sensitive to hyper parameters such as batch size and hidden dimensions [22]. Also, they are not specific to language processing; they are only generic sequence learning machines, which lack task specificity [23].

3.5 Convolutional Neural Networks

However, in deep learning approaches such as convolutional neural networks (CNNs) [24], deep hierarchical representations can be obtained. The main idea of CNNs is to perform feature extraction and classification as one task. Typically, there will be many successive layers of filter convolutions and feature pooling. The convolution filters are learned from data with the intention of eliminating handcrafted features. These learned filters are then used to perform the convolution.

For the application of CNNs in NLP, semantically related terms are close in word embedding space, which enhances the classification performance by keeping the deep knowledge of relationship between terms. A study by Johnson and Zhang [25] used one dimensional CNNs directly with sentences represented as one-hot vectors as inputs. This has the overhead of dealing with sparse features inherent to the data. Typically, this is not well-suited for convolution networks wherein dense data is preferred. To mitigate this issue, others have used Word2Vec vectors as tunable inputs into their CNN [26], but this may cause a decrement in classification performance if the Word2Vec vectors are not domain specific. In other studies [23, 27], very deep CNNs (up to 29 paired convolution and pooling layers) have been implemented in sentiment classification by natively learning embedding based on alphanumerical characters. This type of deep CNN is well suited to texts because characters combine to form n-grams, stems, words, phrases, and sentences. Figure 2 demonstrates a part of a text as input into a model where there is a vocabulary size of 5. The embedding layer has an embedding dimension of 3, wherein dropout is performed. This embedding layer can be trained with either random or pre-trained vectors such as Word2Vec or GloVe vectors. The one-dimensional convolution also has three filters on which L2 regularization and dropout is applied before activation with a fully connected dimension of four. The 1-max pooling layer takes only the single best feature per feature map. In the final layer, the text is classified with a score from 1 to 5.

Thus, when using CNNs or LSTM networks in NLP, one must be aware of the advantages and disadvantages relative to the task. Similar classification results may be obtained by the competing methods, but the feature representation may be weak. For instance, it is commonly accepted that a fully connected one hidden layer may learn

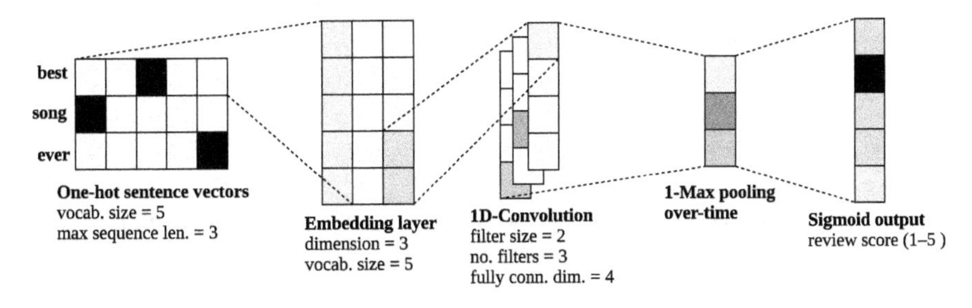

Fig. 2 Example of one dimensional CNN

any real-valued function. However, a hierarchical representation of multiple layers will provide more information regarding the interaction among features, which is represented by the successive layers [23].

3.6 Experimental Results

The described approaches implemented and applied on the "IMDB Movie Reviews" dataset [28] for the task of sentiment analysis. The training data consists of 25,000 movie reviews and another 25,000 reviews were used as testing data. The results illustrated in the Table 1.

In these experiments, the methods are applied only on one dataset and the default parameters and architectures are used, more related results are available in [25, 26, 5].

Table 1 Results of introduced methods applied to IMDB Movie Reviews Dataset

Method	Accuracy	F_score
Bag-of-words—SVM	0.8406	0.8378
Bag-of-bigram—SVM	0.8746	0.8733
Word2Vec (Averaging)—SVM	0.8357	0.8354
Doc2Vec	0.8540	0.8291
GloVe (Averaging)—SVM	0.8304	0.8291
LSTM (100 training epochs)	0.8378	0.8413
CNN (100 training epochs)	0.8686	0.8638

4 Combined Studies

In the field of document representation, a relatively recent study by Kim et al. [29] introduced three approaches that developed the Word2Vec vector of content words. Average pooling, class-specific Gaussian Mixture distribution, and Semantic Space Allocation are three considered approaches. The most successful outcomes were achieved by average pooling. In a Chinese articles classification task, this approach outperformed the LDA method.

A tagging-based document representation method was used in a study by Bernotas et al. [30], the outcome of which was improved clustering by using ontology. This study identified the negative effect of the tagging-based representation on short size documents but offered a better outcome for large scale documents compared to word-based document representation.

Hong [16] subsequently employed two approaches in an effort to further develop the capacity of Paragraph2Vec: the addition of a tensor layer for interaction between paragraph vector and word vectors and addition of a hidden layer. In another study, Hong and Zhao [31] employed the LDA and the Deep Learning-based approaches for the task of anomaly detection in legal texts. This involved first using LDA to extract topics from a EULA corpus, then removing the words in the topics. The next step was to calculate the Word2Vec vector of all the remaining words, enabling a Word2Vec vector for each sentence to be created. The process was completed with the application of agglomerative clustering (which is more successful in comparison to the K-mean clustering method) and LOF (Local Outlier Factor) in order to detect abnormal sentences in the EULA text.

A study by Lao and Jagadeesh [32] presented a classification task in which questions were allocated into 16 legal areas. Bag-of-words, bag-of-bigrams, the TF-IDF technique, and average Word2Vec vectors of questions were used as features and compared to five different classifiers: The Linear Support Vector Machine (SVM); the Logistic Regression; Multi nominal Naïve Bayes; SVM with stochastic gradient descendent; and the one layer Neural Network. The first classifier, the SVM, achieved the best results.

In 2017, Lin et al. [33] used an approach, called self-attentive sentence embedding, which is based on long short-term memory networks with a set of summation weight vectors provided by a self-attention mechanism and a 2-D matrix to represent the embedding. Also in 2017, Kamkarhaghighi and Makrehchi [15] introduced content tree word embedding, which is a framework to improve document representation by Word2Vec and GloVe. In this framework, the vector for each word is updated based on the words around it in the given context.

References

1. C. Wei, S. Luo, X. Ma, H. Ren, J. Zhang, L. Pan, Locally embedding autoencoders: a semi-supervised manifold learning approach of document representation. PLoS ONE **11** (2016)
2. Z.S. Harris, Distributional structure Word **10**, 146–162 (1954)
3. T. Mikolov, K. Chen, G. Corrado, J. Dean, Efficient estimation of word representations in vector space. arXiv:13013781 (2013)
4. J. Pennington, R. Socher, C.D. Manning, Glove: global vectors for word representation, in *EMNLP* (2014), pp. 1532–1543
5. Q.V. Le, T. Mikolov, Distributed representations of sentences and documents. arXiv:14054053 (2014)
6. D.M. Blei, A.Y. Ng, M.I. Jordan, Latent Dirichlet allocation. J. Mach. Learn. Res. **3**, 993–1022 (2003)
7. K. Pearson, LIII. On lines and planes of closest fit to systems of points in space. Lond. Edinb. Dublin Philos. Mag. J. Sci. **2**, 559–572 (1901)
8. S. Deerwester, Improving information retrieval with latent semantic indexing (1988)
9. A. Jung, An introduction to a new data analysis tool: independent component analysis, in *Proceedings of Workshop GK "Nonlinearity"*, Regensburg (2001)
10. E.E. Milios, M.M. Shafiei, S. Wang, R. Zhang, B. Tang, J. Tougas, A systematic study on document representation and dimensionality reduction for text clustering. Technical report (Faculty of Computer Science, Dalhousie University, 2006)
11. J. SzymańSki, Comparative analysis of text representation methods using classification. Cybern. Syst. **45**, 180–199 (2014)
12. E. Gabrilovich, S. Markovitch, Computing semantic relatedness using Wikipedia-based explicit semantic analysis, in *IJcAI* (2007), pp 1606–1611
13. M. Li, P. Vitányi, *An Introduction to Kolmogorov Complexity and Its Applications* (Springer Science & Business Media, 2009)
14. A.L. Maas, R.E. Daly, P.T. Pham, D. Huang, A.Y. Ng, C. Potts, Learning word vectors for sentiment analysis, in *Proceedings of the 49th Annual Meeting of the Association for Computational Linguistics: Human Language Technologies*, vol. 1 (Association for Computational Linguistics, 2011), pp. 142–150
15. M. Kamkarhaghighi, M. Makrehchi, Content tree word embedding for document representation. Expert Syst. Appl. **90**, 241–249 (2017)
16. S. Hong, Improving Paragraph2Vec (2016)
17. S. Hochreiter, J. Schmidhuber, Long short-term memory. Neural Comput. **9**, 1735–1780 (1997)
18. A. Graves, J. Schmidhuber, Framewise phoneme classification with bidirectional LSTM and other neural network architectures. Neural Netw. **18**, 602–610 (2005)
19. K. Greff, R.K. Srivastava, J. Koutník, B.R. Steunebrink, J. Schmidhuber, LSTM: a search space Odyssey. IEEE Trans. Neural Netw. Learn. Syst. (2016)
20. M. Sundermeyer, R. Schlüter, H. Ney, LSTM neural networks for language modeling, in *Thirteenth Annual Conference of the International Speech Communication Association* (2012)
21. I. Sutskever, O. Vinyals, Q.V. Le, Sequence to sequence learning with neural networks, in *Advances in Neural Information Processing Systems* (2014), pp. 3104–3112
22. R. Pascanu, T. Mikolov, Y. Bengio, On the difficulty of training recurrent neural networks, in *International Conference on Machine Learning* (2013), pp. 1310–1318
23. A. Conneau, H. Schwenk, L. Barrault, Y. Lecun, Very deep convolutional networks for natural language processing. arXiv:160601781 (2016)
24. Y. LeCun, L. Bottou, Y. Bengio, P. Haffner, Gradient-based learning applied to document recognition. Proc. IEEE **86**, 2278–2324 (1998)
25. R. Johnson, T. Zhang, Effective use of word order for text categorization with convolutional neural networks. arXiv:14121058 (2014)
26. Y. Kim, Convolutional neural networks for sentence classification. arXiv:14085882 (2014)
27. X. Zhang, J. Zhao, Y. LeCun, Character-level convolutional networks for text classification, in *Advances in Neural Information Processing Systems* (2015), pp. 649–657

28. Kaggle, Bag of Words Meets Bags of Popcorn, vol. 2016 (2015)
29. H.K. Kim, H. Kim, S. Cho, Distributed representation of documents with explicit explanatory features (2014)
30. M. Bernotas, K. Karklius, R. Laurutis, A. Slotkienė, The peculiarities of the text document representation, using ontology and tagging-based clustering technique. Inf. Technol. Control **36** (2015)
31. Y. Hong, T. Zhao, Automatic Hilghter of Lengthy Legal Documents (2015)
32. B. Lao, K. Jagadeesh, Classifying legal questions into topic areas using machine learning (2014)
33. Z. Lin, M. Feng, C.Nd. Santos, M. Yu, B. Xiang, B. Zhou, Y. Bengio, A structured self-attentive sentence embedding. arXiv:170303130 (2017)

Applications of Deep Learning in Medical Imaging

Sanjit Maitra, Ratul Ghosh and Kuntal Ghosh

Abstract Deep Learning techniques have recently been widely used for medical image analysis, which has shown encouraging results especially for large datasets. In particular, convolutional neural network has shown better capabilities to segment and/or classify medical images like ultrasound and CT scan images in comparison to previously used conventional machine learning techniques. This chapter includes applications of deep learning techniques in two different image modalities used in medical image analysis domain. The application of convolutional neural network in medical images is shown using ultrasound images to segment a collection of nerves known as Brachial Plexus. Deep learning technique is also applied to classify different stages of diabetic retinopathy using color fundus retinal photography.

Keywords Deep learning · Medical image analysis · Convolutional neural network · Brachial plexus segmentation · Diabetic retinopathy detection

1 Introduction

Application of convolutional neural network for pattern recognition is about three decades old using backpropagation network for handwritten zip code recognition for U.S. postal service [1]. As the computation power of the systems increased, the

S. Maitra (✉)
Theoretical and Applied Sciences Unit, Indian Statistical Institute, North-East Centre, Tezpur 784028, Assam, India
e-mail: sanjit@isine.ac.in

R. Ghosh
Department of Electronics and Communication Technology, Indian Institute of Information Technology, Allahabad, Uttar Pradesh 211011, India
e-mail: ratulghoshr@gmail.com

K. Ghosh
Machine Intelligence Unit & Center for Soft Computing Research, Indian Statistical Institute, Kolkata 700108, West Bengal, India
e-mail: kuntal@isical.ac.in

V. E. Balas et al. (eds.), *Handbook of Deep Learning Applications*,
Smart Innovation, Systems and Technologies 136,
https://doi.org/10.1007/978-3-030-11479-4_6

size of the network grew such that it could handle multiple layers thereby handling complex machine learning tasks. Convolutional Neural Network (CNN) was used in medical image analysis in 1995 to detect lung nodule from radiographs of chest and microcalcifications from digital mammograms [2]. These were originally done with small dataset having limited number of hidden layers. In 2012, CNN was applied to a large dataset of about 1.2 million images for a 1000 class classification problem [3]. This was the first time where it was being shown that how five convolutional layers, 650 thousand neurons can achieve a considerable high classification accuracy in comparison to other conventional machine learning classification algorithms. This opened the gates for Deep CNN for image classification and segmentation in varied fields.

Applications of CNN in Medical Imaging, a very recently developed research domain, can be broadly separated into two categories, viz. classification and segmentation problems. In case of classification, from a given set of labelled images, the model is trained where it establishes a functional relationship between the features from the input images and its corresponding class. For example, Anthimopoulos et al. performed classification using CNN on CT scan lung images to classify image patches into 7 classes [4]. This includes healthy tissue and six different interstitial lung disease patterns. CNN is also applied for classification of images of other modalities like ultrasound imaging for identification of thyroid nodules as malignant or benign [5], chest X-ray for classification of view orientation [6] and many more. Segmentation of medical images is also a very important task to identify organs, lesions or substructures of organs for analysis. Examples of CNN in medical image analysis include segmentation of brain tumors from MRI images [7], pancreas segmentation using CT scan images [8], segmentation of neuronal membrane using images from electron microscope [9]. CNN is widely used nowadays in medical image analysis domain on different modalities of images.

2 Convolutional Neural Network

Convolutional Neural Network (CNN) consists of multilayer architecture that performs multilevel feature extraction [2]. Additional layers of abstract features are then mapped to get the desired output. Before the use of CNN, researchers used to connect every neuron in the first hidden layer to each and every node of the input for computation [3]. This increases the number of parameters to be optimized in the training procedure. In case of CNN, the input image is convolved with a set of kernels. The convolution operation forces the hidden neurons in the first layer to have a less number of connections to the input nodes. Instead of usual fully connected architecture, CNN can be considered as a locally connected network because of the convolution function with kernel size much smaller than the input image size. A basic convolution operation is illustrated in Fig. 1. A simple 2×2 kernel is convolved with a 3×3 image and the result is shown. For the particular example, when the filter position is as shown by the red square on the image (Fig. 1b), the pixel values of the image are multiplied with the value of the kernel on the corresponding

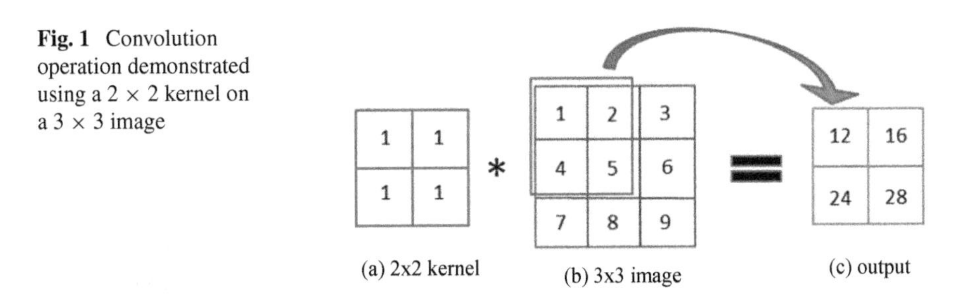

Fig. 1 Convolution operation demonstrated using a 2 × 2 kernel on a 3 × 3 image

position. In order to get the first value of the output, the computation is (1 × 1) + (2 × 1) + (4 × 1) + (5 × 1) which equals to 12. Then the kernel is moved to the next pixel to get the next value of the output and this keeps on repeating until this weighted sum is computed throughout the image. These locally connected networks obtained through convolution actually mimic the visual cortex neurons of human brain having local receptive field [10]. Input to the convolution layer is $\mathbf{m} \times \mathbf{n} \times \mathbf{r}$ dimensional input image. In case of ordinary color image, the value of $\mathbf{r}$ is 3. The convolutional layer will have $\mathbf{k}$ kernels/filters of size $\mathbf{t} \times \mathbf{t} \times \mathbf{s}$, where $\mathbf{s}$ is the number of bands of the kernel that should be less than equal to the value of $\mathbf{r}$ of the original image. These $\mathbf{k}$ filters will produce $\mathbf{k}$ feature maps by extracting locally connected features.

Each convolution operation is followed by an activation function where point wise non-linear transform is applied to each of the components in the extracted feature maps. For example, let's consider a kernel that extracts horizontal edges from the input image (e.g. Sobel operator) [11]. This transformation gets activated only for horizontal edges greater than a threshold. The convolution operation is repeated at different layers to extract additional abstract features.

Output from convolutional layers can be mathematically written as

$$\mathbf{y}_{\mathrm{k}} = \sigma\left(\mathbf{w}_{\mathrm{k}-1} * \mathbf{x}_{\mathrm{k}-1} + \mathbf{b}_{\mathrm{k}-1}\right) \tag{1}$$

where, $\mathbf{y}_{\mathrm{k}}$ is the output from the kth convolutional layer, σ the activation function, $\mathbf{w}_{\mathrm{k}-1}$ is the kernel used for convolution, $\mathbf{x}_{\mathrm{k}-1}$ is the feature map from the previous $(\mathrm{k}-1)^{\mathrm{th}}$ layer and $\mathbf{b}_{\mathrm{k}-1}$ is the bias added after the feature map is generated.

Each convolutional operation with non-linear transformation is followed by a pooling operation [3]. The convolutional feature map is divided into disjoint blocks and pooling operation is performed over the feature values in each of the blocks. Average and max operations are the most used pooling functions. Figure 2 shows how pooling operations are computed. The convolutional output (Fig. 2a) is divided into 2 × 2 blocks shown as blue dotted squares.

Max pooling (Fig. 2b) takes in the maximum feature value from the 2 × 2 blocks which are then used as inputs to the higher layers. Similarly, average pooling computes the average of the feature values from the disjoint blocks. The output from pooling layer induces shift invariant property by down-sampling the input feature maps [12]. This combination of convolution, non-linear transformation and pooling

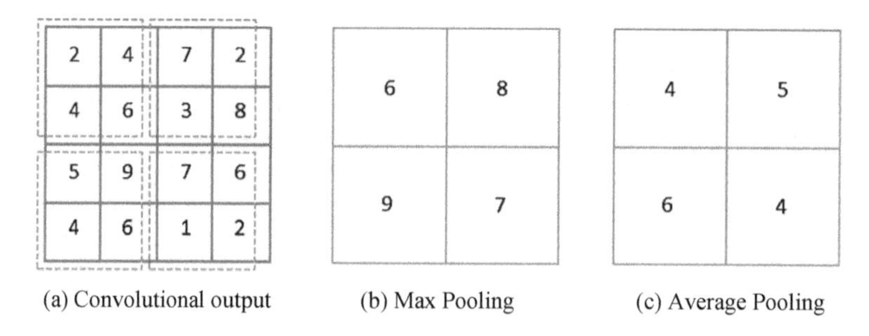

(a) Convolutional output (b) Max Pooling (c) Average Pooling

Fig. 2 Two types of pooling operation on convolutional output

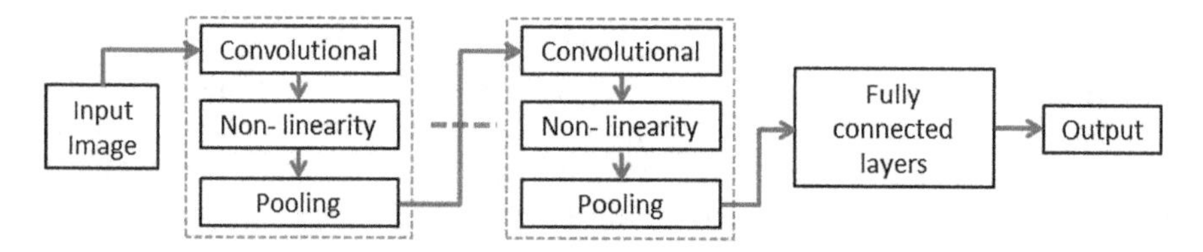

Fig. 3 Architecture of standard CNN

layers are repeated to extract features at different levels or scales which are then usually followed by fully connected conventional neural network layers for the classification task [3]. Block diagram of standard CNN used for classification is given in Fig. 3.

Training of the CNN is usually performed using the gradient descent algorithm which is widely used in training multi-layer feedforward networks for decades [13]. Stochastic gradient descent [14] is currently gaining popularity specially for training CNN due to faster convergence and less chances of getting trapped in local minima. In case of gradient descent algorithm, the cost or error function is computed for the entire training set and the weights are updated by gradient descent. Stochastic gradient descent considers the training samples to be arriving one at a time. So the cost function is computed for each sample of training data on arrival and the weights are updated using gradient descent [13, 14]. Instead of considering one at a time arrival of training samples, the training dataset is also divided into batches, such that the computation of the cost function is done in batches to update the weights through gradient descent. This is known as mini-batch gradient descent algorithm [15]. Depending on the optimized choice of the batch size, mini-batch gradient descent can give faster convergence than gradient descent and stochastic gradient descent [13–15].

Backpropagation method is usually used to compute the gradient of the cost/error function in terms of the weights for gradient descent [16]. The weights of the hidden units are randomly initialized and inputs from the training sample move forward through the hidden layer of the network to finally give an expected value of the output. This is compared with the actual output class of the corresponding input to get the error function. In case of backpropagation, error information flows backwards through the network to find the gradient of the error in terms of the weights of the nodes in the layers [16]. In case of backpropagation, the gradient in one layer is a product of the gradient from the preceding layer and the derivative of the activation function [13].

Application of CNN in two different modes of medical imaging is discussed in the following section that includes segmentation of ultrasound images to identify a collection of nerves known a brachial plexus and classification problem based on severity stages of diabetic retinopathy using color fundus images of the retina.

3 Segmentation of Brachial Plexus from Ultrasound Images

Brachial Plexus (BP) is a collection of nerves that goes to the ribs and armpit from the spinal cord through the neck [17] (Fig. 4). These are responsible for the effect of sensation in the arm and also control the muscles of the hand, wrist, shoulder and elbow [18]. Any injury in the BP portions can result in loss of feeling in the arm and hand area. Severe injury can permanently disable the arm, hand, shoulder, and elbow. The treatment procedure includes surgical methods that require time to recover [17]. Minor injuries can be cured with therapy by keeping the joints flexible. The usual process of examining the progress of the treatment procedure is by checking the presence of sensation in hand, shoulder, elbow area and the strength of muscles. Tests like MRI or ultrasound scan is done to check the extent of damage and finding nerve blocks [19].

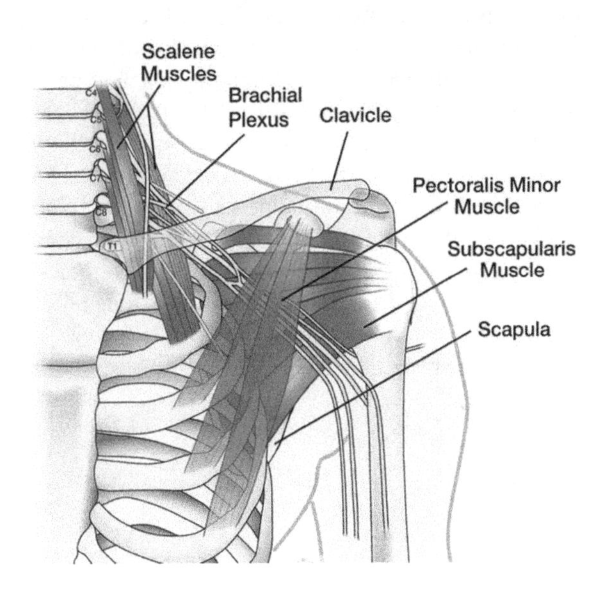

Fig. 4 Brachial Plexus. Image from [17]

In recent years, ultrasound imaging of BP portion has become an important tool to analyze the extent of damage, location of lesions, and identification of abnormal growth (tumoral) of tissue in the area [19]. Use of ultrasound imaging in BP is advantageous as well as challenging because it is very difficult to identify the portions due to its complex structure and obstruction due to the presence of the collar bone. It has been shown that using ultrasound imaging one can identify and track the interscalene and supraclavicular blocks through specific locations where the BP portions are located superficially [20]. Experts, during the ultrasound analysis procedure, usually use some typical anatomical landmarks to identify the difficult-to-detect BP portion regions [19]. This task of segmenting the ultrasound images to identify the BP portions is discussed using CNN where multiple levels of abstract features are extracted to locate the BP portions.

3.1 Dataset

The dataset used for demonstration of CNN on ultrasound images for nerve segmentation is from "Kaggle Ultrasound Nerve Segmentation" competition [21]. The training images are manually annotated by ultra-sonographers to mark the location of the BP portions in the ultrasound image. Sample of image with manual annotations (shown in red) is given in Fig. 5. It can be seen that identification of the region is difficult with different structures throughout the image. The training dataset contains 120 images per 47 patients to capture the BP portions from different orientations and angles with mask images specifying the location of BP portions by experts. These mask images are from the manual annotations provided to train the network. Sample of the mask images are shown in Fig. 6. These masks are actually binary images with BP portions represented as white pixels and black elsewhere. Each image size is 420×580 with a total of 5635 training images with corresponding mask images and 5508 test images without the annotations. As these images are taken from different locations and orientations, there are about 47% (2649 images) of the training images that do not contain BP portions and their corresponding mask images are completely black. The task here is to segment the ultrasound images such that the BP portions can be identified using the training images along with labelled annotations provided, marking the location of BP. In this case, the output should include classification of the pixels of ultrasound images to identify the collection of nerves in the images. It is basically a two class problem where the BP regions identified should be marked as white and the rest of the image should be black (similar to the mask images provided for training).

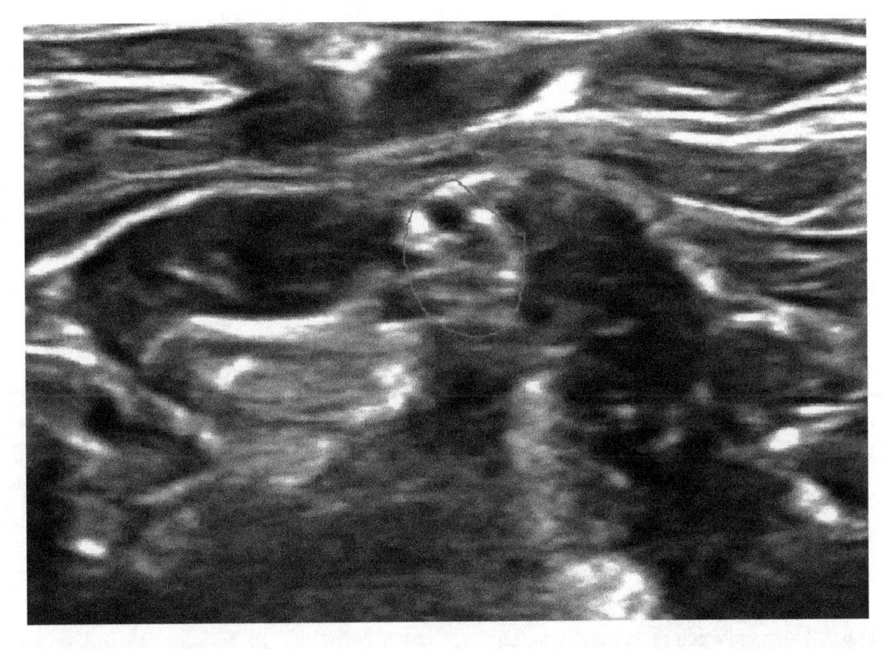

Fig. 5 Sample of ultrasound image with manual annotation (in red) specifying the location of the BP portion

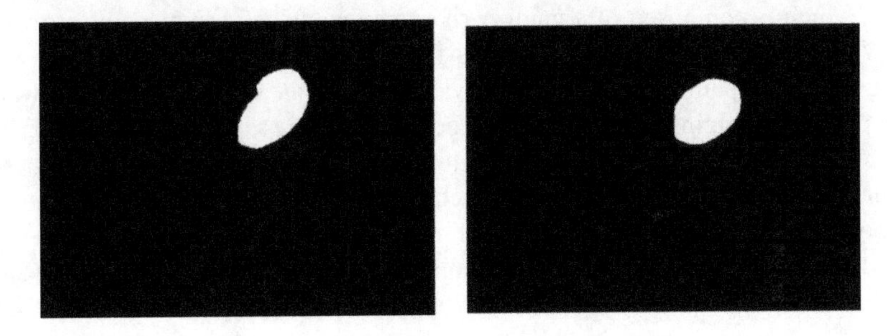

Fig. 6 Sample of the mask images representing the annotations specifying the location of the BP portions

3.2 Network Architecture

The U-Net architecture [22] of CNN for segmentation of medical images is one of the most popular techniques that won the ISBI cell tracking challenge [23] in 2015 by a significant margin. The basic architecture of U-Net is shown in Fig. 7. This uses a deep CNN method that takes in the input image and gives the segmentation map as the output, which is one of the crucial tasks in medical image analysis. It has shown better performance in a limited number of labelled training dataset in comparison to other standard convolutional networks. The architecture is similar to other deep

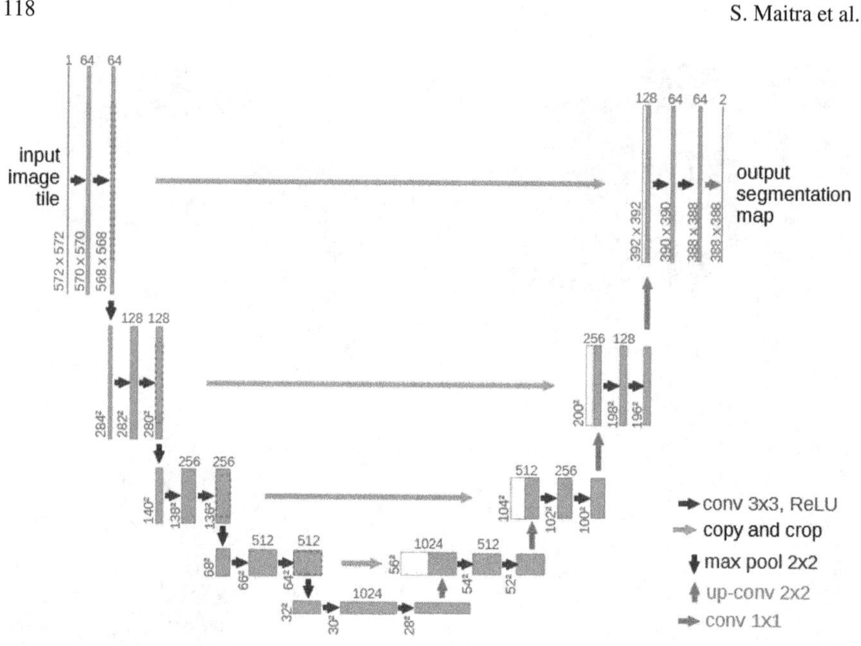

Fig. 7 U-net architecture. Image from [22]

CNN, which is composed of a number of operations which are given as arrows in Fig. 7. The data from the input image is forward propagated through the network along the given paths and finally it provides the output segmentation map where each pixel is belonging to one of the two possible classes which are BP and non-BP. The blue rectangles are multichannel feature maps shown in Fig. 7. The number of feature channels is provided on top of each blue box and size of the feature map on the bottom. For the given example, the dimension of the input image is 572×572 with 1 channel. The convolution operations are followed by a non-linear activation function. The most widely used activation function [24] is Rectified Linear Unit (ReLU).

The ReLU activation function is given as,

$$\sigma_{\text{ReLU}}(\text{x}) = \max(0, \text{x}) \tag{2}$$

where, x is the input and σ_{ReLU} is the activation function which was mentioned as generalized activation function in Eq. 1. The plot of the ReLU function (also known as ramp function) is given in Fig. 8. This shows using ReLU implies that the activation unit is only active when the input is positive and the strength of activation is the magnitude of the input. The function is deactivated for negative values of input.

The next operation followed by convolution and ReLU is the max pooling operation with a 2×2 window and stride of 2. It extracts the maximum activation from the 2×2 window to the next layer feature map. This reduces the dimensions of the feature map. The sequence of convolution and max pooling operation extracts

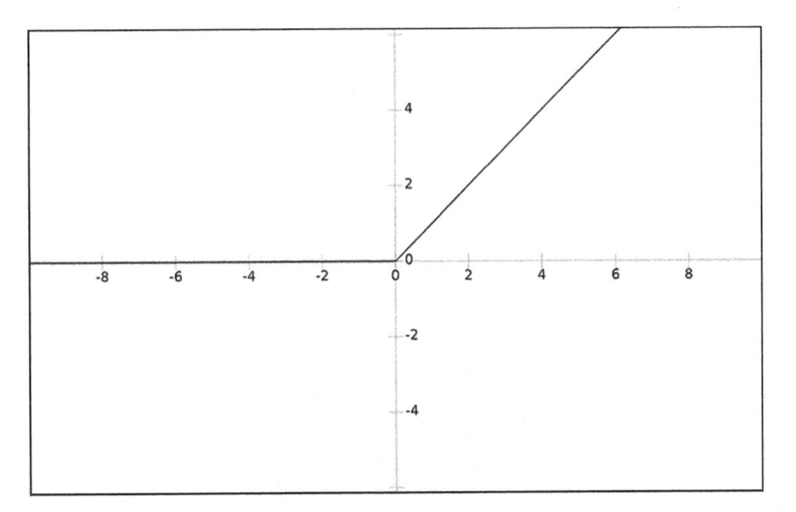

Fig. 8 Rectified Linear Unit as activation function

the features starting from a low level to higher levels. This is used in the conventional CNN model. U-Net consists of an additional expansion path to create the final segmentation map instead of fully connected layers shown in Fig. 3. This path consists of up-convolution operations shown as green up arrows followed by linking with the corresponding high resolution features shown as grey horizontal arrows. Up-convolution with a 2×2 kernel is mapping each feature vector to a 2×2 output feature map followed by the ReLU activation function. The output segmentation map dimension is actually smaller than the input image due to the unpadded convolution operations. Extrapolation by mirroring the input image is done for predicting the pixel values in the border region [22].

The basic idea of U-net comes from training larger and deep network using VGG network architecture [25]. The most important characteristics of the network is passing the input image through stack of 3×3 convolutional layers which mimics the visual cortex neurons of human brain having local receptive field. VGG network has shown greater classification accuracy due to greater depth of the network.

The U-net architecture is used to segment the ultrasound images with 3×3 convolution followed by activation using ReLU. Max pooling in 2×2 window and 2×2 up-convolution.

Along with the basic U-net architecture, a modified version of the network is also applied that uses the inception architecture introduced in 2015 [26, 27]. The main idea of inception model is that instead of performing fixed 3×3 convolutions throughout the network, perform 1×1, 3×3, 5×5 (total of three) convolution within the same module of the network. The outputs are concatenated and forwarded to the next layer. Instead of a single convolution, a combination of 1×1 convolution, 1×1 followed by 3×3, 1×1 followed by 5×5 and 3×3 max pooling followed by 1×1 convolution are used. The basic block diagram of the module is given in

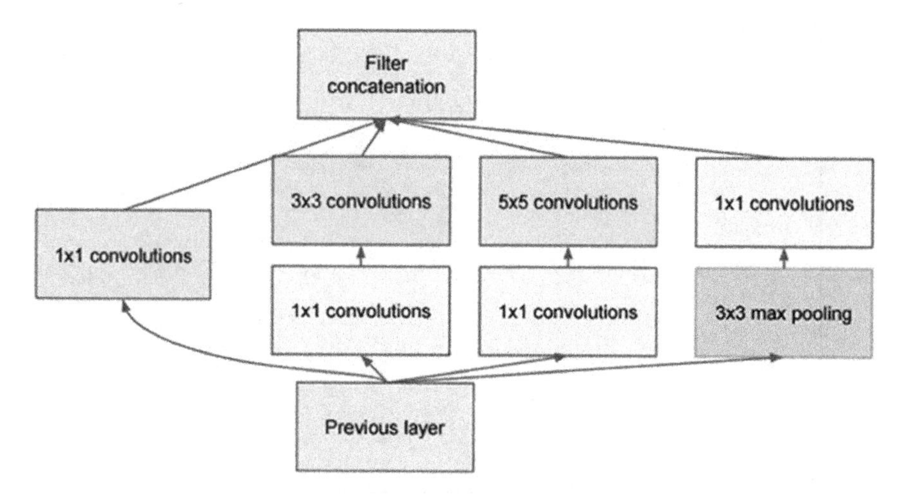

Fig. 9 Inception module. Image from [29]

Fig. 9. In this case, multi-level features are extracted at the same time within the same module and the results are merged using concatenation. This merging of information adds up all the information like size, texture, color, location at different scales at the same level. It is already shown that inception model can lead to comparatively lesser computational cost in comparison to other deep learning network architectures [27]. Modified version of the U-net architecture is applied using the inception module replacing the conventional 3×3 convolution of VGG network architectures.

3.3 Data Augmentation

Increasing number of training sample is required to boost up the performance of deep networks and incorporate robustness with required invariance properties [22]. Deep networks need a large variety of datasets for training to develop the ability to identify complex structures within the images. Standard data augmentation techniques [22, 25, 27] were applied on the ultrasound image dataset like flipping x and y of the images, random zoom, elastic transformation by moving the pixels around based on distortion. Same augmentation techniques were applied to mask of the training dataset images that represent the annotations of the BP region.

3.4 Results

The network is trained using stochastic gradient descent algorithm [15] discussed before in Sect. 2. From the 5635 training images, 500 images were kept for validation of the training and 5135 images were used for training the network. The image shown

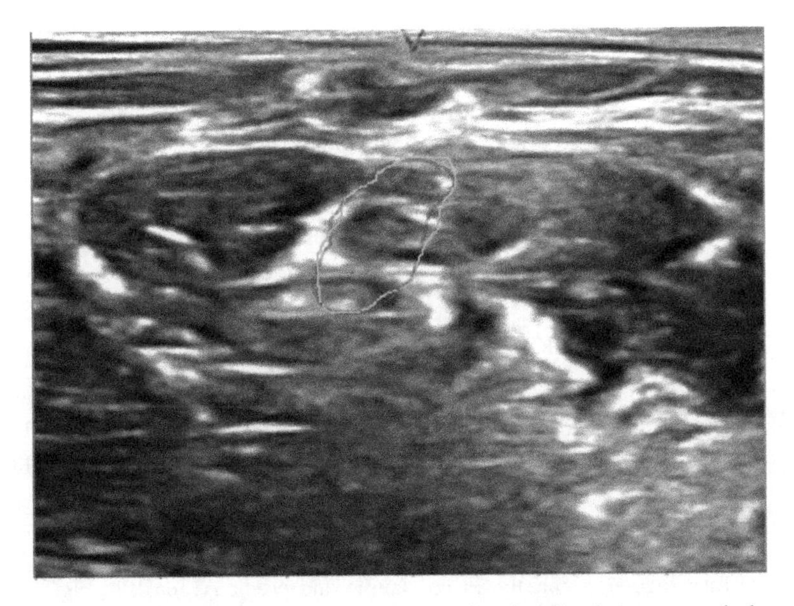

Fig. 10 Sample image with actual and predicted annotations in red and green respectively

in Fig. 10 is a sample of validation image after training with the actual and predicted annotations in red and green respectively. Actual and predicted mask of a sample validation image is given in Fig. 11. There are many metrics for validating medical image segmentation results [28]. Dice coefficient is one of the popular metric used in medical image segmentation that directly compares the overlap of the segmented annotations.

The dice coefficient between the actual and predicted annotations can be computed using the relation

$$Dice\,(a, b) = (2 * (a \cap b))/(|a| + |b|) \tag{3}$$

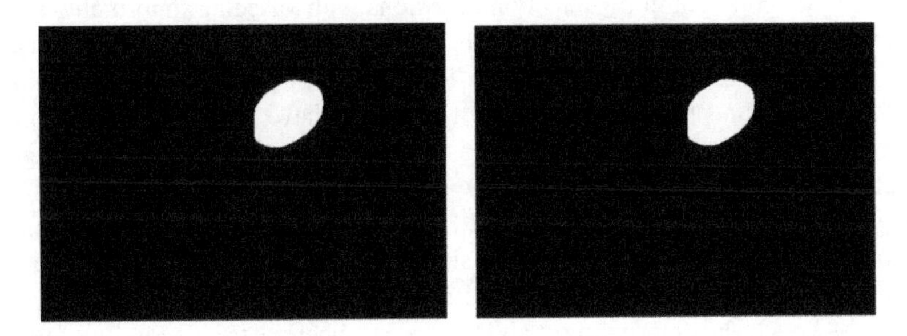

Fig. 11 Sample mask image in the left and the corresponding predicted mask in the right

where, a and b are the actual and predicted annotations, $a \cap b$ is the number of pixels having value as 1 in both a and b, and $|a|$, $|b|$ are the total number of pixels having value 1 in a and b respectively. In the ultrasound images, the annotations are binary images with BP portions represented as white pixels (pixel value 1) and black (pixel value 0) elsewhere. The value of the dice coefficient ranges from 0 to 1. Zero for non-overlapping masks and perfect score of 1 for perfectly matching segmentations. There were about 2649 images of the training dataset that do not contain BP potions and their corresponding mask images are black. For them, if the mask images are black, i.e., all pixel values are zero for both the actual and predicted masks, the dice coefficient is considered as 1.

Using the original U-net architecture, the mean dice coefficient was 0.53 while using the modified architecture with inception module instead of VGG net, the mean dice coefficient of the validation dataset was 0.702. For the given problem, calculation of dice co-efficient is a good measure to compare the segmentation result with the actual annotations.

It can be clearly seen that using the inception module instead of standard 3×3 convolution operation throughout the layers of the network, increases the segmentation performance significantly. From sample and predicted mask in Fig. 11 one can compare the actual and predicted annotations and conclude how this deep learning network has performed to identify the collection of nerves from the complex structures in the ultrasound images.

4 Classification of Diabetic Retinopathy Stages from Color Fundus Retinal Images

Diabetic Retinopathy (DR) is an eye disease that affects the retina of the patients as a prolonged effect of diabetes [29]. Current figures tell that about 350 million people worldwide are affected by DR. Countries like USA, China and India reported the highest number of diabetes mellitus cases which can lead to high numbers of DR affected peoples [30]. The identification of DR is usually performed through color fundus retinal photograph by detection of lesions with vascular abnormalities [29]. Trained analysts are required to identify the lesions from the fundus images and infer about the severity of DR in the patients. Rural areas of India usually have a large number of patients suffering from diabetes that can have long term effects on the retina [30]. Early stage detection is necessary to decrease the spread of damage on the retina, which otherwise can lead to severe visual impairment.

Color fundus images are usually used to capture the condition of the interior surface of the eye, especially the retina using fundus camera. Automated method of classification of the color fundus retinal images based on the severity of damage is necessary. This task is more important in the areas where there is scarcity of trained technician to detect the lesions manually. The task of classification of the color fundus images based on the severity of DR is discussed using CNN.

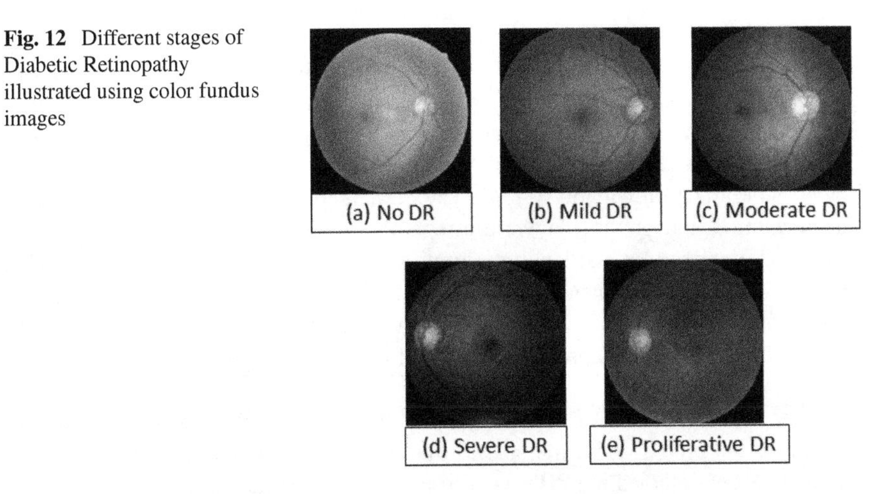

Fig. 12 Different stages of Diabetic Retinopathy illustrated using color fundus images

4.1 Dataset

The dataset used to demonstrate the classification using CNN is from "Kaggle Diabetic Retinopathy Detection" competition [31]. The color fundus images of the dataset are labelled into one of the four classes based on the severity of the DR. The four stages of DR are shown in Fig. 12 along with normal negative DR. There were about 35,000 images which are labelled into one of these five classes for training a deep network. About 73% of the images are from No-DR class (negative DR). Proliferative DR is the highest stage of severity that can lead to permanent blindness. Only 708 images were present that were labelled as Proliferative DR class.

4.2 Network Architecture

The CNN applied to this dataset is the conventional deep network as mentioned in Fig. 3. The combination of convolution, non-linear transformation and pooling layers are repeated to extract additional levels of abstract features. The convolution kernel size was 3×3 following the standard VGG network architecture [25]. Each convolution was followed by a non-linear activation layer that incorporates the non-linearity in the model. Leaky ReLU function was used as activation function which is a modified version of the ReLU function described before. Instead of output of zero when the input is negative, Leaky ReLU gives an output of 0.01 times the input implying the leakage in the function for negative values of the input. This modified ReLU function has shown higher efficiency in comparison to standard sigmoid function for some applications [32]. The activation layer is followed by max pooling with a kernel size of 2×2.

Dropout is used as a regularization method to tackle the problem of overfitting [33]. During the training of the network, each unit is retained with a probability **p**. While testing, the weights of the units are scaled by a factor of **p** thereby forcing the network to extract more information from the active neurons at each phase. The probability **p** was taken as 0.5 which showed better predicting performance and higher generalization capability [34]. The combination of convolution, non-linear transformation and pooling layers are repeated in seven layers followed by fully connected layers and classification output layer using softmax function [35]. This function squeezes n-dimensional arbitrary valued vector to n-dimensional real valued vector with values between 0 and 1 such that the elements of the vector adds up to 1 [35]. This is required to predict the class of the input image at the output layer of the network.

To tackle the problem of class imbalance, data augmentation was performed by rotation of the images from the classes that were having fewer images. Stochastic gradient descent algorithm [15] was used for training the network.

4.3 Results

About 10% of the images were randomly selected as validation dataset. Confusion matrix is a good way to visualize the performance of any classification algorithm. The confusion matrix for the validation dataset is given in Table 1.

The data imbalance is also reflected in the validation dataset where one can find that about 70% of the images are from No DR (class 0). The algorithm performs well in predicting images from the No DR class with an accuracy of 95%. Maximum confusion was found to be between No DR (class 0), mild DR (class 1) and moderate DR (class 2). A total of 85% accuracy was achieved in this five class classification case. This might not reflect the actual capability of the model to classify the images due to the high imbalance of class 0 images in the dataset. We restructured our dataset to include 100 images per class. The corresponding confusion matrix is given in Table 2. This drastically reduces the accuracy of the model to 65%, mainly due to the fact that images from class 1 were predicted as class 0. Instead of classifying into 5 classes based on severity, one can perform a 2-class classification to separate as Non-

Table 1 Confusion matrix of the validation dataset

Predicted class						
Actual class		**0**	**1**	**2**	**3**	**4**
	0	**2186**	47	29	31	7
	1	84	**79**	34	3	0
	2	197	68	**210**	18	7
	3	5	2	17	**51**	15
	4	6	0	8	37	**49**

Table 2 Confusion matrix with 100 images per class

Predicted class						
Actual class		**0**	**1**	**2**	**3**	**4**
	0	**96**	3	1	0	0
	1	48	**46**	4	2	0
	2	23	3	**51**	17	6
	3	3	0	7	**63**	27
	4	2	0	3	24	**71**

Severe and Severe class. For the given dataset, the two class problem achieved an accuracy of 97%. Quadratic weighted kappa is computed that measure the agreement between actual and predicted class of the images by the model. The final model for 5-class classification problem has a kappa score of 0.74.

5 Conclusions

This chapter presents a basic introduction of convolutional neural network, which is gaining popularity for analysis of medical images. Preliminary concepts of convolution, pooling and non-linear operations are discussed that constitutes the building blocks of the CNN. The basic architecture of conventional CNN is explained. The two major applications of machine learning, i.e., image segmentation and classification which are currently very important in medical image analysis are demonstrated. Two completely different modes of imaging are used to demonstrate the application of deep learning. Image segmentation required a modified version of U-Net architecture to segment the collection of nerves from the ultrasound images. More conventional CNN architecture is explained that was used to classify the color fundus images based on the severity of diabetic retinopathy. Deep learning is already having a great impact on the medical imaging domain. The importance of deep learning in medical imaging is sure to increase in the days to come.

Acknowledgements The authors would like to thank Kaggle for making the ultrasound nerve segmentation and diabetic retinopathy detection datasets publicly available. Thanks to California Healthcare Foundation for sponsoring the diabetic retinopathy detection competition and EyePacs for providing the retinal images.

References

1. Y. LeCun, B. Boser, J.S. Denker, D. Henderson, R.E. Howard, W. Hubbard, L.D. Jackel, Backpropagation applied to handwritten zip code recognition. Neural Comput. **1**(4), 541–551 (1989)

2. S.C.B. Lo, H.P. Chan, J.S. Lin, H. Li, M.T. Freedman, S.K. Mun, Artificial convolution neural network for medical image pattern recognition. Neural Netw. **8**(7), 1201–1214 (1995)
3. A.I. Krizhevsky, S.G. Hinton, Imagenet classification with deep convolutional neural networks. Adv. Neural. Inf. Process. Syst. **25**, 1106–1114 (2012)
4. M. Anthimopoulos, S. Christodoulidis, L. Ebner, A. Christe, S. Mougiakakou, Lung pattern classification for interstitial lung diseases using a deep convolutional neural network. IEEE Trans. Med. Imaging **35**(5), 1207–1216 (2016)
5. T. Liu, S. Xie, J. Yu, L. Niu, W. Sun, Classification of thyroid nodules in ultrasound images using deep model based transfer learning and hybrid features, in *IEEE International Conference on Acoustics, Speech and Signal Processing (ICASSP)*, New Orleans, USA, 2017 (2017), pp. 919–923
6. A. Rajkomar, S. Lingam, A.G. Taylor, High-throughput classification of radiographs using deep convolutional neural networks. J. Digit. Imaging **30**(1), 95–101 (2017)
7. S. Pereira, A. Pinto, V. Alves, C.A. Silva, Brain tumor segmentation using convolutional neural networks in MRI images. IEEE Trans. Med. Imaging **35**(5), 1240–1251 (2016)
8. H.R. Roth, A. Farag, L. Lu, E.B. Turkbey, R.M. Summers, Deep convolutional networks for pancreas segmentation in CT imaging. Proc. SPIE Medical Imaging pp. 94–131 (2015)
9. D. Ciresan, A. Giusti, L.M. Gambardella, J. Schmidhuber, Deep neural networks segment neuronal membranes in electron microscopy images, in *Advances in Neural Information Processing Systems*, vol. 25 (Red Hook, NY, 2012), pp. 2843–2851
10. D.A. Pollen, S.F. Ronner, Visual cortical neurons as localized spatial frequency filters. IEEE Trans. Syst. Man Cybern. **13**(5), 907–916 (1983)
11. I. Pitas, A.N. Venetsanopoulos, Edge detectors based on nonlinear filters. IEEE Trans. Pattern Anal. Mach. Intell. **4**, 538–550 (1986)
12. D. Scherer, A. Müller, S. Behnke, Evaluation of pooling operations in convolutional architectures for object recognition, in *Artificial Neural Networks—ICANN 2010*. Lecture Notes in Computer Science, ed. by K. Diamantaras, W. Duch, L.S. Iliadis, vol. 6354 (Springer, Berlin, Heidelberg, 2010)
13. H. Guo, S.B. Gelfand, Analysis of gradient descent learning algorithms for multilayer feedforward neural networks. IEEE Trans. Circuits Syst. **38**(8), 883–894 (1991)
14. Y. Jia, E. Shelhamer, J. Donahue, S. Karayev, J. Long, R. Girshick, S. Guadarrama, T. Darrell, Caffe: convolutional architecture for fast feature embedding. arXiv:1408.5093 (2014)
15. M. Li, T. Zhang, Y. Chen, A. Smola, Efficient mini-batch training for stochastic optimization, in *ACM SIGKDD Conference on Knowledge Discovery and Data Mining* (2014)
16. A. Von Lehmen, E.G. Paek, P.F. Liao, A. Marrakchi, J.S. Patel, Factors influencing learning by backpropagation, in *IEEE International Conference on Neural Networks*, San Diego, CA, USA, vol. 1 (1988), pp. 335–341
17. http://www.assh.org/handcare/hand-arm-injuries/Brachial-Plexus-Injury#prettyPhoto
18. K. Saladin, *Anatomy and Physiology*, 7 edn. (McGraw Hill, New York 2015), pp. 489–491
19. F. Lapegue, M. Faruch-Bilfeld, X. Demondion, C. Apredoaei, M.A. Bayol, H. Artico, H. Chiavassa-Gandois, J.J. Railhac, N. Sans, Ultrasonography of the brachial plexus, normal appearance and practical applications. Diagn. Interv. Imaging **95**(3), 259–275 (2014)
20. A. Perlas, V.W.S. Chan, M. Simons, Brachial plexus examination and localization using ultrasound and electrical stimulation: a volunteer study. Anesthes **99**(2), 429–435 (2003)
21. Ultrasound Nerve Segmentation Challenge, https://www.kaggle.com/c/ultrasound-nerve-segmentation/data (2016)
22. O. Ronneberger, P. Fischer, T. Brox, U-Net: convolutional networks for biomedical image segmentation. arXiv:1505.04597 [cs] (2015)
23. Cell Tracking Challenge, http://www.codesolorzano.com/Challenges/CTC/Welcome.html (2017)
24. H. Ide, T. Kurita, Improvement of learning for CNN with ReLU activation by sparse regularization, in *International Joint Conference on Neural Networks*, Anchorage, AK, USA (2017), pp. 2684–2691

25. K. Simonyan, A. Zisserman, Very deep convolutional networks for large-scale image recognition. arXiv:1409.1556 (2014)
26. C. Szegedy, W. Liu, Y. Jia, P. Sermanet, S. Reed, D. Anguelov, D. Erhan, V. Vanhoucke, A. Rabinovich, Going deeper with convolutions, in *Proceedings of the IEEE Conference on Computer Vision and Pattern Recognition* (2015), pp. 1–9
27. C. Szegedy, V. Vanhoucke, S. Ioffe, J. Shlens, Z. Wojna, Rethinking the inception architecture for computer vision, in *IEEE Conference on Computer Vision and Pattern Recognition (CVPR)*, Las Vegas (2016), pp. 2818–2826
28. A.A. Taha, A. Hanbury, Metrics for evaluating 3D medical image segmentation: analysis selection and tool. BMC Med. Imaging **15**(1), 15–29 (2015)
29. R. Williams, M. Airey, H. Baxter, J. Forrester, T. Kennedy-Martin, A. Girach, Epidemiology of diabetic retinopathy and macular oedema: a systematic review. Eye **18**(10), 963–983 (2004)
30. J. Cornwall, S.A. Kaveeshwar, The current state of diabetes mellitus in India. Australas. Med. J. 45–48 (2014)
31. Diabetic Retinopathy Detection Challenge, https://www.kaggle.com/c/diabetic-retinopathy-detection (2015)
32. K. He, X. Zhang, S. Ren, J. Sun, Delving deep into rectifiers: surpassing human-level performance on ImageNet classification. arXiv:1502.01852 (2015)
33. N. Srivastava, G.E. Hinton, A. Krizhevsky, I. Sutskever, R. Salakhutdinov, Dropout: a simple way to prevent neural networks from overfitting. J. Mach. Learn. Res. **15**(1), 1929–1958 (2014)
34. P. Baldi, P.J. Sadowski, Understanding dropout, in Advances in Neural Information Processing Systems, ed. by C.J.C. Burges, L. Bottou, M. Welling, Z. Ghahramani, K.Q. Weinberger, vol. 26 (2013), pp. 2814–2822
35. C. Bishop, *Pattern Recognition and Machine Learning* (Information Science and Statistics) (Springer, New York, 2006)

Deep Learning for Marine Species Recognition

Lian Xu, Mohammed Bennamoun, Senjian An, Ferdous Sohel and Farid Boussaid

Abstract Research on marine species recognition is an important part of the actions for the protection of the ocean environment. It is also an under-exploited application area in the computer vision community. However, with the developments of deep learning, there has been an increasing interest about this topic. In this chapter, we present a comprehensive review of the computer vision techniques for marine species recognition, mainly from the perspectives of both classification and detection. In particular, we focus on capturing the evolution of various deep learning techniques in this area. We further compare the contemporary deep learning techniques with traditional machine learning techniques, and discuss the complementary issues between these two approaches. This chapter examines the attributes and challenges of a number of popular marine species datasets (which involve coral, kelp, plankton and fish) on recognition tasks. In the end, we highlight a few potential future application areas of deep learning in marine image analysis such as segmentation and enhancement of image quality.

Keywords Marine species · Deep learning · Convolutional neural networks (CNNs) · Image recognition · Object detection

---

L. Xu · M. Bennamoun (✉) · F. Boussaid
The University of Western Australia, Perth, Australia
e-mail: mohammed.bennamoun@uwa.edu.au

L. Xu
e-mail: lian.xu@research.uwa.edu.au

F. Boussaid
e-mail: farid.boussaid@uwa.edu.au

S. An
Curtin University, Perth, Australia
e-mail: s.an@curtin.edu.au

F. Sohel
Murdoch University, Perth, Australia
e-mail: F.Sohel@murdoch.edu.au

V. E. Balas et al. (eds.), *Handbook of Deep Learning Applications*,
Smart Innovation, Systems and Technologies 136,
https://doi.org/10.1007/978-3-030-11479-4_7

1 Introduction

Oceans are the largest ecosystems on Earth. They generate approximately 50% of the O_2 that we breathe and more than 97% of our water. Besides, they provide us with 1/6 of the animal protein that we consume. They are also home to a myriad of compounds that could well constitute promising medicines to cure cancer and other diseases. Furthermore, oceans mitigate the influence of climate change by absorbing CO_2 from the atmosphere. As such, the world's oceans play a vital role in sustaining life on Earth. With natural and anthropogenic effects (e.g., global warming, climate change, overfishing, and pollution) posing severe threats (e.g., coral bleaching, rising sea temperatures, and changing species distributions) to marine habitats and biodiversity, it has become essential to understand the underlying causes of changes in the ecological processes, for example, to help inform environmental governmental policies.

Surveys conducted between 2001 and 2015 revealed that, driven by decades of ocean warming, Australian temperate reef communities lost their defining kelp forests and became dominated by persistent seaweed turfs and other species that are characteristic of the subtropical and tropical waters. These species, if not controlled, would devastate lucrative fishing and tourism industries that are worth billions of dollars per year. They will also have catastrophic consequences for the thousands of endemic species which are supported by the kelp forests of Australia's Great Southern Reef [1]. A survey over a four-year (2010–2013) period showed that coral cover declined from 73 to 59%, while macroalgal cover increased from 11 to 24% at the Houtman Abrolhos Islands, a high-latitude reef system in Western Australia.

Such significant changes can lead to a regime shift with profound ecological social and economic consequences [2]. Thus, marine environmental protection has aroused great attention from all circles of the society, and it also has led to a large number of collaborative and interdisciplinary programs. Decades ago, marine raw survey data was collected using satellite, aerial and shipborne sensors or diver held cameras and towed video sleds. The resulting images were of low-quality, since it was difficult to precisely control the position and altitude of the acquisition systems. In recent years, technological developments in underwater imagery acquisition have enabled the development of high-definition cameras connected to Remotely Operated Vehicles (ROVs), and Autonomous Underwater Vehicles (AUVs), which can operate continuously at precise depths and across spatial scales with geolocation. Such acquisition systems can capture sensor data and stereo images several times a second. Furthermore, scientists can use the acquired data to produce a 3D visual map of the surveyed sites [3, 4]. However, with the amount of the collected marine imagery growing exponentially, the manual annotation of raw data has become an extremely laborious task for expert analysts. Consequently, only a small fraction of the collected data can be annotated, which obviously affects the reliability and accuracy of the conducted studies [5]. The development of an automated annotation system for marine data recognition analysis, would address this limitation.

When it comes to object recognition, the ImageNet Large Scale Visual Recognition Challenge (ILSVRC) is the foremost global annual competition, in which research teams demonstrate their achievements in classifying and detecting objects and scenes from ImageNet. ImageNet is currently the world's largest dataset for image recognition. It contains more than 15 millions high-definition labeled images of 22,000 classes in total. The dataset for ILSVRC competition is only a subset of ImageNet. It comprises 1000 classes with around 1000 samples per class. Since 2012, with the increasing popularity of deep learning, we have witnessed a dramatic increase in its performance, which was accompanied by the growing depth of convolutional neural networks (CNNs). Specifically, in 2012, Krizhevsky et al. [6] trained an 8-layer convolutional neural network and achieved top-1 and top-5 error rates of 37.5% and 16.4% respectively. In 2014, the runner-up Simonyan and Zisserman [7] achieved a top-5 error rate of 7.3% by using networks of 19 layers, which are usually referred to as "VGG-19", while Google Inception Net, with its 22 layers, earned the first prize at a top-5 error rate of 6.7%. However, a degradation in performance occurred when the network depth increased. In 2015, He et al. [8] solved this problem by proposing the residual network, which consists of up to 152 layers and achieves 3.57% error on the ImageNet dataset. Traditional neural networks (e.g., multilayer perceptron) accomplish "shallow learning" and have been so far unable to match the performance of deep learning techniques when following the ILSVRC protocols.

Due to its remarkable performance in several visual important recognition tasks, such as image classification and object detection, deep learning has been widely used in many application fields. In this chapter, we will discuss how deep learning techniques can help automate the process of marine species recognition. Recall that object recognition involves: (i) object classification, which focuses on assigning a label to a patch or an image of an object. (ii) object detection deals with the detection of the locations and classes of the possible objects in images and videos. In the following two sections, we will review and discuss prior works which are related to the classification and detection of marine species, respectively. Section 4 will describe future prospects, followed by the conclusion of this chapter.

2 Deep Learning for Marine Species Classification

In this section, we illustrate the uses of deep learning techniques in improving marine species classification. Image classification techniques are often used for mapping benthic species habitat. Unlike marine mammals and fish with clear contours, benthic species, such as corals, algae, and seagrass, face a challenging obstacle of ambiguous class boundaries, making recognition more difficult. Moreover, common annotation techniques, such as images labels and bounding boxes, are inappropriate for benthic species images. Instead, marine ecologists rely on random point sampling as shown in Fig. 1. Generally, small patches centered around each point annotation are extracted from an image for classification.

Fig. 1 Point annotation. Boundaries of benthic species are often ambiguous. Common annotation techniques, such as images labels and bounding boxes, are inappropriate for benthic species images. Instead, marine ecologists use point annotation to label a number of chosen locations in an image. Image from the EFC dataset [37]

2.1 Marine Species Classification Based on Deep Convolutional Neural Network (CNN) Features

Deep convolutional neural networks perform extremely well when a large amount of labeled training data is available. However, many visual recognition challenges have tasks with insufficient training examples. Without enough training data, deep architectures, such as AlexNet and VGGNet, will be likely to overfit the training data, since a large number of parameters need to be learnt in these architecture. Fortunately, deep CNN features have been found to be universal representations for various applications. Donahue et al. [9] demonstrated that, by leveraging a deep convolutional architecture trained on a large labeled dataset, one can learn features that are sufficiently powerful and generic to a spectrum of visual recognition tasks by using simple linear classifiers. It has been shown that this can reliably outperform approaches based on sophisticated multi-kernel learning techniques and traditional hand-crafted features. Their experimental results of feature visualization are consistent with the common deep learning knowledge that the initial layers learn low-level features and the latter layers learn semantic or high-level features. The work by Razavian et al. [10] confirms and extends the results in [9] and strongly suggests that features obtained from deep learning with convolutional networks can be the primary candidates for most visual recognition tasks. In marine species classification, examples of use of deep CNN features include:

- Jaeger et al. [11] presented a benchmark dataset, called the Croatian Fish Dataset which contains 794 images of 12 fish species, for fine-grained visual classification (FGVC) of fish species in unrestricted natural environments. They used the activations of the 7th hidden layer of AlexNet pre-trained on ImageNet as features, and a linear SVM for classification. They achieved an accuracy of 66.78% on their dataset.
- Mahmood et al. [12] used the parameters of the first fully connected layer of a pre-trained VGGNet as features and a Multi-Layer Perceptron (MLP) network for classification in a "corals versus non-corals" classification problem. They evaluated their proposed method on the Western Australia (WA) subset of the Australian benthic dataset (Benthoz15) [13], which contains 407,968 labels of almost 40 classes. This WA subset consists of 237,500 annotated points collected from 2011 to 2013. Their results were shown that the highest accuracy of 97% was achieved when the training and testing data were from the same year. This result revealed that the recognition performance was largely affected by the variations between testing data and training data which was caused by the environmental changes cross years. Moreover, they pointed out that most of their misclassifications occurred because of ambiguous class boundaries.
- In another study, Mahmood et al. [14] proposed "ResFeats", i.e., features extracted from the pre-trained deep residual networks. By using the outputs of the last residual unit of the 5th convolutional layer of a 152-layer ResNet as features, and a PCA-SVM classifier, they achieved the state-of-the-art accuracy of 80.8% on the MLC dataset, which corresponds to a 6.8% gain over the baseline performance of [15] and 2.9% over that of [16].

From the above examples, we can observe that deep features generalized well for the case of marine images. It is true when dealing with simple binary classification problems. As a matter of fact, when the number and complexity of classes increases, the performance degrades (as expected). This is because, typical deep models, such as AlexNet, VGGNet and ResNet (from which the deep features are learned), were trained on ImageNet, a dataset where most of the marine species are rare or unseen. Subtle inter-class differences of marine species can be hardly captured by those deep features.

2.2 Marine Species Classification Based on Hybrid Features

In this section, we will introduce hybrid features which are constructed by combining traditional hand-crafted features and deep CNN features. We analyze the strengths and weaknesses of each kind of features, and further discuss how these hybrid features affect the recognition performance.

2.2.1 Traditional Methods Based on Hand-Crafted Features

Before feature learning techniques became prevalent, feature engineering was used in a wide range of visual recognition tasks. The process of feature engineering is complex, time-consuming and requires expert knowledge. However, the hand-crafted features can explicitly describe the texture, color or shape of objects. Some descriptors, such as HOG [17] and SIFT [18], produced notable performance gains a decade ago. The quality and quantity of these features have a significant influence on the prediction models. Generally, a combination of hand-crafted features yield better results than using a single one. Different combinations of features are usually selected based on the type of data and the recognition tasks. Prior marine species works based on some popular hand-crafted features are discussed in the following:

- Marcos et al. [19] encoded color information from a normalized chromaticity coordinate (NCC) space using histograms of mean values, combined with Local Binary Pattern (LBP) texture descriptors. They used linear discriminant analysis (LDA) for classification. When training and testing on a dataset of 138 images with around 40 labels per image, captured at a depth of 1–3 m, their proposed method achieved a binary classification accuracy ranging from 60 to 77%. The performance was shown to degrade as the depth increases, because of the resulting color degradation and poorer image quality. Besides, their results indicated that the proposed features cannot provide sufficient discriminative information for coral reef habitat classification.
- In the work of Pizarro et al. [20], the bag of SIFT features approach is applied to encode the gray scale information. Moreover, histograms of NCC and Gabor wavelet responses are used to represent the color and texture features, respectively. The proposed algorithm was evaluated on a dataset with 453 images of 8 classes, in which 35 images per class were available for training while the rest was used for testing. The reported results demonstrate noticeable improvements when color and texture information are included. However, because an entire image is required to be classified as one class, heterogeneity within the image cannot be classified or quantified.
- Stokes and Deane [21] introduced their discrimination metrics, which involves the encoding of the color information with RGB histograms, and texture features with discrete cosine transform (DCT). For their classification scheme, the probability density-weighted mean distance (PDWMD) is calculated between the sample patch and all the patches in the benthic library, which is generated by manually selecting over 3,000 images from 18 classes. This method was tested on 16 quadrat images. The reported results show their automated routine is more time-efficient and has a comparable performance to the manual classification. However, it is not trivial to determine the weights of the color and texture features when combining them. Manually trying different options usually leads to suboptimal results. In addition, the performance is largely limited by the size of the library.

- Beijbom et al. [15] first introduced a benchmarking marine dataset, called the Moorea Labeled Corals (MLC), which has 400,000 labels of 9 classes. In their work, a powerful dictionary-based texture descriptor (texton) was proposed, which encodes information in each channel of the LAB space using Maximum Response (MR) as feature representation. Their results were shown to achieve accuracies between 67 and 83%. This proposed algorithm is regarded as a strong baseline on the MLC dataset.
- Bewley et al. [22] presented a solution to a binary classification problem of kelp, i.e., Kelp versus "Other". The proposed approach uses a single SVM classifier with Radial Basis Function (RBF) kernel applied to a large AUV dataset with 62,900 labels. The performance of different descriptors, (e.g., raw pixels, PCA, GLCM, LBP and HOG), extracted from patches of various scales, was compared. In addition, they assess whether the color information is useful for recognition by extending some descriptors (Raw, PCA and GLCM) to work in the RGB space. Their results revealed that a color extension of any descriptor at any scale could provide superior results to its grayscale counterpart.
- In [23], the authors of [22] extended the classification problem to 19 classes following a taxonomical hierarchy using the same dataset collected under the IMOS program [24]. They investigated the performance of PCA, LBP and feature learning (FL) with logistic regression (LR) classifiers. Their reported results show that a simple LBP representation achieves the best performance, resulting in an F1-score of 80.2% at the root node and around 85% at the highest level of the hierarchy.

In summary, texture-based features, including intensity histogram statistics, gray-level co-occurrence matrix (GLCM) [25–27] statistics, Gabor wavelet response statistics [28], and local binary pattern (LBP) [29], are the most commonly used hand-crafted features for marine benthic species classification. For underwater benthic images, the color degradation caused by the physical properties of the water medium and ambiguous boundaries of benthic species reduce the reliability of individual color or shape based features. However, advanced color-based features, such as the opponent angle and hue color histograms [30], could still be used as complementary information for marine species classification.

2.2.2 State-of-the-Art Methods Based on Hybrid Features

Deep CNN features have shown their strong discriminative power and transferability on many datasets. Several studies has revealed that a deep learning architecture could capture semantic knowledge that is implicit within image features. However, carefully selected handcrafted features can provide explicit physical descriptions of certain objects. Therefore, when dealing with challenging datasets of uncommon objects, such as rare marine species, it is necessary to develop more sophisticated methods based on hybrid features. Several research groups have considered such methods for marine species recognition tasks and achieved an improved performance:

Mahmood et al. [16] combined color and texture based features, texton features from [15], and CNN features extracted from a pre-trained VGGNet to train a 2-layer MLP classifier. Under the same experimental settings, the proposed model achieves accuracies of 77.9, 70.1, 84.5%, outperforming the approach in [15] (the corresponding accuracies are 74.3, 67.3, 83.1%). It is worth mentioning that the problem of ambiguous class boundaries of corals is partially alleviated by extracting features at multiple scales. In addition, the class imbalance problem of the MLC dataset was taken into account by downsampling the training data and assigning a weight, which is inversely proportional to the downsampling rate, to the cost function. However, there is still room for their accuracy to be improved.

Zheng et al. [31] investigated suitable ways to combine CNN features extracted from the penultimate layer of pre-trained AlexNet and ten different types of hand-crafted features. The minimal-redundancy maximal-relevance (mRMR) [32] was used to select suitable feature combinations from all feature candidates. Given that different features may fit different kernels, multiple kernel learning (MKL) [33] was applied before using the one-versus-rest SVM classifiers. The proposed method was evaluated on the same dataset with [34], i.e., the Taiwan sea fish dataset, which has 27,370 fish images of 23 classes from the Fish4Knowledge project. Their reported results show that CNN features outperform handcrafted features, but are inferior to hybrid features, which achieve a top accuracy of around 97.91%. However, this method is highly time-consuming because it uses MKL with the large computational complexity. Moreover, high false negatives indicate the detrimental effects of a large imbalance in data. This highlights the importance of resolving this class imbalance problem for future research.

Blanchet et al. [35] used three state-of-the-art feature representations: completed local pattern binary (CLBP) with hue and opponent angle histograms [36], textons [15] and DeCAF [7]. Instead of combining those features by simply concatenating them, they trained a one-versus-one SVM, with RBF kernel, for each feature representation. Then, a fusion function is used to aggregate the normalized outputs from the three classifiers. The final prediction corresponds to the one with the maximum fused score. Experiments were conducted on a dataset with 75,195 labeled patches of 4 classes. The best achieved accuracy was 78.7%. Moreover, the prediction accuracy was shown to be further enhanced by introducing rejection thresholding to the fused scores so as to eliminate ambiguous points. Note that, in this work, species appearing in small quantity were filtered out. However, this is not an advisable approach to deal with the class imbalance problem, as it somehow discards potentially useful information.

Deep CNN features and hand-crafted features have been shown to describe different aspects of natural images. As a general rule, the discriminative ability of a certain feature may act differently on various datasets. It is possible that these two kinds of features can complement each other in some respects. Furthermore, an increased feature dimension due to the combination of multiple features may lead to increased computation loads. Generally, for a given dataset, deep CNN features could be used as the first option for recognition tasks, while hybrid features may be used to improve performance.

2.3 *End-to-End Training for Marine Species Classification with Deep Convolutional Neural Networks*

Classical machine learning methods implement classification tasks in two separate steps: features extraction and classification. The performance largely depends on the quality of the feature. Decades ago, machine learning scientists spent much time designing "good" features. Hence, the previous machine learning approach had a more suitable name of "feature engineering". In recent years, researchers have developed machine learning techniques which allow a system to automatically capture "optimal" features from raw data. This is referred to as "feature learning". As a typical supervised feature learning method, neural networks, which employ end-to-end training, can learn multi-level features of inputs from the outputs of different hidden layers. With the number of layers increasing, deep neural networks achieved a better recognition rate, which led to the emergence of "deep learning".

End-to-end pattern refers to an approach in which a neural network accepts inputs from one end and produces outputs at the other end, to solve multi-stage problems with only one training. The learning process optimizes the network weights by considering the inputs and outputs directly, i.e., all the weights of the network are trainable to map the inputs to their corresponding outputs. However, implementing an end-to-end training for a visual task is challenging, since it requires a proper design of network, enough training data, and high-performance GPUs. For marine species classification, there are few examples of previously reported end-to-end training:

- CNNs were first used for the automated annotation of underwater images in [37]. Another contribution of this work was a newly proposed dataset, the Eliat Fluorescence Coral (EFC) dataset, which comprises 212 image-pairs of reflectance and fluorescence images with 200 point annotations per image. They investigated two methods of training CNNs on that data, with the same network structure of LeNet [38]. The first method consists of an end-to-end training on the registered images with five color channels, three from the reflectance image and two from the fluorescence image, respectively. The second approach firstly trains two separate CNNs on each image type and then uses an SVM classifier to consolidate the outputs of the two CNNs. The results show that the latter achieves a better performance, with a 90.5% accuracy. Theoretically, the five-channel network should be more effective than the second approach, as it uses all image information. Two possible factors, including inadequate training data and an imperfect registration quality, may explain these results. Moreover, additional experimental results also demonstrate that adding fluorescence information can improve the accuracy of automated annotation.
- Li and Cui [39] addressed a plankton classification problem with 30,336 grayscale images of various sizes from 121 classes. A top-5 accuracy of 95.8% was achieved by training an end-to-end network based on the structure of deep residual networks (ResNet). Specifically, by comparing the performance of training ResNet and VGG-19 net, ResNet outperformed VGG net in terms of both training time and

testing accuracy. Their reported results confirm the significance of the depth of convolutional neural networks and that deep residual networks can generalize well in the case of plankton classification.

- Khan et al. [40] proposed a cost-sensitive deep neural network which can automatically learn robust feature representations for both the majority and minority classes. The pre-trained model of VGG net (configuration D) was used with two extra fully connected layers which were added before the output layer. The last two fully connected layers were initialized by random weights. The full network was trained with their proposed cost functions. Their method was evaluated on MLC dataset and achieved an improvement of 0.9 and 1.3% compared with the baseline method of [15] (which achieved 74.3 and 67.3%).

End-to-end training reduces the need for pre-processing and post-processing, and the results largely rely on self-learning models based on the data. Therefore, the outputs can better fit the tasks. Existing underwater datasets are mostly too small to train an end-to-end deep network. Therefore, it may be a future option to create large-scale labelled marine datasets for achieving higher performance in deep learning.

3 Deep Learning for Marine Species Detection

Object detection aims at detecting instances of objects of certain classes in images or videos. Popular applications of object detection include tracking objects, video surveillance, people counting, and anomaly detection. In marine environments, underwater video monitoring has been widely used for marine video surveillance in the recent years. This approach is non-invasive and provides sufficient research material. However, it is labor intensive and prone to error, since a massive quantity of video data needs to be manually analyzed. Therefore, it is important to develop an automated video-based marine species detection system, which can be used in a range of applications, such as the estimation of the abundance of certain fish and the study of the behavior and interactions between marine animals.

An object detection system needs to achieve: localization and classification. The object detectors first look for objects and then use classifiers to test for the presence of objects of certain classes. The traditional process of object detection comprises three main steps: (i) region selection, which is often achieved by exhaustive methods, such as sliding windows; (ii) feature extraction, in which the commonly used feature extractors are SIFT and HOG [17]; and (iii) classification, with SVM and Adaboost [41] being the primary classifiers. However, there are two major problems in traditional object detection. First, the sliding window based region search is computationally expensive and provides a large number of redundant windows. Second, handcrafted features have poor robustness to various changes of appearance of objects. In this section, current top performing object detectors for PASCAL and ImageNet datasets will be introduced, as well as their applications to marine species detection.

An improved detection performance is often the result of more complex classifiers, which leads to dramatically increased computation time. Until recently, this trade off between computational tractability and high detection quality has been overcome by the use of "detection proposals". In the LifeCLEF 2015 fish task evaluation campaign, the approach of [42] achieved the best performance in this video-based fish recognition task with a sequence of simple but effective techniques. The approach involves four steps: pre-processing, detection, recognition and post-processing. In the pre-processing phase, the foreground image is segmented from the background with a median value method and then smoothed using a bilateral filter. Subsequently, a selective search is used to extract the candidate fish object window from the foreground image. In the recognition stage, GoogleNet pre-trained on ImageNet is used and fine-tuned using the fish training data. Finally, the outputs of the recognition module are refined by considering temporally connected video segments. Although the pre-processing and post-processing steps clearly help improve the final performance, this approach is more likely to achieve a better performance if more advanced and complex techniques are used.

Girshick et al. [43] presented "R-CNN" as a solution to object detection problems. This approach involves the use of detection proposals and deep learning based classifiers. Region proposals, which correspond to possible locations of objects in the image, were extracted using methods based on texture, edge and color information, such as selective search [44], edge boxes [45], or binarized normed gradients (BING) [46]. These methods largely reduce the number of region windows and maintain a high recall rate. AlexNet was used to extract features of each region proposal. These extracted features were then fed into SVMs for classification. The last step involves the use of a bounding-box regression to refine the locations of the region proposals, making the resulting windows closer to the ground-truth ones.

SPP-Net [47] was proposed to increase the running speed of R-CNN. Instead of extracting CNN features in each region proposal separately, SPP-Net takes the whole image as an input to a CNN and generates a feature map, in which region proposals are projected from the original image. A spatial pyramid pooling layer is added before the fully connected layers to transfer feature maps of arbitrary size into a fixed-length feature vector. SPP-Net runs much faster than R-CNN for object detection.

Combining the advantages of both R-CNN and SPP-Net, Girshick et al. [48] proposed an improved algorithm: fast R-CNN. Compared with the framework of R-CNN, a region of interest (RoI) layer is added after the last convolutional layer. This layer replaces the max pooling layer, and a multi-task loss is adopted in the fast R-CNN. This is achieved by replacing the last fully connected layer with two sibling layers, one of which is a softmax layer for classification (instead of SVM classifiers used in R-CNN), while the other integrates the bounding box regression in the CNN training. Li et al. [49] applied the fast R-CNN framework to deal with a more challenging fish dataset of poorer-quality images compared to ImageNet and VOC datasets. Their experimental results show that they achieve a better performance in detection precision and speed over two popular approaches built on HOG-based Deformable Part Models (DPM) and R-CNN, respectively.

In the case of region-based CNN frameworks, the quality of the generated proposals directly affects the accuracy of object detection. It is thus required to find a more efficient way to extract less and higher-quality region proposals. Faster R-CNN [50] was developed with a region proposal network (RPN) incorporated into a fast R-CNN. This approach achieved the generation of region proposals by sharing convolutional features with detection network, making region candidates cost-free. RPN is a two-layer network, with an intermediate layer mapped from a spatial $n \times n$ sliding window of the input feature map from the last convolutional layer, and the other layer consisting of two sibling layers outputting the probability of object or not-object and the coordinates for each region proposal, respectively. To generate region proposals, an $n \times n$ window is slided over all the locations in the feature map and the central position of each sliding window of the feature map is projected to a point in the input image. Each of k anchors centered at that point is associated with a scale and aspect ratio. Therefore, a feature map of size $w \times h$ can relate to $k \times w \times h$ anchors, which are generated proposals. In the pipeline of faster R-CNN, alternating optimization is used to ensure that RPN and the fast R-CNN are trained with shared convolutional features. Faster R-CNN unifies the process of region proposal generation with classification by CNN, achieving a comparable accuracy but a faster running speed than fast R-CNN. Li et al. [51] adopted the faster R-CNN architecture to improve the results in [49], demonstrating 15.1% higher Mean Average Precision (mAP) over DPM baseline and a $3\times$ boost in speed compared to fast R-CNN for fish detection. Moreover, their results has shown that the high-quality region proposals generated by RPN not only contributed to a higher fish detection precision over the general proposal generation methods, but also improved the segmentation performance.

In addition to the series of region proposal based deep learning methods, more recent works have reported more advanced end-to-end object detection techniques, such as YOLO [52] and SSD [53]. These methods will have a strong research potential in the area of marine species detection.

4 Future Prospects

Although deep learning has achieved breathtaking performance in many aspects of information processing, a number of challenges still remain. First, the training of deep neural networks is computationally expensive. It is not easy to train deep networks as this requires large datasets and a huge computing power. Moreover, it critically depends on the expertise of the uses for the parameter tuning and it often converges slowly to an optimal solution, making it a time-consuming process. Second, future deep architectures should take into account sensory data transformations, such as various geometric and photometric transformations, to further improve the recognition accuracy. Third, there is a need to develop unsupervised feature learning approaches. Since the acquired datasets become larger and larger, it is unreasonable and unrealistic to label most of the data. Based on the latest advances of deep learning, we will next introduce two techniques that could be applied to marine image recognition systems to improve their overall performance.

4.1 Deep Learning to Improve Image Quality

The average accuracy of underwater image recognition is far behind the accuracy of ground image recognition. This is mainly due to the poor quality of underwater images, which often suffers from low contrast, blurring, and color degradation. These degradation factors are due to the physical properties of the water medium and they are not present in images taken on the ground. Several works, have shown good effects and improved accuracy, when using simple image enhancement techniques to improve the image contrast [15, 16]. This is a good indication that improved enhancement techniques would achieve a better performance.

Recently, a number of works [54–56], have used deep learning methods to improve image quality with ground images. For example, in [54], an end-to-end mapping is achieved between low-resolution images and high-resolution ones. A lightweight structure was proposed for their deep convolutional network, which demonstrates a state-of-the-art restoration quality as well as a fast speed for practical on-line usage. Sun et al. [55] proposed a deep learning approach to address the problem of complex non-uniform motion blur, which is usually caused by camera rotations or object motion. Moreover, Schuler et al. [56] designed an end-to-end trainable model for blind image deconvolution, which can be adapted to blurry images with strong noise and enable a competitive performance with respect to both image quality and runtime.

Overall, deep learning-based approaches have achieved better performance in addressing image quality problems of normal images, when compared to state-of-the-art handcrafted ones. For marine images, there is also a need for developing such techniques in the future research.

4.2 Deep Learning in Segmentation

Traditional convolutional neural network structures, such as AlexNet, are suitable for image-level classification and regression tasks whereby the final prediction of an input image is a numerical description. For example, the output of AlexNet for ImageNet dataset is a 1000-dimensional vector. This output vector represents the probabilities of each class, to which the input image belongs. There is high demand for image pixel-level classification in some application scenarios. This includes semantic image segmentation, which requires classification results corresponding to each pixel location of an input image, and edge detection, which is equivalent to binary classification (edge or not edge) for each pixel.

As to the problems of semantic segmentation and edge detection, the classical approach is to cut a patch centered around each pixel, and then to train classifiers using features extracted from the patch. In the testing phase, the learned model is used to classify the patch around each pixel and assign a label to that pixel. This is the idea of DeepContour [57] for image contour detection. However, this process which requires the classification of each pixel one by one is time-consuming. Moreover, the classification results are limited by the patches, making it difficult to model contextual information, which thus impacts on the algorithm performance. Long et al. [58] proposed fully connected networks (FCNs) for pixel-level classification, which efficiently address the problem of semantic segmentation. Unlike traditional CNNs, in which convolutional layers are followed by fully connected layers yielding a fixed-length feature vector for classification, FCNs accept arbitrary-size input images. In addition, the deconvolutional layers are used to up-sample the feature map of the last convolutional layer, outputting an input-size matrix of predictions for each pixel, while retaining the spatial information of the original input image.

Segmentation techniques are often used in maritime video surveillance to separate moving objects (such as fish) from the static background. However, these techniques are rarely used for marine image classification. Specifically, benthic data is annotated by random point annotation, since the boundaries of benthic species are often ambiguous. Therefore, the common annotation methods that are adopted with ground datasets, such as image labels and bounding boxes, cannot provide the required level of details. For automated point annotation, which is similar to pixel-level classification, the goal is to assign labels to a fixed number of randomly chosen points in the image. Similarly to the methods mentioned above, the feature of each point for training is extracted from a patch centered around that point. As a result, there are some problems which are inherent to the patch size selection. Although it is inappropriate to identify a point relying on the feature extracted from any single patch, a multiple scale approach would also not be adequate because it leads to a redundant computation of features in the patches [15]. In future research, FCNs should be used to address this problem, which should result in an improved accuracy and efficient automated annotation.

5 Conclusion

This chapter has covered a broad literature review of deep learning techniques for marine species recognition. Image classification for marine species is the main focus of this chapter, as it is the basis of other visual tasks and the key tool for creating marine species distribution maps. As the core of image classification, feature extraction and feature combination methods specific for marine species have been presented in detail. The pros and cons of the deep features and handcrafted features for marine species have been discussed. In addition, this chapter has explored results for marine species detection on videos. Marine data with their unique characteristics such as color degradation and large variations in morphologies, and increasing quantity, has been providing computer vision field a great opportunity yet a big challenge. We also argue that in future, challenges in marine data (e.g., poor image quality and ambiguous boundaries) can be addressed by deep learning based image quality enhancement and semantic segmentation techniques.

Acknowledgements This research was partially supported by China Scholarship Council funds (CSC, 201607565016) and Australian Research Council Grants (DP150104251 and DE120102960).

References

1. T. Wernberg, S. Bennett, R.C. Babcock et al., Climate-driven regime shift of a temperate marine ecosystem. Science **353**(6295), 169–172 (2016)
2. T.C. Bridge, R. Ferrari, M. Bryson et al., Variable responses of benthic communities to anomalously warm sea temperatures on a high-latitude coral reef. PLoS ONE **9**(11), e113079 (2014)
3. H. Singh, R. Armstrong, G. Gilbes et al., Imaging coral I: imaging coral habitats with the SeaBED AUV. Subsurf. Sens. Technol. Appl. **5**(1), 25–42 (2004)
4. J.W. Nicholson, A.J. Healey, The present state of autonomous underwater vehicle (AUV) applications and technologies. Mar. Technol. Soc. J. **42**(1), 44–51 (2008)
5. O. Beijbom, P.J. Edmunds, C. Roelfsema et al., Towards automated annotation of benthic survey images: variability of human experts and operational modes of automation. PLoS ONE **10**(7), e0130312 (2015)
6. A. Krizhevsky, I. Sutskever, G.E. Hinton, ImageNet classification with deep convolutional neural networks, in *Advances in Neural Information Processing Systems* (2012), pp. 1097–1105
7. K. Simonyan, A. Zisserman, Very deep convolutional networks for large-scale image recognition. arXiv:1409.1556 (2014)
8. K. He, X. Zhang, S. Ren et al., Deep residual learning for image recognition, in *Proceedings of the IEEE Conference on Computer Vision and Pattern Recognition* (2016), pp. 770–778
9. J. Donahue, Y. Jia, O. Vinyals et al., DeCAF: a deep convolutional activation feature for generic visual recognition, in *Proceedings of the 31st International Conference on Machine Learning (ICML)*, Beijing, China, vol. 32, June 2014, pp. 647–655
10. S. Razavian, H. Azizpour, J. Sullivan et al., CNN features off-the-shelf: an astounding baseline for recognition, in *Proceedings of the IEEE Conference on Computer Vision and Pattern Recognition Workshops* (2014), pp. 806–813
11. J. Jaeger, M. Simon, J. Denzler et al., Croatian Fish dataset: fine-grained classification of fish species in their natural habitat (2015), pp. 1–7. http://dx.doi.org/10.5244/C.29.MVAB.6
12. A. Mahmood, M. Bennamoun, S. An et al., Automatic annotation of coral reefs using deep learning, in *Proceedings of OCEANS 16*, Monterey, California, USA, Sept 2016, pp. 17–23
13. M. Bewley, A. Friedman, R. Ferrari et al., Australian seafloor survey data, with images and expert annotations. Sci. Data **2** (2015)
14. A. Mahmood, M. Bennamoun, S. An et al., ResFeats: residual network based features for image classification. arXiv:1611.06656 (2016)
15. O. Beijbom, P.J. Edmunds, D.I. Kline et al., Automated annotation of coral reef survey images, in *Proceedings of IEEE Conference on Computer Vision and Pattern Recognition (CVPR)*, Providence, Rhode Island, June 2012, pp. 16–21
16. A. Mahmood, M. Bennamoun, S. An et al., Coral classification with hybrid feature representations, in *Proceedings of IEEE International Conference on Image Processing (ICIP)*, Phoenix, Arizona, USA, Sept 2016, pp. 25–28
17. N. Dalal, B. Triggs. Histograms of oriented gradients for human detection, in *IEEE Computer Society Conference on Computer Vision and Pattern Recognition, 2005. CVPR 2005*, vol. 1 (IEEE, 2005), pp. 886–893

18. D.G. Lowe, Object recognition from local scale-invariant features, in *The Proceedings of the Seventh IEEE International Conference on Computer Vision*, 1999, vol. 2 (IEEE, 1999), pp. 1150–1157
19. M. Marcos, S. Angeli, L. David et al., Automated Benthic counting of living and non-living components in Ngedarrak Reef, Palau via Subsurface Underwater video. Environ. Monit. Assess. **125**(1), 177–184 (2008)
20. A. Pizarro, P. Rigby, M. Johnson-Roberson et al., Towards image-based marine habitat classification, in *Proceedings of OCEANS 08*, Quebec City, QC, Canada, Sept 2008, pp. 15–18
21. M.D. Stokes, G.B. Deane, Automated processing of coral reef benthic images. Limnol. Oceanogr.: Methods **7**(2), 157–168 (2009)
22. M. Bewley, B. Douillard, N. Nourani-Vatani et al., Automated species detection: an experimental approach to kelp detection from sea-floor AUV images, in *Proceedings of Australasian Conference on Robotics and Automation* (2012)
23. M. Bewley, N. Nourani-Vatani, D. Rao et al., Hierarchical Classification in AUV imagery, in *Springer Tracts in Advanced Robotics*, vol. 105, Jan 2015, pp. 3–16
24. IMOS: integrated marine observing system, Sept 2013. http://www.imos.org.au
25. R.M. Haralick, K. Shanmugam, I. Dinstein, Textural features for image Classification. IEEE Trans. Syst. Man Cybern. **SMC-3**(6), 610–621 (1973)
26. L. Soh, C. Tsatsoulis, Texture analysis of SAR sea ice imagery using gray level co-occurrence matrices. IEEE Trans. Geosci. Remote Sens. **37**(2), 780–795 (1999)
27. D.A. Clausi, An analysis of co-occurrence texture statistics as a function of grey level quantization. Can. J. Remote Sens. **28**(1), 45–62 (2002)
28. G.M. Haley, B.S. Manjunath, Rotation-invariant texture classification using a complete space-frequency model. IEEE Trans. Image Process. **8**(2), 255–269 (1999)
29. Z. Guo, L. Zhang, A completed modeling of local binary pattern operator for texture classification. IEEE Trans. Image Process. **19**(6), 1657–1663 (2010)
30. J. Van de Weijer, C. Schmid, Coloring local feature extraction, in *Proceedings of the 9th European Conference on Computer Vision (ECCV 06)*, Graz, Austria, May 2006, pp. 334–438
31. Z. Chao, J.C. Principe, B. Ouyang, Marine animal classification using combined CNN and hand-designed image features, in *OCEANS'15 MTS/IEEE*, Washington (IEEE, 2015), pp. 1–6
32. H. Peng, F. Long, C. Ding, Feature selection based on mutual information criteria of max-dependency, max-relevance, and min-redundancy. IEEE Trans. Pattern Anal. Mach. Intell. **27**(8), 1226–1238 (2005)
33. C. Zheng, Jose C. Principe, B. Ouyang, Group feature selection in image classification with multiple kernel learning, in *2015 International Joint Conference on Neural Networks (IJCNN)* (IEEE, 2015), pp. 1–5
34. H. Qin, X. Li, J. Liang et al., DeepFish: accurate underwater live fish recognition with a deep architecture. Neurocomputing **187**, 49–58 (2016)
35. J.N. Blanchet et al., Automated annotation of corals in natural scene images using multiple texture representations. PeerJ Preprints **4**, e2026v2 (2016)
36. A.S.M. Shihavuddin, N. Gracias, R. Garcia et al., Image-based coral reef classification and thematic mapping. Remote Sens. **5**(4), 1809–1841 (2013)
37. O. Beijbom, T. Treibitz, D.I. Kline et al., Improving automated annotation of benthic survey images using wide-band fluorescence. Sci. Rep. **6** (2016)
38. Y. LeCun, Y. Bengio, Convolutional networks for images, speech, and time series, in *The Handbook of Brain Theory and Neural Networks*, vol. 3361, no. 10 (1995)
39. X. Li, Z. Cui, Deep residual networks for plankton classification, in *OCEANS 2016 MTS/IEEE*, Monterey, Sept 2016, pp. 1–4
40. S.H. Khan, M. Hayat, M. Bennamoun et al., Cost-sensitive learning of deep feature representations from imbalanced data. IEEE Trans. Neural Netw. Learn. Syst. (2017) (in press)
41. P. Viola, M. Jones, Rapid object detection using a boosted cascade of simple features, in *Proceedings of the IEEE Computer Society Conference on Computer Vision and Pattern Recognition (CVPR)*, vol. 1 (2001), pp. I–I

42. S. Choi, Fish identification in underwater video with deep convolutional neural network: SNUMedinfo at LifeCLEF fish task 2015, in *CLEF* (Working Notes) (2015)
43. R. Girshick, J. Donahue, T. Darrell et al., Rich feature hierarchies for accurate object detection and semantic segmentation, in *Proceedings of the IEEE Conference on Computer Vision and Pattern Recognition*, pp. 580–587, 2014
44. J.R. Uijlings, K.E. Van De Sande, T. Gevers et al., Selective search for object recognition. Int. J. Comput. Vis. **104**(2), 154–171 (2013)
45. C. Zitnick, P. Dollr, Edge boxes: locating object proposals from edges, in *European Conference on Computer Vision* (Springer International Publishing, 2014), pp. 391–405
46. M. Cheng, Z. Zhang, W. Lin et al., BING: binarized normed gradients for objectness estimation at 300 fps, in *Proceedings of the IEEE Conference on Computer Vision and Pattern Recognition* (2014), pp. 3286–3293
47. K. He, X. Zhang, S. Ren et al., Spatial pyramid pooling in deep convolutional networks for visual recognition, in *European Conference on Computer Vision* (Springer International Publishing, 2014), pp. 346–361
48. R. Girshick, Fast R-CNN, in *Proceedings of the IEEE International Conference on Computer Vision* (2015), pp. 1440–1448
49. X. Li, M. Shang, H. Qin et al., Fast accurate fish detection and recognition of underwater images with Fast R-CNN, in *OCEANS'15 MTS/IEEE*, Washington, Oct 2015, pp. 1–5
50. S. Ren, K. He, R. Girshick et al., Faster R-CNN: towards real-time object detection with region proposal networks, in *Advances in Neural Information Processing Systems* (2015), pp. 91–99
51. X. Li, M. Shang, J. Hao et al., Accelerating fish detection and recognition by sharing CNNs with objectness learning, in *OCEANS 2016*-Shanghai, 10 Apr 2016 (IEEE, 2016), pp. 1–5
52. J. Redmon, S. Divvala, R. Girshick et al., You only look once: unified, real-time object detection, in *Proceedings of the IEEE Conference on Computer Vision and Pattern Recognition* (2016), pp. 779–788
53. W. Liu, D. Anguelov, D. Erhan et al., SSD: single shot multibox detector, in *European Conference on Computer vision* (Springer, Cham, 2016), pp. 21–37
54. C. Dong, C.L. Chen, K. He et al., Image super-resolution using deep convolutional networks. IEEE Trans. Pattern Anal. Mach. Intell. **38**(2), 295–307 (2016)
55. J. Sun, W. Cao, Z. Xu et al., Learning a convolutional neural network for non-uniform motion blur removal, in *Proceedings of the IEEE Conference on Computer Vision and Pattern Recognition* (2015), pp. 769–777
56. C.J. Schuler, M. Hirsch, S. Harmeling et al., Learning to deblur. IEEE Trans. Pattern Anal. Mach. Intell. **38**(7), 1439–1451 (2016)
57. W. Shen, X. Wang, Y. Wang et al., Deepcontour: a deep convolutional feature learned by positive-sharing loss for contour detection, in *Proceedings of the IEEE Conference on Computer Vision and Pattern Recognition* (2015), pp. 3982–3991
58. J. Long, E. Shelhamer, T. Darrell, Fully convolutional networks for semantic segmentation, in *Proceedings of the IEEE Conference on Computer Vision and Pattern Recognition* (2015), pp. 3431–3440

Deep Molecular Representation in Cheminformatics

Peng Jiang, Serkan Saydam, Hamed Lamei Ramandi, Alan Crosky and Mojtaba Maghrebi

Abstract Quantum-chemical descriptors are powerful predictors of discovering and designing new materials of desired properties. Wave-function-based methods are often employed to calculate quantum-chemical descriptors, which are time consuming. Recently, machine learning models have been used for predicting quantum-chemical descriptors because of their computational advantages. However, it is difficult to generate a proper molecular representation for training. This work reviews recent molecular representation techniques and then employs variational autoencoders to encode Bag-of-Bond molecular representation. The encoded representation reduce the dimensionality of features and extract the essential information through a deep neural network structure. Results on a benchmark dataset show that the deep encoded molecular representation outperforms Bag-of-Bond representations in predicting electronic quantum-chemical descriptors.

P. Jiang · A. Crosky
School of Materials Science and Engineering, UNSW Australia, Sydney, NSW, Australia
e-mail: peng.jiang@student.unsw.edu.au

A. Crosky
e-mail: a.crosky@unsw.edu.au

S. Saydam · H. L. Ramandi
School of Mining Engineering, UNSW Australia, Sydney, NSW, Australia
e-mail: s.saydam@unsw.edu.au

H. L. Ramandi
e-mail: h.lameiramandi@unsw.edu.au

M. Maghrebi (✉)
School of Civil and Environmental Engineering, UNSW Australia, Sydney, NSW, Australia
e-mail: maghrebi@unsw.edu.au

M. Maghrebi
Department of Civil Engineering, Ferdowsi University of Mashhad, Mashhad, Razavi Khorasan, Iran

V. E. Balas et al. (eds.), *Handbook of Deep Learning Applications*,
Smart Innovation, Systems and Technologies 136,
https://doi.org/10.1007/978-3-030-11479-4_8

1 Introduction

For centuries, chemists have designed and performed time-consuming experiments to synthesize new materials of desired properties. With the development of computers, Cheminformatics, a new discipline, has been established and developed. Cheminformatics applies state-of-the-art informational and computer techniques to solve a wide range of problems in fields of Materials and Chemistry such as storing, indexing and searching. Moreover, Cheminformatics has absorbed machine learning-based and data mining-based techniques to learn the latent knowledge in molecules. The knowledge can be used by chemists and materials engineers for guiding the development of new materials of desired properties.

Quantitative structure-activity relationship (QSAR) is a typical application of Cheminformatics in predicting properties of materials. For example, QASR models have been widely used in drug discovery [1–3] and biomedical study [4, 5] because of their moderate computation cost and advantages in accelerating development and testing. Recently, researchers have begun to use QSAR models to solve problems in different fields. Gómez-Bombarelli et al. [6] developed a data-driven QSAR model and demonstrated its effectiveness in designing organic light-emitting diodes. Camacho-Mendoza et al. [7] showed the reliability of QSAR model in predicting the inhibition performance of organic corrosion inhibitors. Li et al. [8] proved the effectiveness of QSAR model in predicting performances of corrosion inhibitors with simulations and experiments, and they also showed that the QSAR-based similarity analysis can be employed for developing novel corrosion inhibitors.

To conduct QSAR analysis, a set of features that represent the material is often needed. Ideal molecular features convey all of the essential information required for predicting its properties so that the relationship can be well approximated by mapping these features to the properties of the corresponding material. Most of the works employing QSAR models for predicting materials properties or discovering new compounds have used quantum-chemical descriptors as the features [9, 10]. Quantum-chemical descriptors, also known as quantum mechanical properties, deliver simple molecular information that can provide insight into the electronic and thermochemical structures of molecules. Quantum-chemical descriptors that are widely used in QSAR studies include atomic charges,molecular orbital energies, frontier orbital densities, dipole moments and polarity indices. Karelson et al. [9] gave detailed descriptions of quantum-chemical descriptors and applications. The correlations between quantum-chemical descriptors and different properties such as biological activities [11], reactivity of organic compounds [12] and octanol/water partition coefficients [13] have been established and demonstrated.

In general, most of quantum-chemical descriptors are obtained from computational chemical methods, which are based on classical molecular force fields and quantum-chemical methods. Semiempirical quantum-chemical methods based on simplification and approximation of the molecular orbital theory (SCF MO), have also been extensively used. Recently, Density functional theory (DFT) has been becoming one of the most popular methods for calculating quantum-chemical descriptors. DFT is an Ab initio quantum-chemical method based solely on quantum mechanics. In comparison to experiments, computational chemistry methods are easy to perform and less costly. Furthermore, calculations can be performed on any molecules, even those that do not exist. This offers valuable opportunities to explore the huge unknown chemical compound space (CCS). However, calculations could be very costly in terms of the amount of time required. DFT can take up to several days to perform the calculations for one molecule, which is not an economic means given the billions of molecules in CCS.

To accelerate the calculation of quantum-chemical descriptors, a few researchers have tried to use machine learning methods [14, 15]. Though less common in physical chemistry, they demonstrated that it is possible to estimate the quantum-chemical descriptors of molecules based on the correspondence principle in quantum mechanics and statistical mechanics. Ramakrishnan and von Lilienfeld [16] pointed out that the success of machine learning methods in producing quantum-chemical descriptors depends on whether four conditions can be met: (i) there is a rigorous relationship between the system and the quantum properties; (ii) the representations carry enough information needed to reconstruct the systems' Hamiltonian; (iii) the training set and test set need to be from the same distribution; (iv) the training set is large enough. Due to the computational advantage and good transferability in machine learning, more works on machine learning-based quantum-chemical studies have been published [16, 17]. However, the four conditions determining the success of machine learning still act as the main challenges. The most difficult one, to our knowledge, is the representations of molecules. These representations are often required to be fixed-size vectors, which are used as the input features in machine learning models. The elements, length and bonds of molecules in CCS vary dramatically, which add difficulties in developing a proper representation for machine learning models.

This work is structured as follows: Sect. 2 reviews the recent findings in developing molecular representations; Sect. 3 introduces the deep learning techniques used in producing a deep molecular representation, followed by Sects. 4 and 5, where the database and the model structure are discussed. Section 6 gives the simulation results and discussion; the conclusions that can be drawn from this work is in Sect. 7.

2 Molecular Representation

As mentioned previously, a proper representation of molecules is critical for machine learning models and often requires feature engineering techniques to improve the calculation performance. In this section, several popular molecular representation methods are reviewed.

2.1 Molecular Fingerprint

Molecular fingerprint is an approach for generating a string of binary numbers (0, 1) that shows the existence of a specific function group [18]. One simple interpretation of fingerprints is a series of binary numbers. For a given length vector, "1" in one location of the vector means a predefined function group exists in this compound, and "0" means the absence of this group. On advantage of this representation method is that these binary vectors can be effectively processed by computers. In addition, researchers can customize the predefined vectors based on different scenarios. Fingerprint requires little reconstruction of the original information compared to many other representations. This helps in improving the calculation efficiencies and more importantly, similarity tests among molecules can be conducted easily with fingerprint descriptors. However, the searching and designing of molecule fingerprint might be difficult and time-consuming for some large and complex polymers.

Merck [19] sponsored a competition in 2012 on predicting activities of different molecules. This challenge involved 15 datasets of different sizes (2000–50000 molecules) with thousands of descriptors as the features (fingerprints). The first prize went to a team of researchers who used an ensemble of several machine learning methods, including the state-of-art deep learning, gradient boosting machine, and Gaussian process. Their model achieved a mean R^2 of 0.49 over the 15 datasets, which was a remarkable improvement compared to the Merck internal benchmark (0.42) [20]. Inspired by their work, Unterthiner et al. [21] applied deep learning in the dataset from the ChEMBL database. They compared the performance of deep learning to seven target prediction models, demonstrating that deep learning outperformed other methods regarding the area under receiver operating characteristic (ROC) curve, i.e. one of the useful criteria for assessing the classification performance.

2.2 Coulomb Matrix

Among all of the available molecular representation methods, fingerprints is one of the most popular methods because of its simplicity and low computational cost. However, to generate a complete fingerprint, a large feature database is required. Moreover, the length of fingerprint might be too long. Coulomb Matrix [14, 15], which considers only the structural information of molecules, has been proposed to produce a relatively short and simple representation. Let Z be the nuclear charge and R be the atomic position in 3D space of each atom in one molecule. With Z and R as input, the quantum-chemical descriptors can be estimated through first principles-based DFT methods. Using the same information as input features, machine learning-based models can provide an alternative to time-consuming DFT calculations when sufficient samples are available. Coulomb Matrix serves as a means of feature engineering that combines both Z and R. Specifically, for a molecule of N atoms, Coulomb Matrix is a symmetric matrix of $N \times N$. Each element M in a Coulomb Matrix is defined as:

$$M_{ij} = \begin{cases} 0.5Z_i^{2.4}, & \text{if } i = j \\ \frac{Z_i Z_j}{|R_i - R_j|}, & \text{if } i \neq j \end{cases} \tag{1}$$

In a Coulomb Matrix, the diagonal elements correspond to the estimation of potential energy of the free atom, while the off-diagonal elements encode the Coulomb interaction between two atoms. Due to its differentiability with respect to the atomic position and nuclear charge, Coulomb Matrix is a proper candidate for the input features used by Machine learning.

Although Coulomb Matrix adequately summarizes the essential information that is needed for DFT calculation, it fails to represent the unique ordering of a molecule. In other words, a molecule can be represented by multiple Coulomb matrices by introducing a permutation in the rows and columns. Several solutions to this problem have been proposed [22]. A solution is using a sorted Coulomb Matrix based on the norms of each row. But this method removes the differentiability of a Coulomb Matrix. Another solution is using a randomly sorted Coulomb Matrices and letting the machine learning algorithms to learn the invariance from the permuted matrices. Results [22] showed that the latter solution outperforms the former solution by the predicted accuracy of atomization energies.

2.3 *Bag of Bonds*

To compromise the invariance property of Coulomb Matrix, Bag of Bonds (BOB) was proposed [23]. BOB is inspired by Bag of Words (BOW), which is one of the natural language processing (NLP) techniques. Similar to Cheminformatics, one of the biggest challenges in the field of NLP is representing texts with fixed-size vectors so that algorithms can extract and process the true information. BOW addresses this through tokenization, counting and normalization. Tokenization is the process of breaking the texts into possible tokens. The tokens can be words, phrases or any predefined elements. Then the occurrences of tokens in each text is counted. Finally, normalization is performed to assign weights to tokens for obtaining the numerical feature vectors. BOB adopts the three key processes from BOW and produces fixed-size feature vectors for molecules. Specifically, the elements in Coulomb Matrix are handled as tokens. These elements are sorted by the combination of nuclear charges and filled into one bag. Therefore, each token in one bag carries the spatial information and the nuclear charges.

Though Coulomb Matrix and BOB reduce the feature length dramatically compared to molecular fingerprints, they generate fixed-size features by simply padding. This makes the features of some molecules very sparse if a complex molecule is involved in the database. Besides the three molecular representation methods, there are several interesting methods including scattering transforms [24], atomic distances [25] and graph models [26, 27]. The choice of a proper representation is a trade-off among computational efficiency, information complexity and accessibility. A comparison of molecular representations was reviewed in [28].

3 Theoretical Background

3.1 Deep Neural Networks

Deep neural networks (DNN) is an advanced version of conventional artificial neural networks (ANN). An ANN is a multiple-layer network consisting of neurons that process the raw input information using an activation function $f(z)$. A neuron is associated with a default bias term b and weights from neurons w in the front layer. The input features go through each layer of ANN, and then a neuron processes the information in the following form:

$$f(z) = f(\sum_{i=1}^{N} w_i x_i + b) \tag{2}$$

A typical ANN contains three types of layers for different tasks: (i) the input layer, which is built at the bottom, is for receiving the raw input information; (ii) the output layer, which is built at the top, is for generating predictions and (iii) the hidden layers that are built in the middle are for reconstructing the raw information. Normally, DNN involves two or more hidden layers. In addition, several special architectures are designed for DNN to process the complex relationship in the raw input such as spatial and temporal information [29]. DNN has obtained a number of notable successes in image processing [30], speech recognition [31] and natural language processing [32].

3.2 Variance Autoencoders

Proposed by Kingma and Welling [33], VAE is a deep learning generative architecture that attempts to construct a probability distribution of latent variables. An autoencoder is an algorithm that reproduces original data by encoding and decoding. A simple autoencoder consists of an encoder and a decoder. Let x be the original data. An encoder transforms original data into a hidden representation z through a function f: $z = f(x)$, while a decoder projects the hidden representation from the encoder into r through function g: $r = g(f(x))$, where r should be close enough to x so the raw information can be reserved after the encoding-decoding process. Figure 1 shows a typical structure of VAE.

Since the final output of the decoder is only an approximation of the input, the feature transformation by encoding is considered. This type of feature transformation is the essential and effective information extracted from the original data [33]. Another remarkable ability of VAE is that it produces new data points through sampling from the distribution in the latent space which enables obtaining the unknown samples similar to the input. This ability makes VAE a generative model. Generative models is known as a powerful tool for health diagnosis due to following justifications. The first reason is that labeling observations, which is costly and time consuming, is

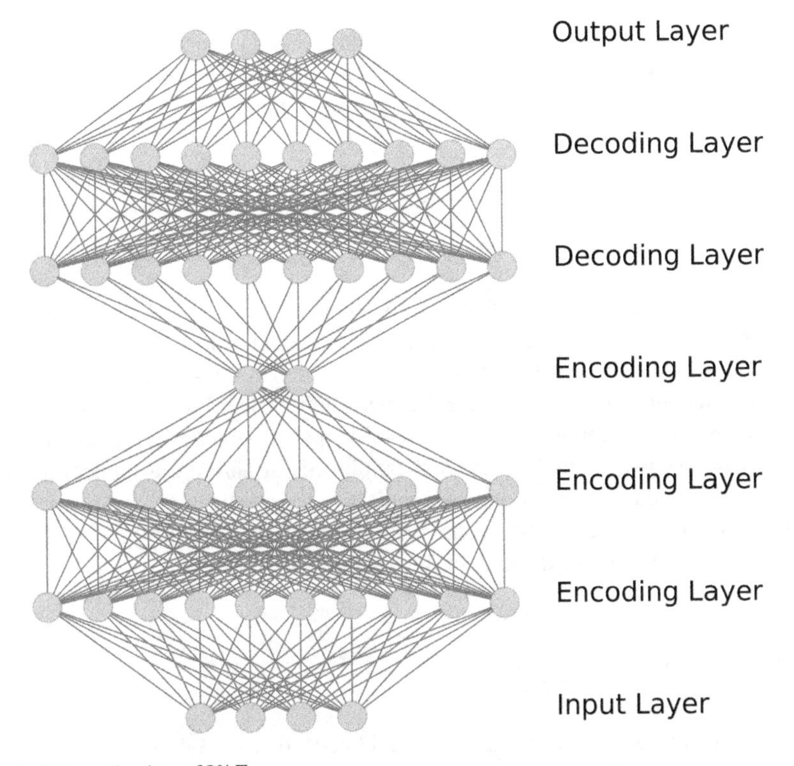

Fig. 1 Schematic view of VAE

eliminated using generative models. This is performed through allowing for observing the similar data points by providing a few labeled samples. The second reason is that for a generative model, architectures of the model, latent distribution and tuning hyperparameters are modularized. This means that prior knowledge or experience can be easily integrated in this model.

VAE learns to produce a probabilistic distribution of original data by encoding, which is actually an approximation on a true posterior density $p_\theta(z \mid x) = p_\theta(x \mid z)p_\theta(z)/p_\theta(x)$. VAE tries to infer a variational approximation $q_\phi(z \mid x)$ to the intractable true posterior. A metric, Kullback-Leibler (KL) divergence (D_{KL}) is then introduced to measure the degree of similarity between $q_\phi(z \mid x)$ and $p_\theta(z \mid x)$. To obtain an optimal model, i.e. the optimal weights for the encoder and the decoder networks, with the lowest generalization error, the marginal likelihood needs to be maximized. This is achieved by summation of the log-likelihood across data points:

$$\log p_\theta(x) = \mathscr{L}(\theta, \phi; x) + D_{KL}(q_\phi(z \mid x) \parallel p_\theta(z \mid x)) \tag{3}$$

where $\mathscr{L}(\theta, \phi; x)$ is the variational lower bound. Since D_{KL} is non-negative, the variational lower bound can be rewritten as:

$$\mathscr{L}(\theta, \phi; x) = \mathbb{E}_{q_\phi(z|x)}[\log p_\theta(x \mid z)] - D_{KL}(q_\phi(z \mid x) \parallel p_\theta(z \mid x)) \tag{4}$$

The first right hand side (RHS) term is the expected reconstruction error that is used to force the decoder to reconstruct the data. The second RHS term, KL divergence, measures how close the produced approximation is to the true posterior, which can be viewed as a regularizer.

For each input sample x, a set of hidden representations z is generated by sampling from $q_\phi(z \mid x)$. VAE utilize the reparameterization trick to sample z:

$$z = \mu + \sigma \odot \epsilon \tag{5}$$

where μ and σ are the mean value and the standard deviation of the approximate posterior $q_\phi(z \mid x)$. $\odot$ denotes an element-wise product. ϵ is a Gaussian noise $\epsilon \sim \mathcal{N}(0, I)$. This process enables transforming z from a random drawn value to a deterministic value with noise. Since the mean and standard deviation are obtained from the encoder's inference procedure, backpropagation with respect to θ through the variational lower bound function is employed. In addition to generating new samples and dimension reduction, VAE can also visualize the high-dimensional data by projecting it into a low-dimensional latent space. We use this property to discover the structural similarity of the data in the following case study.

4 Database

In this study, QM9 set [34] is used as a benchmark dataset. QM9 set is a subset of GDB-13 set. GDB-13 was published by Reymond et al. [35], which contains 166 billion organic small molecules made of CHONF. The atom distribution is shown in

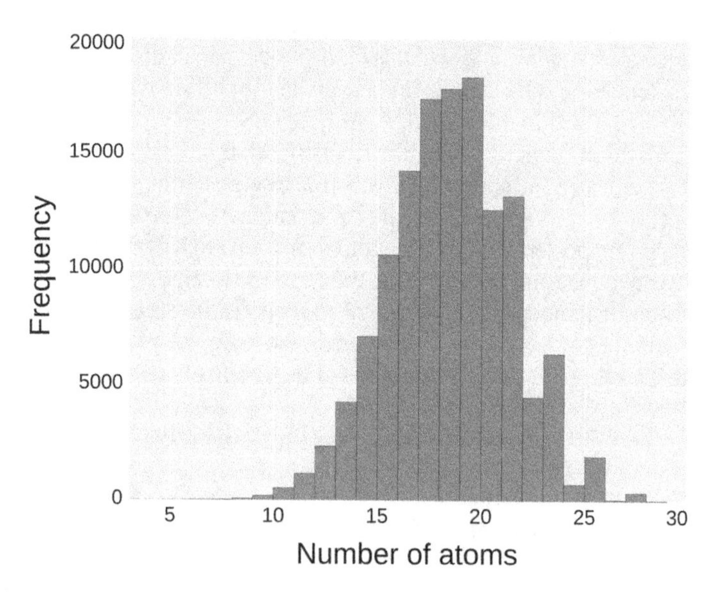

Fig. 2 Atom distributions of QM9 set

Table 1 Description of the calculated properties in QM9 set

No.	Property	Unit	Description
1	μ	Debye	Dipole moment
2	α	a_0^3	Isotropic polarizability
3	E_{HOMO}	Hartree	Energy of highest occupied molecular orbital (HOMO)
4	E_{LUMO}	Hartree	Energy of lowest occupied molecular orbital (LUMO)
5	E_{gap}	Hartree	Gap, difference between LUMO and HOMO

Fig. 2. Most of the molecules in QM9 set have 18 atoms. The geometric and quantum chemical properties were obtained for smallest 133885 (around 134 k) molecules in GDB-9 from DFT quantum chemistry calculations at B3LYP/6-31G(2df,p) level of theory. The electronic quantum-chemical descriptors are chosen as targets in this work, which are provided in Table 1. Properties such as gap energy, E_{HOMO} and E_{LUMO} have proven to be powerful features for predicting performances of organic corrosion inhibitors [7] and semiconductors [36].

5 Model

BOB is used as the method for producing the molecular representations for QM9 dataset in this study. As mentioned in Sect. 2, BOB extracts the elements in Coulomb Matrix and sorts it according to their magnitudes. Therefore, elements in BOB basically carry information about the positions of atoms and nuclear charges. However, Hansen et al. [23] found that the inclusion of the diagonal elements from the Coulomb Matrix contribute little to the improvement of machine learning-based predictions. Hence, only the off-diagonal elements that carry interatomic interactions are used for BOB molecular representations. Since BOB simply produce fixed-size representations by padding, this makes some molecules' features pretty sparse if a complex molecule is involved in the database. Moreover, this might introduce redundancy and lead to a poor prediction performance. Inspired by Rafael's work [6], VAE is used for reconstructing BOB molecular representations in this study to address the problem, and then a three-layer conventional ANN is used to predict several quantum-chemical descriptors. All of the models are built with Keras [37] and Tensorflow [38]. The employed VAE consists of an encoder and a decoder. The encoder has two 1D convolutional layers and one dense layer (fully connected layer) of dimension 200. The decoder has a dense layer of dimension 200 for connecting the encoder's last dense layer, followed by two recurrent unit networks of dimension 666. With the encoded representations as the input, a conventional baseline ANN model is built for fitting the properties. This ANN contains only one hidden layer of dimension 200 and an output layer of dimension 1. The raw dataset is divided into a training set (80%) and a test set (20%). The training set is used for training VAE-ANN and ANN, and the the test set is used for assessing its out-of-sample performance.

6 Simulation Results and Discussion

During the training for producing VAE-based molecular representations, 20% of the training data (raw BOB features) is used for validation, and the mean absolute error (MAE) is employed as an error measure. The MAE loss for training and validation can be seen in Fig. 3. After almost 80 iterations, both training loss and validation loss converge although an obvious fluctuation is observed for the validation set. The two curves are close to each other, demonstrating that training set and validation set are from the same distribution.

Table 2 shows the MAE and the corresponding standard deviation (prediction errors) of the predicted quantum-chemical descriptors for VAE-ANN and the baseline ANN model. Results show that VAE-ANN outperforms the baseline model regarding the MAE for all of the electronic quantum-chemical descriptors. This is because of the deep extraction ability of VAE that removes the redundancies while keeping the essential information from the raw data.

By adjusting the dimensions for the latent space in VAE's encoder from 200 to 2, the encoded representations is visualized and in Fig. 4. It is clearly observed that the 2D molecular representations are clustered according to E_{LUMO} values. This allows discovering new molecules of desired properties by sampling from the clusters of similar properties. Then VAE can employ the decoder to reconstruct the new molecule's BOB molecular representation.

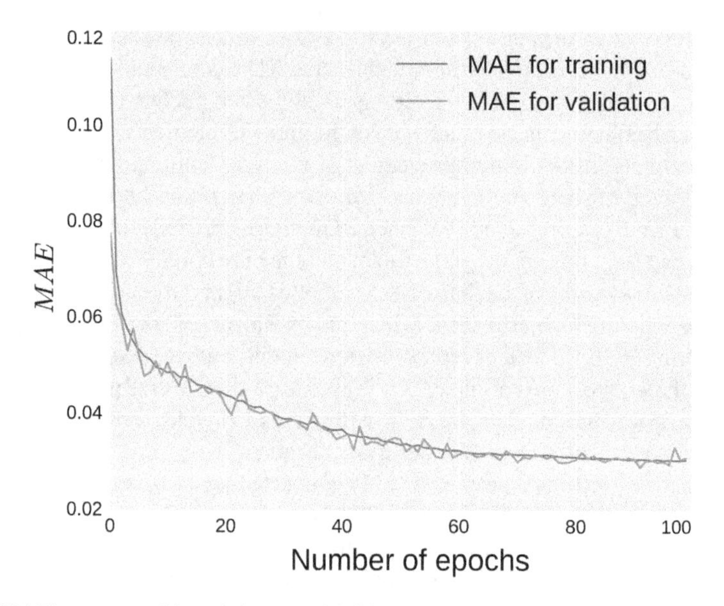

Fig. 3 MAE loss curves of the training set and validation set during producing VAE-based molecular representations

Table 2 MAE of electronic quantum-chemical descriptors in the test set for ANN using VAE encoded molecular representations

No.	Property	VAE-ANN	ANN
1	μ (Debye)	0.6443 (0.9200)	0.9312 (1.2864)
2	α (a_0^3)	0.9870 (1.5538)	2.8997 (4.0793)
3	E_{HOMO} (Hartree)	0.0084 (0.0114)	0.0141 (0.0193)
4	E_{LUMO} (Hartree)	0.0131 (0.0176)	0.0268 (0.0338)
5	E_{gap} (Hartree)	0.0154 (0.0202)	0.0278 (0.0352)

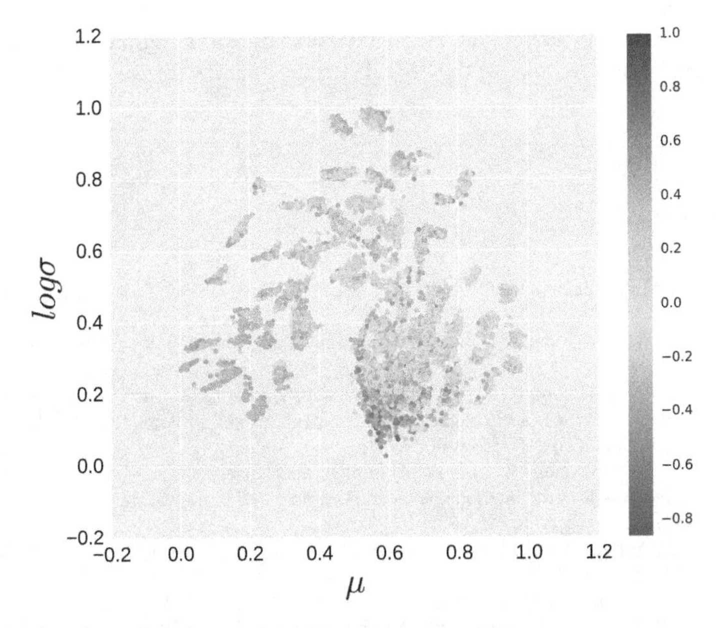

Fig. 4 VAE-based molecular representations in 2D latent space. The color represents the normalized actual E_{LUMO} values. It is clear that the molecular representations are clustered by the corresponding E_{LUMO} values

7 Conclusion

In this work the applications of machine learning in Cheminformatics are outlined together with the background of quantum-chemical descriptors in discovering and designing new materials of desired properties. Currently DFT methods used in producing quantum-chemical descriptors is costly in terms of calculation time given the billions of molecules in CCS. Due to the computation advantages, machine learning methods have been becoming popular in producing quantum-chemical descriptors. However, a reliable fixed-size molecular representation used as the input features is essential for determining the prediction performances. Representations such as molecular fingerprints, Coulomb Matrix and BOB are popular molecular representations used in industry and academia. However, due to the variable lengths of molecules in CCS, many of these representations can only generate fixed-size representation by simple padding. This makes molecular representations very sparse; notably, this might introduce redundancies. To produce a more reliable molecular representation, this work employs VAE to encode BOB features and reduces the dimensions from 666 to 200. Tests on QM9 set show that the prediction performance using the encoded molecular representation on several electronic quantum-chemical descriptors are improved compared to a baseline ANN model using the raw BOB descriptors. In addition, VAE enables producing new BOB by simply sampling from the latent space so that new molecules of desired properties can be found in clusters of molecules of similar properties. In future work, more detailed studies are recommended to further investigate the methods for transforming the calculated BOB features back into formats like the SMILE strings to allow for validation of the new molecules.

References

1. M.A. Lill, Multi-dimensional QSAR in drug discovery. Drug Discov. Today **12**(23), 1013–1017 (2007)
2. Hugo Kubinyi, QSAR and 3D QSAR in drug design Part 1: methodology. Drug Discov. Today **2**(11), 457–467 (1997)
3. D. Qi-Shi, R.-B. Huang, K.-C. Chou, Recent advances in QSAR and their applications in predicting the activities of chemical molecules, peptides and proteins for drug design. Curr. Protein Pept. Sci. **9**(3), 248–259 (2008)
4. C. Hansch, D. Hoekman, A. Leo, D. Weininger, C.D. Selassie et al., Chem-bioinformatics: comparative QSAR at the interface between chemistry and biology. Chem. Rev. **102**(3), 783–812 (2002)
5. C. Hansch, A. Leo, D. Hoekman, Albert Leo, *Exploring QSAR*, vol. 631 (American Chemical Society, Washington, DC, 1995)
6. R. Gómez-Bombarelli, D. Duvenaud, J. Hernández-Lobato, J. Aguilera-Iparraguirre, T.D. Hirzel, R.P. Adams, A. Aspuru-Guzik, Automatic chemical design using a data-driven continuous representation of molecules. arXiv:1610.02415 (2016)
7. R.L. Camacho-Mendoza, E. Gutierrez-Moreno, E. Guzman-Percastegui, E. Aquino-Torres, J. Cruz-Borbolla, J.A. Rodriguez-Avila et al., Density functional theory and electrochemical studies: structure–efficiency relationship on corrosion inhibition. J. Chem. Inf. Model. **55**(11), 2391–2402 (2015)
8. L. Li, X. Zhang, S. Gong, Hongxia Zhao, Yang Bai, Qianshu Li, Lin Ji, The discussion of descriptors for the QSAR model and molecular dynamics simulation of benzimidazole derivatives as corrosion inhibitors. Corros. Sci. **99**, 76–88 (2015)
9. M. Karelson, V.S. Lobanov, A.R. Katritzky, Quantum-chemical descriptors in QSAR/QSPR studies. Chem. Rev. **96**(3), 1027–1044 (1996)
10. Z. Zhang, N. Tian, L. Wu, L. Zhang, Inhibition of the corrosion of carbon steel in HCL solution by methionine and its derivatives. Corros. Sci. **98**, 438–449 (2015)
11. C. Gnerre, M. Catto, F. Leonetti, P. Weber, P.-A. Carrupt, C. Altomare et al., Inhibition of monoamine oxidases by functionalized coumarin derivatives: biological activities, QSARs, and 3D-QSARs. J. Med. Chem. **43**(25), 4747–4758 (2000)
12. G. Schüürmann, QSAR analysis of the acute fish toxicity of organic phosphorothionates using theoretically derived molecular descriptors. Environ. Toxicol. Chem. **9**(4), 417–428 (1990)
13. Ramon Carbó-Dorca, Stochastic transformation of quantum similarity matrices and their use in quantum QSAR (QQSAR) models. Int. J. Quantum Chem. **79**(3), 163–177 (2000)

14. M. Rupp, A. Tkatchenko, K.-R. Müller, O.A. Von Lilienfeld, Fast and accurate modeling of molecular atomization energies with machine learning. Phys. Rev. Lett. **108**(5), 058301 (2012)
15. K. Hansen, G. Montavon, F. Biegler, S. Fazli, M. Rupp, M. Scheffler et al., Assessment and validation of machine learning methods for predicting molecular atomization energies. J. Chem. Theory Comput. **9**(8), 3404–3419 (2013)
16. R. Ramakrishnan, O.A. von Lilienfeld, Machine learning, quantum mechanics, and chemical compound space. arXiv:1510.07512 (2015)
17. E. Gutiérrez, Development of a predictive model for corrosion inhibition of carbon steel by imidazole and benzimidazole derivatives. Corros. Sci. **108**, 23–35 (2016)
18. P.D Lyne, Structure-based virtual screening: an overview. Drug Discov. Today **7**(20), 1047–1055 (2002)
19. G.E. Dahl, N. Jaitly, R. Salakhutdinov, Multi-task neural networks for QSAR predictions. arXiv:1406.1231 (2014)
20. J. Ma, R.P. Sheridan, A. Liaw, G.E. Dahl, V. Svetnik, Deep neural nets as a method for quantitative structure–activity relationships. J. Chem. Inf. Model. **55**(2), 263–274 (2015)
21. T. Unterthiner, A. Mayr, M. Steijaert, J.K. Wegner, H. Ceulemans, S. Hochreiter, Deep learning as an opportunity in virtual screening
22. G. Montavon, K. Hansen, S. Fazli, M. Rupp, F. Biegler, A. Ziehe et al., Learning invariant representations of molecules for atomization energy prediction, in *Advances in Neural Information Processing Systems* (2012), pp. 440–448
23. K. Hansen, F. Biegler, R. Ramakrishnan, W. Pronobis, O.A Von Lilienfeld, K.-R. Müller et al., Machine learning predictions of molecular properties: accurate many-body potentials and nonlocality in chemical space. J. Phys. Chem. Lett. **6**(12), 2326 (2015)
24. M. Hirn, N. Poilvert, S. Mallat, Quantum energy regression using scattering transforms. arXiv:1502.02077 (2015)
25. K.T. Schütt, F. Arbabzadah, S. Chmiela, K.R. Müller, A. Tkatchenko. Quantum-chemical insights from deep tensor neural networks. Nat. Commun. **8** (2017)
26. A. Lusci, G. Pollastri, P. Baldi, Deep architectures and deep learning in chemoinformatics: the prediction of aqueous solubility for drug-like molecules. J. Chem. Inf. Model. **53**(7), 1563 (2013)
27. Predicting activities without computing descriptors: graph machines for QSAR §
28. C.R. Collins, G.J. Gordon, O.A. von Lilienfeld, D.J. Yaron, Constant size molecular descriptors for use with machine learning. arXiv:1701.0664 (2017)
29. Y. LeCun, Y. Bengio, Geoffrey Hinton, Deep learning. Nature **521**(7553), 436–444 (2015)
30. D. Ciregan, U. Meier, J. Schmidhuber, Multi-column deep neural networks for image classification, in *2012 IEEE Conference on Computer Vision and Pattern Recognition (CVPR)* (IEEE, 2012), pp. 3642–3649
31. L. Deng, J. Li, J.-T. Huang, K. Yao, D. Yu, F. Seide et al., Recent advances in deep learning for speech research at microsoft, in *2013 IEEE International Conference on Acoustics, Speech and Signal Processing* (IEEE, 2013), pp. 8604–8608
32. R. Collobert and J. Weston, A unified architecture for natural language processing: deep neural networks with multitask learning, in *Proceedings of the 25th International Conference on Machine Learning* (ACM, 2008), pp. 160–167
33. D.P. Kingma, M. Welling, Auto-encoding variational Bayes. arXiv:1312.6114 (2013)
34. R. Ramakrishnan, P.O. Dral, M. Rupp, O.A. Von Lilienfeld, Quantum chemistry structures and properties of 134 kilo molecules. Sci. Data **1** (2014)
35. L. Ruddigkeit, R. Van Deursen, L.C. Blum, J.-L. Reymond, Enumeration of 166 billion organic small molecules in the chemical universe database GDB-17. J. Chem. Inf. Model. **52**(11), 2864–2875 (2012)
36. O.G. Mekenyan, G.T. Ankley, G.D. Veith, D.J. Call, QSARs for photoinduced toxicity of aromatic compounds. SAR QSAR Environ. Res. **4**(2–3), 139–145 (1995)
37. F. Chollet, Keras. https://github.com/fchollet/keras (2015)
38. M. Abadi, A. Agarwal, P. Barham, E. Brevdo, Z. Chen, C. Citro et al., TensorFlow: large-scale machine learning on heterogeneous systems. Software available from www.tensorflow.org (2015)

A Brief Survey and an Application of Semantic Image Segmentation for Autonomous Driving

Çağrı Kaymak and Ayşegül Uçar

Abstract Deep learning is a fast-growing machine learning approach to perceive and understand large amounts of data. In this paper, general information about the deep learning approach which is attracted much attention in the field of machine learning is given in recent years and an application about semantic image segmentation is carried out in order to help autonomous driving of autonomous vehicles. This application is implemented with Fully Convolutional Network (FCN) architectures obtained by modifying the Convolutional Neural Network (CNN) architectures based on deep learning. Experimental studies for the application are utilized 4 different FCN architectures named FCN-AlexNet, FCN-8s, FCN-16s and FCN-32s. For the experimental studies, FCNs are first trained separately and validation accuracies of these trained network models on the used dataset is compared. In addition, image segmentation inferences are visualized to take account of how precisely FCN architectures can segment objects.

Keywords Deep learning · Convolutional Neural Network · Fully Convolutional Network · Semantic image segmentation

1 Introduction

With advanced technology, modern camera systems can be placed in many places, from mobile phones to surveillance systems and autonomous vehicles, to obtain very high quality images at low cost [1]. This increases the demand for systems that can interpret and understand these images.

The interpretation of images has been approached in various ways for years. However, the process involving reviewing images to identify objects and assess their

Ç. Kaymak · A. Uçar (✉)
Mechatronics Engineering Department, Firat University, 23119 Elazig, Turkey
e-mail: agulucar@firat.edu.tr

Ç. Kaymak
e-mail: ckaymak@firat.edu.tr

V. E. Balas et al. (eds.), *Handbook of Deep Learning Applications*,
Smart Innovation, Systems and Technologies 136,
https://doi.org/10.1007/978-3-030-11479-4_9

importance is the same [2]. Learning problems from visual information are generally separated into three categories called as image classification [3], object localization and detection [4], and semantic segmentation [5].

Semantic image segmentation is the process of mapping and classifying the natural world for many critical applications such as especially autonomous driving, robotic navigation, localization, and scene understanding. Semantic segmentation, which is a pixel-level labeling for image classification, is an important technique for the scene understanding. Because each pixel is labeled as belonging to a given semantic class. A typical urban scene consists of classes such as street lamp, traffic light, car, pedestrian, barrier and sidewalk.

Autonomous driving will be one of the revolutionary technologies in the near future in terms of the impact on the lives of people living in industrially developed countries [6]. Many research communities have contributed to the development of autonomous driving systems thanks to rapidly the increasing performance of vision-based algorithms such as object detection, road segmentation and recognition of traffic signals. An autonomous vehicle must sense its surroundings and act safely to reach a certain target. Such functionality is carried out by using several types of classifiers.

Approximately up to the end of 2010, the identification of a visual phenomena was constructed as a two-stage problem. The first of these stages is to extract features from the image. Extensive efforts have been made to extract the features as visual descriptors and consequently the descriptors obtained by algorithms such as Scale Invariant Feature Transform (SIFT) [7], Local Binary Patterns (LBP) [8] and Histogram of Oriented Gradients (HOG) [9] have become widely accepted. The second stage includes to use or design classifier. Artificial Neural Networks (ANNs) are one of the most important classifiers. ANNs are not a new approach and its past is based on about 60 years ago. Until the 1990s, ANNs used in various fields did not provide satisfactory achievements on nonlinear systems. Therefore, there are not many studies about ANNs for a certain period. In 2006, Hinton et al. [10] used ANNs in speech recognition problems and achieved successful results. Thus, ANNs have come up again in the scientific world. Henceforth, researchers thought that the ANNs would be the solution to problems in most areas, but they soon realized that it was a wrong idea with various reasons, such as failure in the training of multi-layer ANNs. Then, the researchers turned to new approaches finding the most accurate class boundaries in feature space and input space such as Support Vector Machine (SVM) [11], AdaBoost [12], and Spherical and Elliptical classifiers [13] using the features obtained from the first stage. In addition to over-detailed class models to facilitate the search for completely accurate boundaries, methods of transforming feature space such as Principal Component Analysis (PCA) and kernel mapping have also been developed.

Later, in image recognition competitions such as the ImageNet Large Scale Visual Recognition Competition (ILSVRC), ANN-based systems took the lead and began to get first place every year by making a big difference to other systems. As time progressed, especially through the internet, very large amount of data has begun to be produced and stored in the digital environment. When processing this huge amount of data, Central Processing Units (CPUs) on the computers have been slow. Along with the developments in GPU technology, the computational operations can be performed much faster by using the parallel computing architecture of the graphics processor. With this increase in process power, the use of deeper neural networks has become widespread in practice. By means of this, "Deep Learning" term has emerged as a new approach in the machine learning.

Deep learning is the whole of the methods consisting of ANNs, which has a deep architecture with an increased number of hidden layers. At each layer of this deep architecture, features belonging to the problem is learned and this learned features create an input into an upper layer. This creates a structure from the bottom layer to the top layer, where the features are learned from the simplest to the most complex. It would be useful to analyze the vision system in the human brain to understand this structure. The signals coming to the eyes through nerves are evaluated in a multi-layer hierarchical structure. At the first layer where the signal is coming after the eyes, the local and basic features of the image, such as the edge and corner, are determined. By combining these features, at the next layer, mouth, nose, etc. details and at the subsequent layers, features belonging to the overall of image, such as face, person and location of objects, respectively can also be determined. Convolutional Neural Networks (CNNs) approach, which combines both feature extraction and classification capabilities in computer vision applications, work in this way.

Deep learning brings the success of artificial intelligence applications developed in recent years to very high levels. Deep learning is used in many areas such as computer vision, speech recognition, natural language processing and embedded systems. In the ILSVRC, which has been carried out using huge data sets in recent years, the competitors have been directed to the CNN approaches and achieved great success [14]. Companies such as Google [15], Facebook [16], Microsoft [17] and Baidu [18] have realized the progress in deep learning and carried out studies on this topic with great investments.

A graphical representation of search interest of the "Deep Learning" on the Google search engine in the last 5 years is shown in Fig. 1.

The advancement of CNNs is based on a high amount of labeled data. In general terms, CNNs carry out end-to-end learning by predicting class labels from raw image data by learning millions of parameters, which is more successful than methods based on visual descriptors.

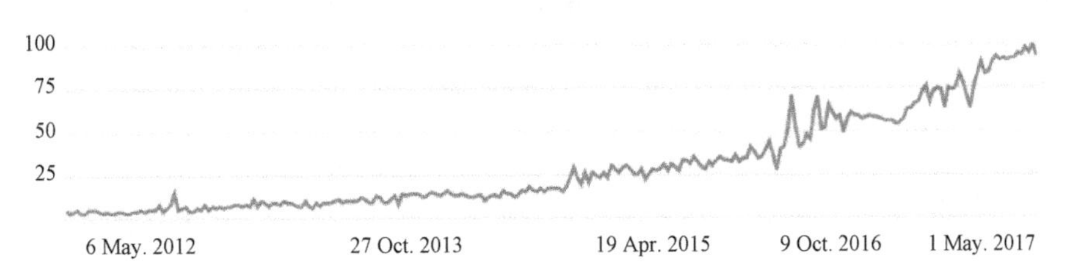

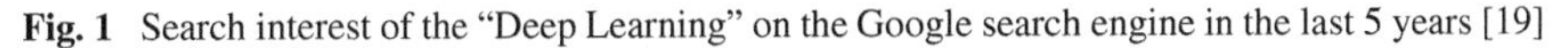

Fig. 1 Search interest of the "Deep Learning" on the Google search engine in the last 5 years [19]

In semantic image segmentation, a large number of studies have recently been conducted [20–32] to overcome of the supervised semantic segmentation using images with pixel-pixel annotations to train the CNNs. Some of the semantic segmentation studies was tried to directly adopt CNN architectures designed for image classification. However, the results were not very satisfactory. Because standard CNN architectures are not suitable to the semantic segmentation due to loss of the spatial position. While on one hand, repeated convolutional and pooling layers reduce the spatial resolution of feature maps, on the other hand, fully connected layers produce class probability values by completely discarding spatial information to produce an output.

In recent years, a lot of network architectures have emerged that are capable of bringing semantic information to a pixel location [20, 21]. First, Long et al. [20] converted the pre-trained VGG-16 [33] CNN architecture for classification into Fully Convolutional Networks (FCNs). For this, they replaced all fully connected layers of the CNN architecture with convolutional layers and added deconvolutional layers that restore the original spatial resolution with the skip connections. Thus, three different versions of FCN called FCN-8s, FCN-16s and FCN-32s were obtained, and the spatial resolution at the output was brought to the spatial resolution of the input image to generate the class probability values for each pixel.

The lack of a deep deconvolutional network trained in a large dataset makes it difficult to completely reconstruct nonlinear structures of object boundaries. Chen et al. [21] have contributed to correct this problem by applying the Conditional Random Field (CRF) method to the output of FCN.

Noh et al. [22] constructed FCN architecture named DeconvNet by using the convolutional layers adopted from the VGG-16 network architecture with the proposed deconvolutional network architecture. DeconvNet performed well on the PASCAL VOC 2012 [21] dataset.

Badrinarayanan et al. [27] proposed a new and practical FCN architecture called SegNet for semantic segmentation. This architecture consists of the encoder, which is the trainable segmentation unit, followed by the corresponding decoder and classifier layer. In the encoder network architecture, 13 convolutional layers in the VGG-16 network architecture were used likewise. They compared the SegNet architecture with the original FCNs [20] and DeconvNet [22] architectures on the SUN RGB-D [34] and CamVid [35] road scene datasets. They provided that the SegNet architecture has fewer trainable parameters than other FCN architectures and is therefore efficient both in terms of memory and computational time. In [28], they implemented the Bayesian SegNet architecture, an extension of SegNet, on the same datasets. The architecture resulted in improving of the boundary lines, increasing the accuracy of prediction, and reducing the number of parameters.

Fourure et al. [29] presented an approach that is enhanced by multiple datasets to improve the semantic segmentation accuracy on the KITTI [36] dataset used for autonomous driving. To take advantage of training data from multiple datasets with different tasks including different label sets, they proposed a new selective loss function that can be integrated into deep networks.

Treml et al. [30] conducted a study to reduce the computational load of embedded systems found in autonomous vehicles for autonomous driving. They designed a network architecture that preserves the accuracy of semantic segmentation while reducing the computational load. This architecture consists of an encoder like SqueezeNet [37], Exponential Linear Unit (ELU) used instead of Rectified Linear Unit (ReLU) activation function and a decoder like SharpMask [38].

Hoffman et al. [31] trained three different FCN architectures in [20] on both GTA5 [39] and SYNTHIA [40] datasets to examine adaptations simulated real images in CityScapes [41] and compared inference performances.

Marmanis et al. [32] proposed a FCN architecture for the semantic segmentation of very high resolution aerial images. They initiated the FCN with learned weight and bias parameters using FCN-PASCAL [20] network model pre-trained on the PASCAL VOC 2012 dataset.

The rest of this paper is organized as follows. In Sect. 2, the necessary concepts for a better understanding of deep learning are introduced in detail. In Sect. 3, firstly, the structures that constitute the CNN architectures are explained, next, information about the training of the CNN and the necessary parameters affecting its performance are given. In Sect. 4, it is explained how to make the conversion from CNN to FCN used in paper by explaining the main differences between image classification and semantic image segmentation applications. The Sect. 5 gives information about the dataset used for the semantic image segmentation application and the experimental results obtained by FCN architectures. In Sect. 6, the paper is concluded.

2 Deep Learning

Deep learning is a fast-growing popular machine learning approach in the artificial intelligence field to create a model for perceiving and understanding large quantities of machines, such as images and sound. Basically, this approach is based on deep architectures, which are the more structurally complex of the ANNs. This deep architecture term refers to ANNs whose number of hidden layers has been increased.

Deep learning algorithms are separated from existing algorithms in machine learning; it needs very high amount of data and hardware with very high computational power that can handle this high data rate. In recent years, the number of labeled images, especially in the field of computer vision, has increased extremely. Deep learning approach has attracted much attention thanks to the great progress in the area of GPU-based parallel computing power. GPUs with thousands of compute cores provide 10–100 times the application performance when processing these data compared to CPUs [42]. Nowadays, deep learning has many application areas, mainly automatic speech recognition, image recognition and natural language processing.

There are many different types of deep learning architecture. Basically, deep learning architectures can be named as in Fig. 2 [43].

In order to be understand better the deep learning term, it is necessary to adopt ANN structures in a good way. For this reason, this information will be given first.

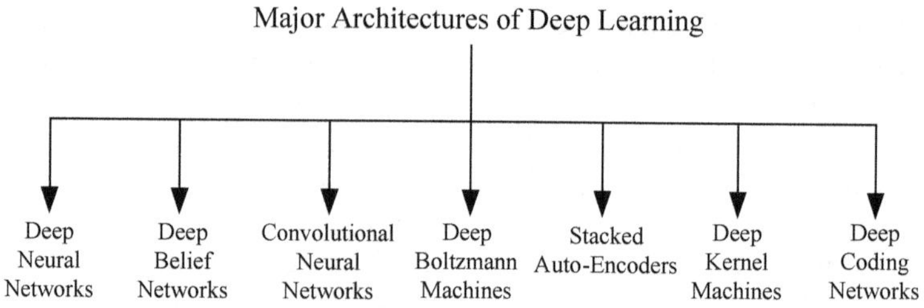

Major Architectures of Deep Learning

Deep Neural Networks | Deep Belief Networks | Convolutional Neural Networks | Deep Boltzmann Machines | Stacked Auto-Encoders | Deep Kernel Machines | Deep Coding Networks

Fig. 2 Major architectures of deep learning

In addition, we will focus on the feedforward ANN because the FCNs are the multi-layer feedforward neural network type from the deep learning architectures that form the basis of our paper.

2.1 *Artificial Neural Networks*

ANNs have been developed in the light of the learning process in the human brain. As the neurons in the biological nervous system connect with each other, the structures defined as artificial neurons in the ANN systems are modeled to be related to each other. ANNs can be used in many areas such as system identification, image and speech recognition, prediction and estimation, failure analysis, medicine, communication, traffic, production management and more.

2.1.1 Neuron

ANNs also have artificial neurons, as biological neural networks are neurons. The neuron in can be called the basic calculation unit in ANN. The neurons can also called node or unit. The structure of an artificial neuron is shown in Fig. 3.

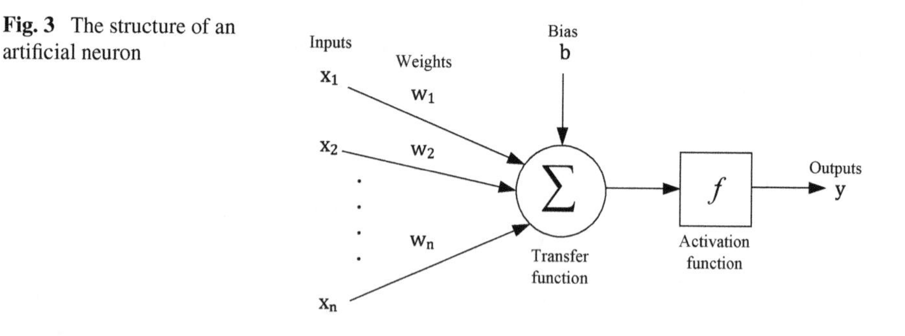

Fig. 3 The structure of an artificial neuron

Inputs are information incoming to a neuron from external world. These are determined by the samples for which the learning of the network is desired. Weights show the importance of information incoming to neurons and their effect on neurons. There are separate weights for each input. For example, W_1 weight in Fig. 3 shows the effect of x_1 input on the neuron. The fact that weights are big or small does not mean that they are important or insignificant. The transfer function calculates the net input incoming to a neuron. Although there are a large number of transfer functions for this, the most commonly used is weighted sum. Each incoming information is summed by multiplying its own weight. The activation function determines the output the neuron will generate in response to this input by processing the net input incoming to the neuron. The generated output is sent to the external world or another neuron. In addition, if desired, the neuron may also send its own output as an input to itself.

The activation function is usually chosen a nonlinear function. The purpose of the activation function is to transfer the nonlinearity to the output of the neuron as in (1). A characteristic of ANNs is nonlinearity, which is due to the nonlinearity of activation functions.

$$y = f\left(\sum_{i=1}^{n} W_i x_i + b\right) \tag{1}$$

The important thing to note when choosing the activation function is that the derivative of the function is easy to calculate. This ensures that the calculations take place quickly.

In the literature, there are many activation functions such as linear, step, sigmoid, hyperbolic tangent (tanh), Rectified Linear Unit (ReLU) and threshold functions. However, sigmoid, tanh and ReLU activation functions are usually used in ANN applications.

The sigmoid activation function, expressed by (2), is a continuous and derivatable function. It is one of the most used activation functions in ANNs. This function generates a value between 0 and 1 for each input value.

$$\sigma(x) = \frac{e^x}{1 + e^x} \tag{2}$$

The tanh activation function, expressed by (3), is similar to the sigmoid activation function. However, the output values range from -1 to 1.

$$\tan h(x) = 2\sigma(2x) - 1 \tag{3}$$

The ReLU activation function, expressed by (4), generates an output with a threshold value of 0 for each of the input values. It has a characteristic as in Fig. 4. Recently, the usage in ANNs has become very popular.

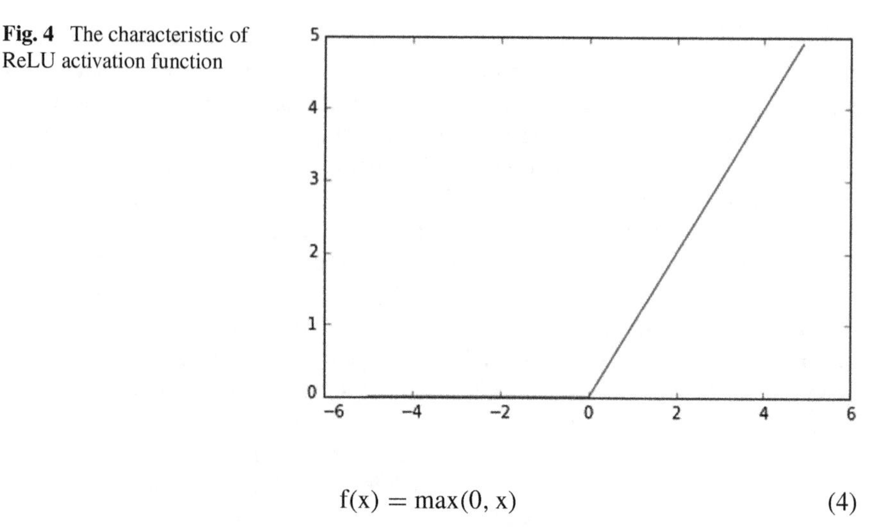

Fig. 4 The characteristic of ReLU activation function

$$f(x) = \max(0, x) \tag{4}$$

2.1.2 Feedforward Neural Network

In the feed forward neural network, the flow of information is only in the forward direction. The neurons in the network are arranged in the form of layers and the outputs of the neurons in a layer are input to the next layer via weights. The feedforward neural networks are basically composed of 3 types of layers. These layers are input, hidden and output layer. The input layer transmits the information incoming from the external world to the neurons in the hidden layer without making any changes. This information is then sequentially processed in the hidden layer/layers, which are not associated with the external world, and the output layer, which transfers the information from the network to the external world, to determine the network output in response to the desired input.

A feedforward neural network may have one or more hidden layers or no hidden layers. If the network does not contain any hidden layers, it is called single-layer perceptron, if it contains one or more hidden layers, it is called multi-layer perceptron.

In Fig. 5, a 3-layer feedforward ANN model is shown as an example.

In a multi-layer feedforward neural network, each neuron is only associated with the next neurons. In other words, there is no connection between the neurons in the same layer.

The term of depth in a multi-layer feedforward neural network is related to the number of hidden layers. As the number of hidden layers of the network increases, the depth increases. In short, a network with multiple hidden layers can be expressed as a deep neural network. Figure 6 shows a deep feedforward ANN model as an example.

The ability to learn from an information source is one of the most important features of ANN. In multi-layer neural networks, learning process takes place by changing weights at each step. Therefore, how weights are determined is important. Since the information is stored in the entire network, the weight value of a neuron does not make sense by itself. The weights on the whole network should get the most appropriate values. The process to achieve these weights is to train the network. In short, the learning of the network occurs by finding the most appropriate values of the weights. In addition, there are a number of considerations to be taken when designing multi-layer neural networks, such as the number of hidden layers in a network, the number of neurons to be found in each hidden layer, the optimal solution for the most reasonable time, and the test of network accuracy [44].

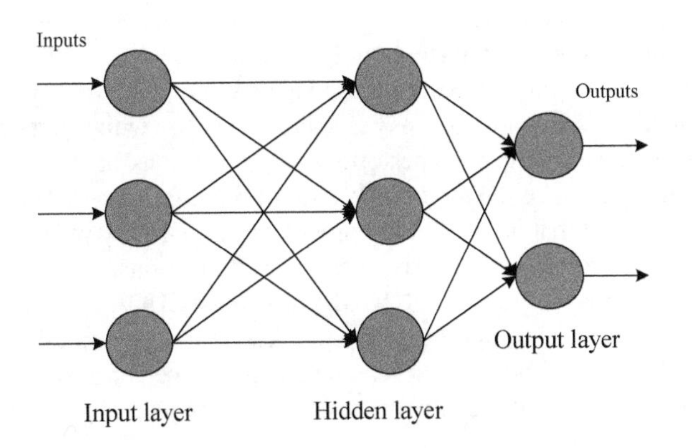

Fig. 5 3-layer feedforward ANN model

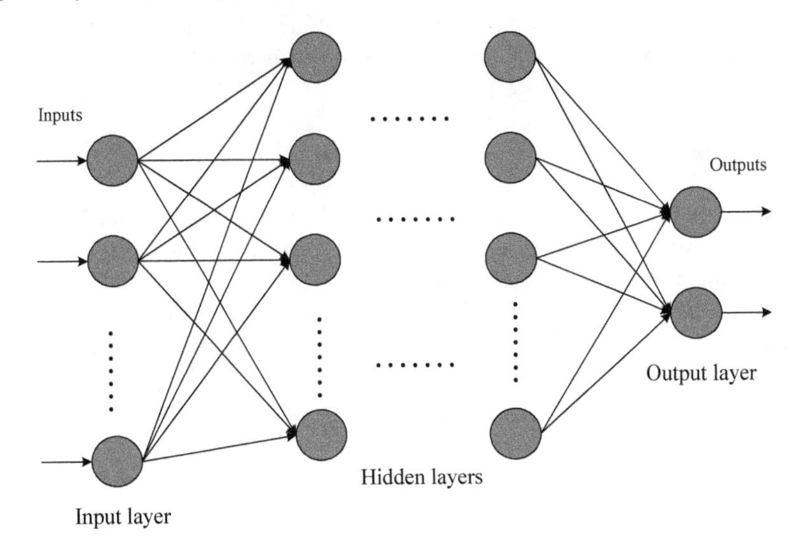

Fig. 6 A deep feedforward ANN model

2.2 *Deep Learning Software Frameworks and Libraries*

The workflow of deep learning is a multi-stage, iterative process. In this process, the data must first be collected and preprocessed if necessary. Large-scale datasets may be needed for to take place a successful deep learning. Nowadays, thanks to the internet and large data sources, datasets have been growing rapidly. The network is trained using these datasets. Existing networks can be used to train the network, or new networks can be developed. The network models created after the training phase must be tested to confirm that they work as expected. Generally, at this point, it is necessary to repeat certain operations to improve the results. These operations include reprocessing the data, arranging the networks, changing the parameters of the networks or solvers, and retesting until the desired result is obtained. The CPUs of these computers are insufficient for these intensive computational processes to be performed in the deep learning process. Because the CPUs with a certain processing capacity and architecture cannot perform many operations at the same time, the training and test phases of the model take a lot of time. Because of this, CPUs have given place to GPUs that allow parallel processing of data. By means of this, deep learning has begun to be used quickly in real life applications.

In the deep learning applications, NVIDIA provides a CUDA extension that allows GPUs to perform parallel computing [45]. CUDA is a parallel computing architecture that uses NVIDIA's GPU power to accelerate computing performance at a high level. CUDA enables the usage of graphics processor cores for general purpose accelerated computing.

There are many popular software frameworks and libraries, especially including Caffe, Torch, Theano, TensorFlow, Keras and DIGITS, for the implementation of deep learning algorithms. Most of them can also run on the GPU.

2.2.1 Caffe

The Caffe deep learning framework, created by Yangqing Jia, is developed by the Berkeley AI Research (BAIR) and community contributors. Caffe was designed to be as fast and modular just like the human brain [46].

Caffe is often preferred in industrial and academic research applications. The most important reason for this is the ability to process data quickly. Caffe can process over 60 million images per day with a single NVIDIA K40 GPU. Caffe is believed to be among the fastest accessible CNN implementations available [46].

2.2.2 Torch

Written in LuaJIT language, Torch is a scientific computing structure that provides extensive support for machine learning algorithms. It is an easy and efficient library because it is written in LuaJIT and uses the C/CUDA application basis [47]. This library, which can use numerical optimization methods, contains various neural networks and energy based models. It is also open source and provides fast and efficient GPU support.

Torch is constantly being developed and is being used by various companies such as Facebook, Google and Twitter.

2.2.3 Theano

Theano is a Python library that effectively identifies, evaluates, and optimizes mathematical expressions containing tensors [48]. Since this library is integrated with NumPy library, it can easily perform intensive mathematical operations. It also offers the option to create dynamic C code, allowing user to evaluate expressions more quickly.

2.2.4 TensorFlow

Tensorflow is an open source deep learning library that performs numerical computations using data flow graphs. This library was developed by Google primarily to conduct research on machine learning and deep neural networks [49]. With its flexible architecture, TensorFlow allows you to deploy the computation to one or more CPUs or GPUs on a server, mobile or desktop device with a single Application Programming Interface (API).

Snapchat, Twitter, Google and eBay, which are popular nowadays, also benefit from TensorFlow.

2.2.5 Keras

Keras is a modular Python library built on TensorFlow and Theano deep learning libraries [50]. These two basic libraries provide the ability to run on the GPU or CPU. By making minor changes in the configuration file of Keras, it is possible to use the TensorFlow or Theano in the background.

Keras is very useful as it simplifies the interface of TensorFlow and Theano libraries, and easier application can be developed than these two libraries. Keras has a very common usage in image processing applications.

2.2.6 DIGITS

In 2015, NVIDIA introduced the CUDA Deep Neural Network library (cuDNN) [51] due to the growing importance of deep neural networks, both in the industrial and academia, and the great role of GPUs. In 2016, Jen-Hsun Huang, NVIDIA CEO and founder, has brought the Deep Learning GPU Training System (DIGITS) into use at the GPU Technology Conference.

DIGITS is a deep learning GPU training system that helps users to develop and test CNNs. This system supports GPU acceleration using cuDNN to greatly reduce training time while visualizing Caffe, Torch and TensorFlow by providing web interface support.

DIGITS supports many educational objectives including image classification, semantic segmentation and object detection. Figure 7 shows the main console window where datasets can be generated from the images and they can be prepared for training. In DIGITS, once a dataset is available, the network model can be configured and training can begin. DIGITS also provides the necessary tools for network optimization. Settings for network configuration can be followed and accuracy can be maximized by changing parameters such as bias, activation functions and layers.

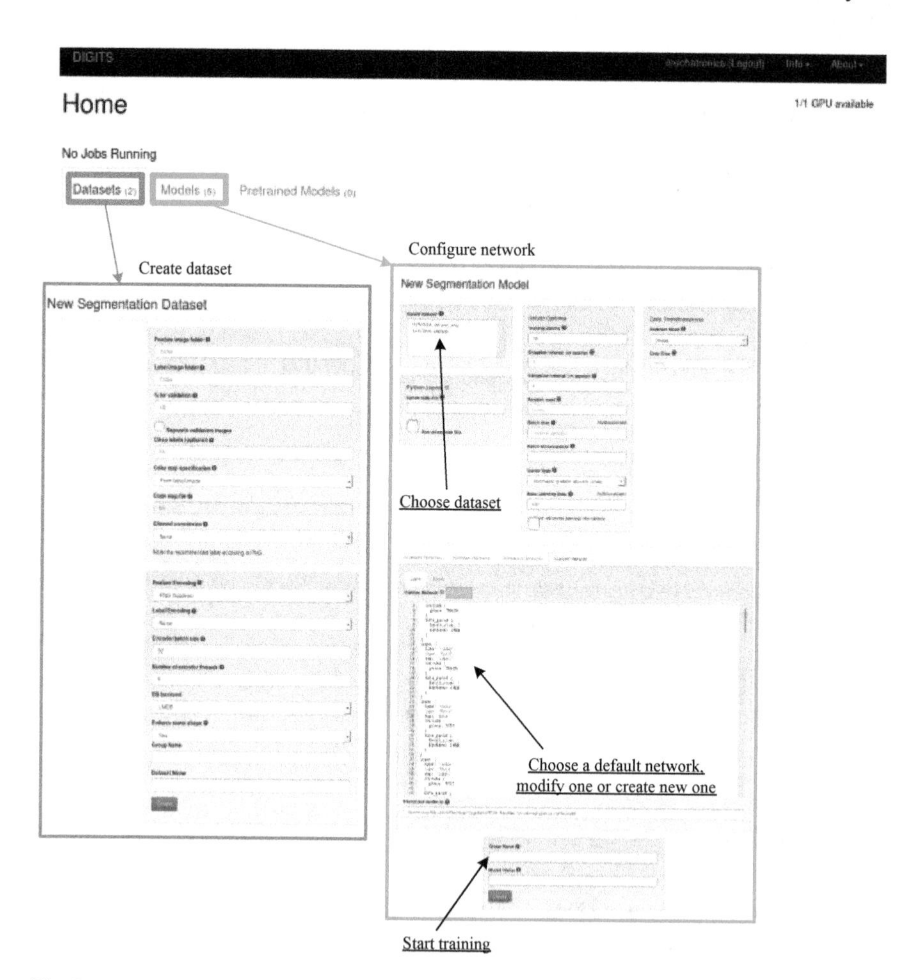

Fig. 7 DIGITS main console

3 Convolutional Neural Networks

CNNs, introduced by LeCun in 1989 for computer vision applications, are a type of multi-layer feedforward-ANN [52]. Nowadays, CNNs become increasingly popular among in-deep learning methods as they can successfully learn models for many computer and visual applications such as object detection, object recognition, and semantic image segmentation.

CNNs can be thought of as classifiers that extract hierarchical features from raw data. In CNN, images are given as input to the network, and learning takes place automatically with a feature hierarchy created without using any feature extractor method.

3.1 Architecture

All neurons in a layer in the feedforward ANNs are connected to all neurons of the next layer. Such connected layers are called fully connected layers and, in addition to fully connected layers in the CNN, convolution is applied to the input image to generate an output. This is caused by the local connection that all regions in the input layer are bound to neurons in the next layer. Thus, the input image is convolved with each learned filter used in this layer to generate different feature maps. The feature maps become more insensitive to rotation and distortion by providing more and more complex generalizations towards higher layers. In addition, the feature maps obtained in the convolutional layer are subjected to the pooling layer in order to perform spatial dimensionality reduction and keeping of important features. A classifier always is the final layer to generate class probability values as an output. The final output from the convolutional and pooling layers is transferred to one or more fully connected layers. Then, the output prediction is obtained by transferring to the classifier layer where the activation functions such as Softmax are used.

A simple CNN architecture is a combination of convolutional, pooling and fully connected layers as in Fig. 8.

3.1.1 Convolutional Layer

The purpose of the convolutional layer, which is the most basic layer of CNNs, is to convolve the input image with learnable filters and extract its features. A feature map is generated with each filter. CNNs draw attention to the fact that when applied to RGB images (images used in this paper), the image is a 3D matrix, and each of the layers is similarly arranged. This is shown in Fig. 9. Each layer of CNN consists of a set of spatial filters of size $d \times h \times w$ that are the spatial dimensions of h and w that appear as volume of the neurons and the number of kernel (or filter) feature channels of d. Each of these filters is subjected to convolution with a corresponding

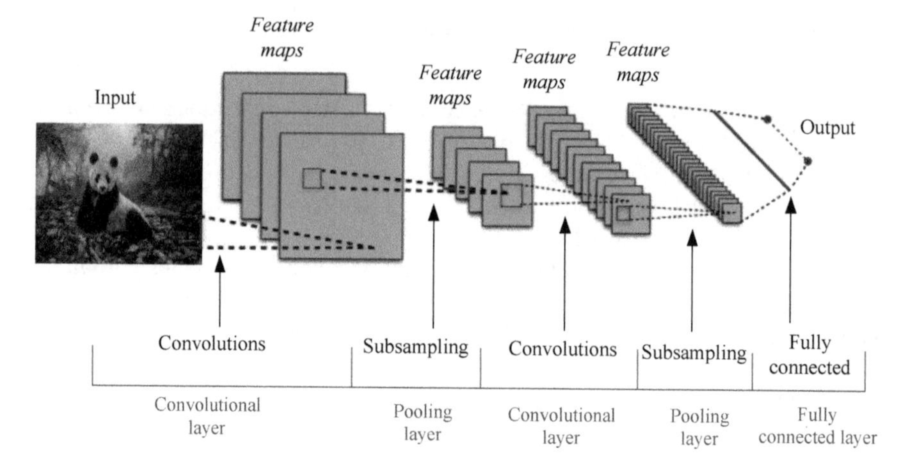

Fig. 8 An example of a simple CNN architecture

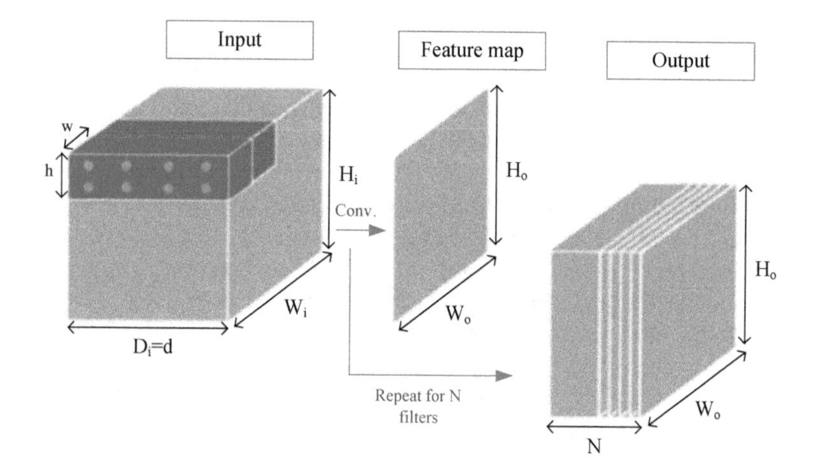

Fig. 9 An illustration of a single convolutional layer [53]

volume of the input image, and slid through the entire image (sized $D_i \times H_i \times W_i$ where H_i, W_i are the spatial dimensions and D_i is the channel number) across its spatial dimensions H_i, W_i. Convolution refers to the sum of element by element multiplication of the neurons in each filter with the corresponding values at the input. Thus, it can be assumed that the first layer in the CNNs is the input image. Based on this, convolution with a single filter in each layer provides a 2-dimensional output with parameters such as stride and padding. This is expressed as a feature map or activation map for a filter in input. At each convolutional layer of the CNNs, N filters are used, each resulting in a feature map. These feature maps are stacked together in a certain volume to obtain the output of a convolutional layer.

A single neuron in a filter of a certain layer can be mapped to connected neurons in all previous layers, following such convolutions. This is called the effective receptive field of the neuron. It is easy to see that convolutions result in very local connections with neurons in the lower layers (closer to the input) with those with smaller receptive fields than in the higher layer. While the lower layers learn to represent small areas of the input, the higher layers learn more specific meanings because they respond to a larger subdivision of the input image. Thus, a feature hierarchy is generated from the local to the global.

The red and blue areas in Fig. 9 represent the two positions of the same filters of size $d \times h \times w$ that are subjected to convolution by sliding through the input volume. Given that the filter size is 2×2, it can be seen that the stride parameter s is 2. For RGB input image, $D_i = d = 3$.

The stride s of a filter is defined as the intervals at which the filter moves in each spatial dimension. p padding corresponds to the number of pixels added to the outer edges of the input. Hence, stride can be considered as an input means of the subsampling [54]. Typically, square filters of the form $h = w = f$ are used. The output volume of such a layer is calculated using Eqs. (5), (6) and (7).

$$D_o = N \tag{5}$$

$$H_o = \frac{H_i - f + 2ps}{+} 1 \tag{6}$$

$$W_o = \frac{W_i - f + 2ps}{+} 1 \tag{7}$$

Figure 10 shows a 3×3 filter to be slid over a 5×5 image matrix representing a binary image. The sliding of the filter is from left to right and continues until the end of the matrix. In this paper, the stride is taken as 1. By sliding filters in order, the process is completed and the final state of the feature map is obtained as shown in Fig. 10.

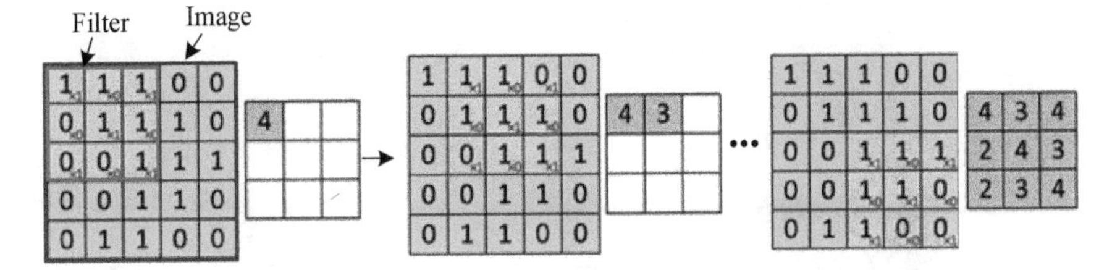

Fig. 10 Image matrix and final state of the feature map [55]

3.1.2 Pooling Layer

Frequently between the convolutional layers, pooling layers are scattered which help to spatially subsample the input features. The pooling layers make the subsampling process of the input image. This is done by sliding a filter over the input image.

The input image (usually non-overlapping) is divided into subregions and each subregion is sampled by non-linear pooling functions. The best-known of these functions are the maximum and average pooling functions. With the maximum pooling function used throughout this paper, the maximum value is returned from each subregion. The average pooling function returns the average value of the subregion. The pooling provides robustness to the network by reducing the amount of translational variance in the image [3]. In addition, unnecessary and redundant features are also discarded, which reduces the network's computational cost and, therefore, makes it more efficient.

The pooling layers also have a stride parameter that provides control over the output sizes. The same equations used for the output size of the convolutional layers can be used for this layer. It can be seen in Fig. 11 that the input volume of $64 \times 224 \times 224$ is subsampled to the volume of $64 \times 112 \times 112$ by 2×2 filters and strides 2. The pooling operation is performed separately for each feature map, and the size of the feature map is reduced as shown.

Figure 12 shows the pooling operation performed with the maximum pooling function by sliding the 2×2 filter, stride 2.

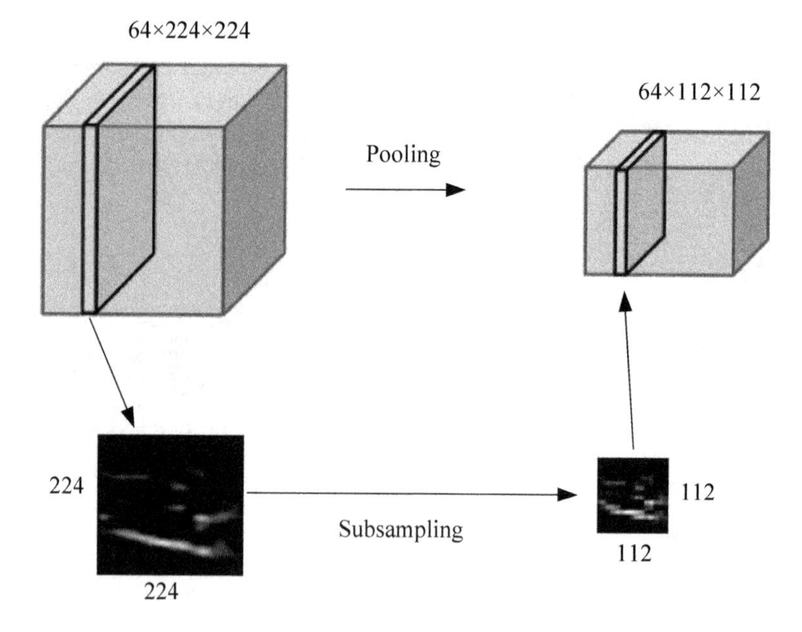

Fig. 11 An example of a subsampling [5]

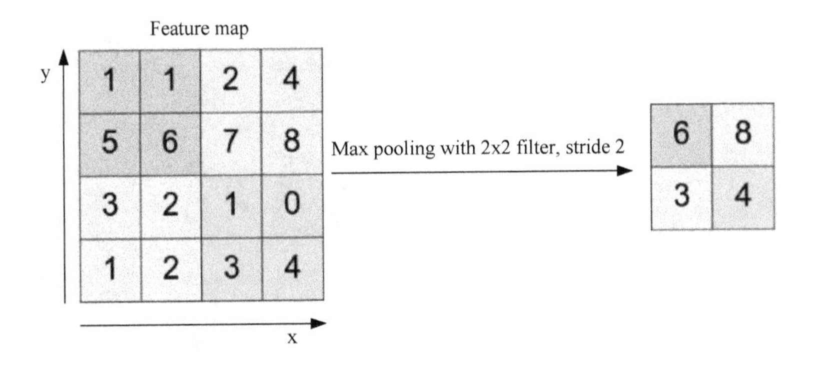

Fig. 12 The pooling operation with the 2 × 2 filter, stride 2 [5]

3.1.3 Rectified Linear Unit Layer

Generally, the outputs of the convolutional layer are fed into activation functions. The nonlinearity layers proposed for this purpose can be composed of functions such as sigmoid, tanh and ReLU. ReLU has been found to be more effective in CNNs and is often preferred [56].

A ReLU layer thresholds negative inputs to 0 and activates the positive inputs as described (8) by passing them unchanged.

$$f(x) = \begin{Bmatrix} x, & x \geq 0 \\ 0, & \text{others} \end{Bmatrix} \tag{8}$$

where; x is the input of the ReLU, and f(x) is the rectified output.

In the ReLU layer, an operation is performed separately for each pixel value. For example, the output of the ReLU is as shown in Fig. 13 if it is considered that the black areas are represented by negative pixels and the white areas are represented by positive pixels in the input feature map.

3.1.4 Fully Connected Layer

After high-level features are extracted with convolutional, pooling and ReLU layers, generally the fully connected layer is placed at the end of the network. The neurons in this layer are completely dependent on all activations in the previous layer. The most important feature of the fully connected layer is that it allows the neurons in this layer to determine which features correspond to which classifications. In short, the fully connected layer can be thought of as the layer that feeds the classifier.

Spatial information is lost as a neuron in the fully connected layer receives activations from all input neurons. This is not desirable in this semantic image segmentation paper where spatial information is very important. One way to get over this situation

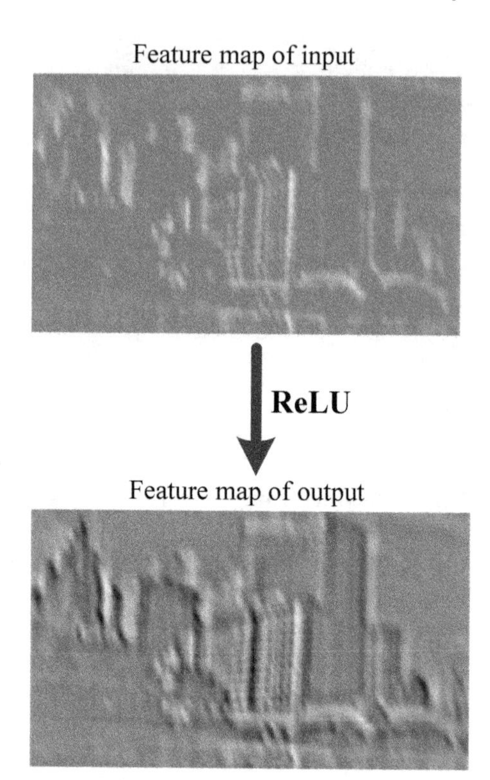

Fig. 13 An example of ReLU operation [57]

is to see the fully connected layer as a corresponding convolutional layer. It is also based on the basis of the FCNs that will be mentioned in the following.

3.1.5 Classifier

A classifier is chosen by considering the problem at hand and the data used. In this paper, the Softmax function is used which allows to predict for a class other than exclusive classes mutually. For the binary class problem, the Softmax function is reduced to a logistic regression. The Softmax function gives the probability value in (9) for a certain input belonging to a certain class c.

$$p_c = \frac{e^{(s_c)}}{\sum_{i=1}^{c} e^{(s_i)}} \tag{9}$$

where; s is the network outputs obtained from previous layers of CNNs for a particular class. For a single input, the sum of all probabilities between classes is always equal to 1. The loss metric is defined as the probability of the negative logarithm of the Softmax function. This is a cross entropy loss.

3.1.6 Regularization

Overfitting over training data is a big issue. Especially when dealing with deep neural networks that the network is strong enough to fit the training set alone is a big problem. The overfitting must be avoided. The methods developed for this are called regularization methods.

Dropout is a simple and effective regularization strategy integrate into the training phase. Dropout, introduced for the first time in [58], is implemented as dropout layers characterized by a probability value. Dropout can be accepted as a reasonable default value of 0.5 proven to be sufficiently effective [58].

3.2 Training

The learning process in CNNs can be divided into 4 basic steps:

1. Forward computation,
2. Error/loss optimization,
3. Backpropagation,
4. Parameter updates.

Forward computation is usually the case where sublayers composed of convolutional or pooling layers are followed by higher fully connected layers. The network returns the class output, which encodes the probability of belonging to a particular class for input. The outputs of the network may be unscaled as in the SVM classifier, or a negative logarithm probability may be obtained as in the Softmax classifier. For semantic segmentation, each pixel in the image is provided with a class output.

The set of class outputs provided by the network should be subjected to optimization processing by adjusting the values of the learned parameters such as weight filters and biases. The uncertainty that occurs in determining which set of parameters is ideal is quantified by the loss function that can be formulated as an optimization problem. For each vector of the class outputs s, the cross entropy loss is calculated as given in (10) for the Softmax classifier.

$$\mathrm{L_i} = -\log\left(\frac{\mathrm{e}^{(\mathrm{s_c})}}{\sum_{i=1}^{C}\mathrm{e}^{(\mathrm{s_i})}}\right) \tag{10}$$

The evaluation function of the cross entropy is calculated by (11).

$$\mathrm{H(p,q)} = -\sum_{x}\mathrm{p(x)logq(x)} \tag{11}$$

where; q is the Softmax function defined in (9), and p is the probability distribution. The total loss is calculated by (12).

$$L = \sum_{i=1}^{N} L_i + \lambda R(W) \quad (12)$$

where; N is the number of training samples, λ is the regularization strength, L is the total loss, and R(W) is the regularization term. To minimize L, the problem is formulated as an optimization step and the loss function is minimized. Thus, the probability is maximized.

Backpropagation is a fundamental concept in learning with neural networks. The purpose of backpropagation is to periodically update the initial weight parameters. The backpropagation of the problem helps to optimize the cost function. The optimization algorithm is generally used to understand the gradient descent and its various types. A simple application of the gradient descent may not work well in a deep network because it is confronted with problems by going and returning around the local optimum. This situation is fixed by the momentum parameter which helps to update the gradient descent, which is necessary to reach an optimal point.

One of the crucial parts of developing neural network architecture is the selection of hyperparameters. The hyperparameters are variables set to specific values before the training process. In [59], a list of the most effective hyperparameters adopted by many researchers for model performance has been proposed. This list includes initial learning rate, mini-batch size, number of training iterations, momentum, number of hidden units, weight initialization, weight decay, regularization strength and more hyperparameters.

3.3 *Some Known CNN Architectures*

LeNet [12], AlexNet [16], VGGNet [33], GoogleNet [60] and ResNet [4] are among the best- known CNN architectures. AlexNet and VGG16, 16-layer (convolutional+fully connected layers) version of VGG-Net, form the basis for the FCN architectures to be addressed in the next section.

Developed by Alex Krizhevsky, Ilya Sutskever, and Geoff Hinton, AlexNet is the first study to make CNN popular in computer vision [16]. AlexNet was presented at the ILSVRC in 2012, and was the first in the competition to perform significantly better than the second and third architectures Even though AlexNet has an architecture similar to LeNet, convolutional layers with deeper and more specific are stacked on top of each other. AlexNet has been trained on more than 1 million high-resolution images containing 1000 different classes.

Developed by Simonian and Zisserman, VGGNet has two versions, called VGG-16 and VGG-19. The VGG-16 architecture was the second in the ILSVRC in 2014, where GoogleNet was first. Recently, ResNet seems to be a very advanced CNN model, but VGG-16 is preferred because of its simple architecture. VGG-16 has also been trained on more than 1 million high-resolution images containing 1000 different classes like AlexNet.

4 Semantic Image Segmentation and Fully Convolutional Networks

Except where it is in the image, there are many situations that need to be learned For example, you should know the locations of the objects around autonomous vehicles so that you can move without hitting anywhere. The process of determining the locations of the objects in the image can be realized by detecting (getting into the bounding box) or segmenting. As mentioned before, an application of semantic image segmentation for the autonomous vehicles will be implemented in this paper.

Image segmentation is the operation of separating images into specific groups that show similarity and labeling each pixel in an image as belonging to a given semantic class. In order for the image segmentation process to be considered successful, it is expected that the objects are independent of the other components in the image, as well as the creating of regions with the same texture and color characteristics. As an example in Fig. 14, each pixel in a group corresponds to the object class as a whole. These classes may represent an urban scene that is important for autonomous vehicles; traffic signs, cars, pedestrians, street lamps or sidewalks.

As mentioned before, AlexNet and similar standard CNNs perform non-spatial prediction. For example, in the image classification, the network output is a single distribution of class probabilities. The CNN must be converted into the FCN in order to achieve a density prediction such as semantic image segmentation. Because, as described in [20], the fully connected layers of the CNNs give information about location. Therefore, in order to be able to carry out a semantic image segmentation, it is necessary to convolve the fully connected layers of the CNNs.

In the conversion to the FCN, the convolution part of the CNN can be completely reused. Fully convolutional versions of existing CNNs predict dense outputs with efficient inference and learning from arbitrary sized inputs. Both learning and inference present the whole visual in a single pass through forward computation and backpropagation, as shown in Fig. 15. The upsampling layers within the network allow learning on the network via pixel-level prediction and subsampling [20].

Fig. 14 A sample of semantic image segmentation [61]

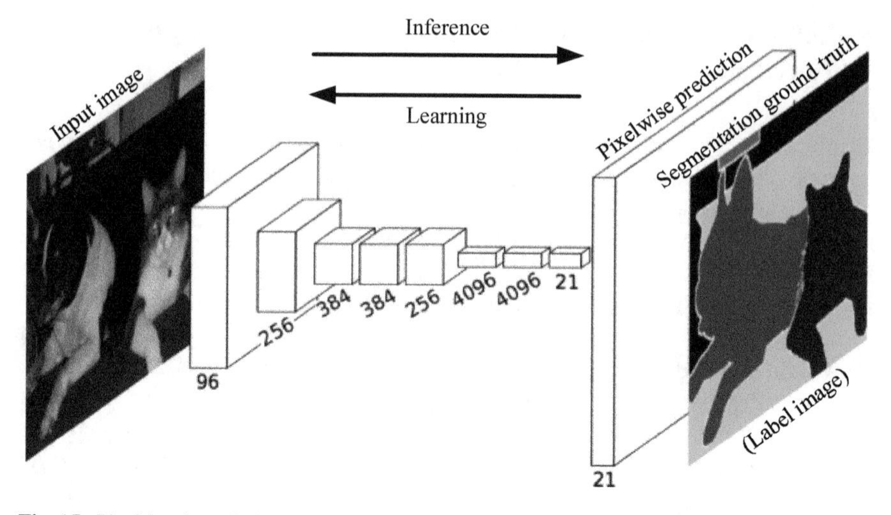

Fig. 15 Pixel-level prediction with FCN [20]

4.1 Semantic Image Segmentation

For semantic image segmentation application for autonomous vehicles, it may be thought that this can overcome its with the thing coming to mind classification network architectures such as AlexNet, GoogleNet and VGG-16. However, the models created by image classification architectures give the output of which class is the dominant object class in the image. That is, the output is a discrete probability distribution. If the image contains more than one object, the desired result cannot be obtained. Considering that classification models such as AlexNet are trained with a data set consisting of more than one million images of only one object in it, it is quite understandable. In addition, location information of the object in the image cannot be obtained with the classification networks. The situation can be understood when they are thought to have never been trained for this aim. The semantic image segmentation eliminates some of these deficiencies. Instead of estimating a single probability distribution for an entire image, the image is divided into several blocks and each block is assigned its own probability distribution. Very commonly, images are divided into pixel-levels and each pixel is classified. For each pixel in the image, the network is trained to predict which class the pixel belongs to. This allows the network not only to identify several object classes in the image but also to determine the location of the objects.

The datasets used for semantic image segmentation are images that are to be segmented and usually consist of label images of the same size as these images. The label image shows the ground truth limits of the image. The shapes in the label images are coded with colors to represent the class of each object. In some of the datasets, especially the SYNTHIA-Rand-CVPR16 dataset used in this paper, the label images

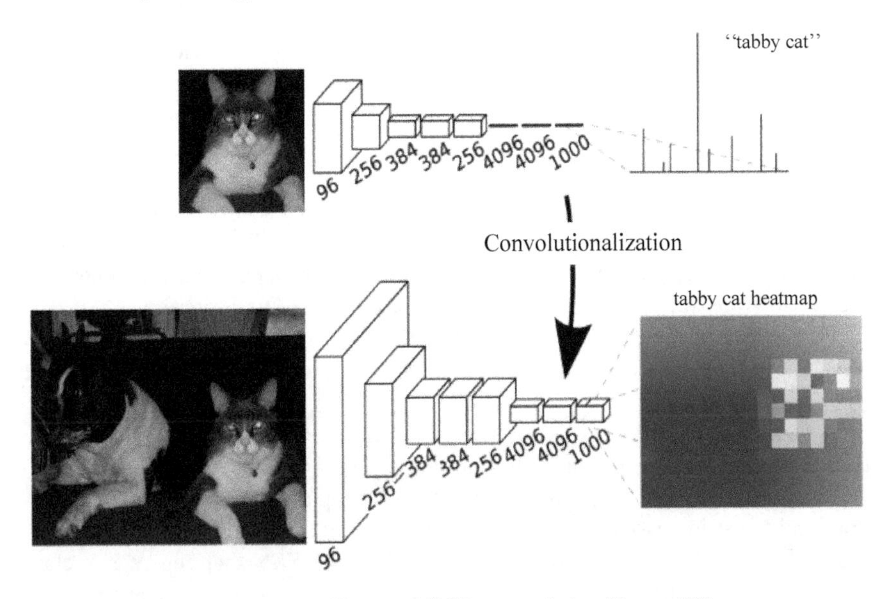

Fig. 16 Converting fully connected layers of CNN to convolutional layers [20]

consist of 24-bit 3 channel RGB images. In this case, the pixels must be indexed by creating color maps with RGB channel conversion.

4.2 Conversion from CNN to FCN

Semantic image segmentation only adds a spatial dimension to the image classification problem. Thus, several minor adjustments are sufficient to convert a classification neural network into a semantic segmentation neural network. It is implemented on AlexNet and VGG-16 architectures using the techniques of conversion to FCN architecture [20], which is necessary for the conventional image segmentation. The conversion to FCN is achieved by converting the fully connected layers of the CNN to convolutional layers. This conversion is shown generally in Fig. 16.

In a fully connected layer, each output neuron calculates the weighted sum of the values in input, while in a convolutional layer, each filter calculates the weighted sum of the values in the receptive field. Although these operations seem to be exactly the same thing, they are the same only when the layer input has the same size as the receptive field. If the input is larger than the receptive field, then the convolutional layer slide input window and calculate another weighted sum. This repeats until the input image is scanned from left to right, from top to bottom. A fully connected layer must be replaced with a corresponding convolutional layer, the size of the filters must be set to the input size of the layer, and as many filters as the neurons in the fully connected layer must be used.

All connected layers in the AlexNet architecture can be converted to corresponding convolutional layers to obtain FCN architecture. This FCN has the same number of learned parameters as the basic CNN and the same computational complexity. Convolutionalization of a basic CNN brings considerable flexibility to the conversion to FCN. The FCN model is no longer limited to work with fixed input size 224×224, as in AlexNet. The FCN can process large images by scanning throughly, such as a sliding window, and the model generates one per 224×224 window rather than generating a single probability distribution for the entire input. Thus, the output of the network has become a tensor in the form of $N \times H \times W$.

Where; N is the number of classes, H is the number of sliding windows (filters) along the vertical axis, and W is the number of sliding windows along the horizontal axis.

In summary, the first significant step in the design of the FCN is completed by adding two spatial dimensions to the exit of the classification network.

The window number depends on the size of the input image, the size of the window, and the stride parameters used between the windows when the input image is scanned, as will be understood during the design of a FCN that generates a class probability distribution per window. Ideally, a semantic image segmentation model should generate a probability distribution per pixel in the image. When the input image passes through the sequential layers of convolutionalized AlexNet, coarse features are extracted. The purpose of semantic image segmentation is to interpolate for these coarse features to reconstruct a fine-tuned classification for each pixel in the input. This can easily be done with deconvolutional layers. The deconvolutional layers perform the inverse operation of their convolutional counterparts. Considering the output of the convolutional layer, the deconvolutional layer finds the input generating the output. As it can be remembered, the stride parameter in the convolutional or pooling layer is a measure of how much the window is to be slid when the input is processed, and how the output is subsampled accordingly. In contrast, the stride parameter in the deconvolutional layer is a measure of how the output is upsampled. The output volume of the deconvolutional layer, $D_o \times H_o \times W_o$, is calculated using the Eqs. (13), (14) and (15).

$$D_o = N \tag{13}$$

$$H_o = s(H_i - 1) + f - 2p \tag{14}$$

$$W_o = s(W_i - 1) + f - 2p \tag{15}$$

where; s stride, p padding, f filter size, H_i and W_i input sizes, and N is the number of channels.

It is important to know how much of the activation of the last convolutional layer in the FCN architecture must be upsampled to obtain an output of the same size as the input image. The upsampling layer added to create FCN-AlexNet is shown to increase the output of the previous convolutional layer by 32 times. This means that

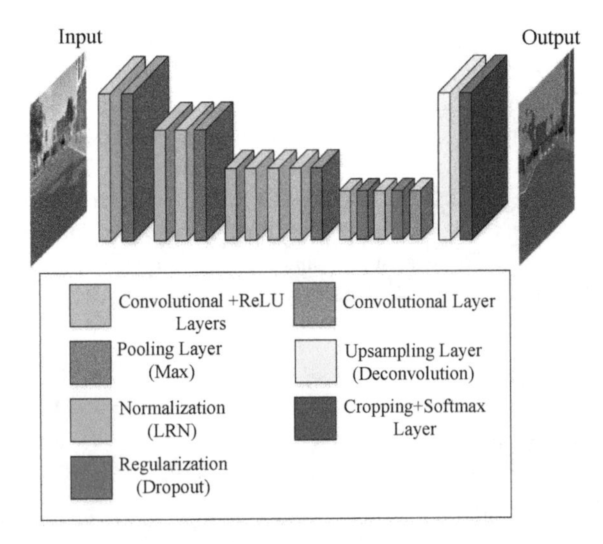

Fig. 17 FCN-AlexNet architecture

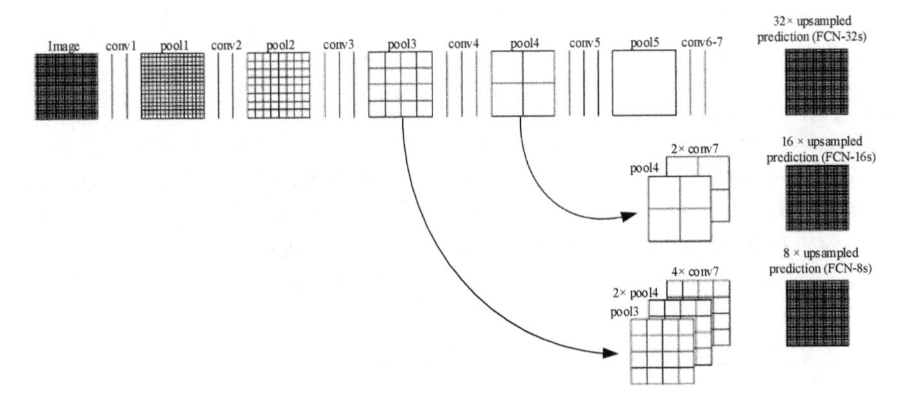

Fig. 18 Visualization of FCN-32s, FCN-16s and FCN-8s architectures [20]

in practice, the network has made a single prediction per 32 × 32 pixel block. This causes the contours of objects in the image to be segmented as rough. Figure 17 shows FCN-AlexNet architecture.

The article in [20] presents the idea of skip architecture for this restriction. The skip connections in this architecture have been added to redirect the outputs of the pooling layers pool3 and pool4 of the FCN architecture derived from VGG-16 directly to the network as shown in Fig. 18. These pooling layers work on low-level features and can capture more fine details.

The FCN architectures proposed in [20] are called FCN-8s, FCN-16s and FCN-32 according to the application of skip connections, converted into corresponding convolutional layers of fully connected layers in VGG-16. The visualization of these architectures is shown in Fig. 18.

In Fig. 18, the pooling layers are shown as grids expressing the spatial density, while the convolutional layers are shown as vertical lines. As it can be seen, the predictions in FCN-32s is upsampled at stride 32 to pixels in a single step without skip connection. The predictions in FCN-16s is combined at stride 16 from both *conv7* and *pool4* layers, allowing to predict finer details when detecting high-level semantic information. FCN-8s provides sharper predictions at stride 8 by adding the *pool3* layer. Thus, FCN-8s architecture allows to make fine-tuned predictions up to 8×8 pixel blocks.

5 Experimental Studies

In this paper, a semantic image segmentation application, which is useful for autonomous vehicles, was performed to observe the performance of the FCNs in semantic image segmentation. Four different popular FCN architectures were used separately for the application: FCN-AlexNet, FCN-8s, FCN-16s and FCN-32s.

The applications were implemented using Caffe framework in DIGITS platform on SYNTHIA-Rand-CVPR16 dataset and the segmentation performances of the used FCN architectures for experimental studies were compared. The studies were carried out on a desktop computer with 4th Generation Intel® Core i5 3.4 GHz processor, 8 GB RAM and NVIDIA GTX Titan X Pascal 12 GB GDDR5X graphics card. Thanks to the CUDA support of the graphics card, the GPU-based parallel computing power has been utilized in the computations required for the application.

5.1 SYNTHIA-Rand-CVPR16 Dataset

The SYNTHIA-Rand-CVPR16 dataset [40] has been generated to support semantic image segmentation in autonomous driving applications. The images in this dataset were created by portraying a virtual city with the Unity development platform [62]. The virtual environment allows them to freely place the desired components in the scene image and generate its semantic annotations without additional effort.

The SYNTHIA-Rand-CVPR16 dataset consists of a 13,407 RGB image with a resolution of 960×720 taken from a virtual camera array randomly moving through the city, limited to the range [1.5 m, 2 m] from the ground. It also consists of ground truth images of the same size as these RGB images. The dataset images, which are taken under different conditions such as night and day, includes 12 object classes: sky, building, road, sidewalk, fence, vegetation, pole, car, sign, pedestrian, cyclist and void.

Sample images from the SYNTHIA-Rand-CVPR16 dataset are shown in Fig. 19. For the paper, a total of 13,407 images of the dataset are used to train the network

Fig. 19 Samples from the SYNTHIA-Rand-CVPR16 dataset: **a** Sample images, **b** Ground truth images with semantic labels

model, and the remaining 2681 (about 20% of the dataset) are used to validate the model.

5.2 *Training and Validations of the Models*

The values of the network parameters determined for the application are given in Table 1. These parameters are generally set to be used for all the used FCN models in the application.

Stochastic Gradient Descent (SGD) as solver type and GPU as solver mode are selected. Epoch is set to 30. An epoch is a single pass through the full training set. Thus, for 10,726 training images, 1 epoch is completed in 10,726 iterations with 1

Table 1 The values of the network parameters

Parameter	Value
Base learning rate	0.0001
Momentum	0.9
Weight decay	10^{-6}
Batch size	1
Gamma	0.1
Maximum iteration	321,780

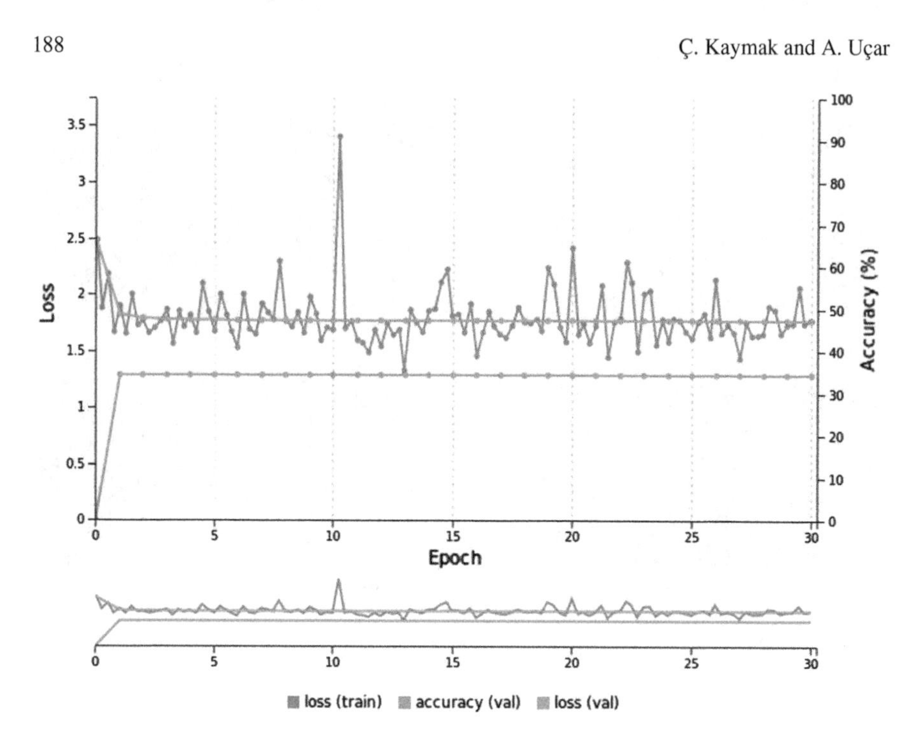

Fig. 20 Training/validation loss and validation accuracy when training FCN-AlexNet using random weight initialization

batch size and it is seen that the number of maximum iteration number for 30 epochs is 321,780, as indicated in Table 1.

Initially, FCN-AlexNet model is trained using random weight initialization in DIGITS and the results in Fig. 20 are obtained.

As shown in Fig. 20, performance is not satisfactory enough. Validation accuracy has reached a stationary point of about 35%. This means that only about 35% of the pixels in the validation set are correctly labeled. The training loss, which indicates that the network is not suitable for the training set, is parallel to the validation loss. When the trained model is tested on sample images in the validation set and visualized in DIGITS, it can be seen in Fig. 21 that the network classifies indiscriminately everything as building.

With the Fig. 21, it is understood that the building is the most representative object class in the SYNTHIA-Rand-CVPR16 dataset, and that the network has learned to achieve approximately 35% accuracy by labeling everything as building.

There are several commonly accepted ways to improve a network that is not suited to the training set [63]. These:

- Increasing the learning rate, and reducing the batch size,
- Increasing the size of the network model,
- Transfer learning.

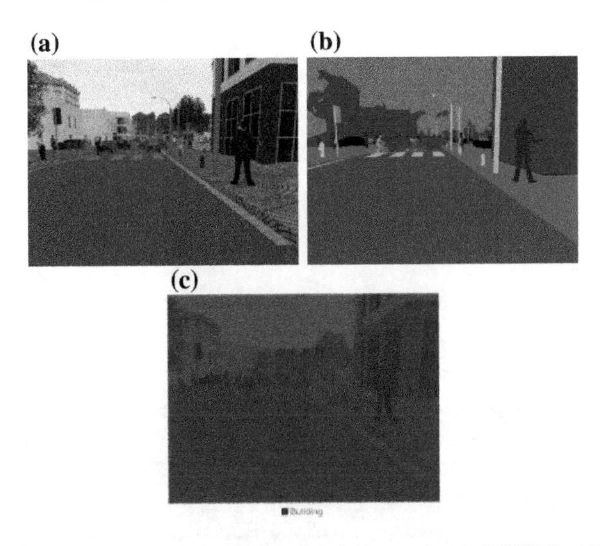

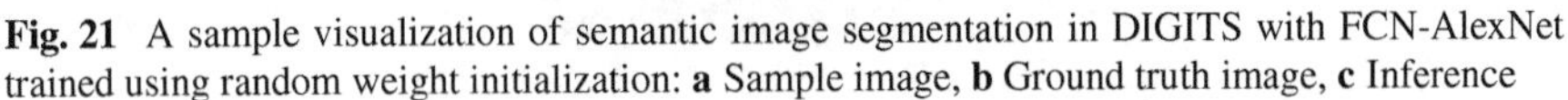

Fig. 21 A sample visualization of semantic image segmentation in DIGITS with FCN-AlexNet trained using random weight initialization: **a** Sample image, **b** Ground truth image, **c** Inference

Information learned by a deep network can be used to improve the performance of another network and this process is very successful for computer vision applications [64]. For this reason, while learning the required models for the application, transfer learning was used.

Recent developments in machine learning and computer vision are first achieved through the use of common criteria. It does not have to start from randomly initialized weights to train a model. Transfer learning is a reuse of information that a network learns in another dataset to improve the performance of another network [65]. A network is trained on any data and gains knowledge from this data, compiled as weights of the network. These weights can be transferred to any network. In other words, instead of training the network from scratch, learned features can be transferred to the network.

Transfer learning is often preferred in the computer vision field, since many low-level features such as line, corner, shape, and texture can be immediately applied to any dataset via CNNs.

Models trained and tested on high-variance standard datasets usually owe their successes to strong features [65]. Transfer learning allows to use a model that learns fairly generalized weights trained on a large dataset such as ImageNet and allows fine-tuning to adapt the situation of the network to be used.

It is very logical to transfer learning from image classification dataset such as ImageNet since the image segmentation has a classification at the pixel level. This process is quite easy using Caffe. However, Caffe cannot automatically carry the weights from AlexNet to FCN-AlexNet because AlexNet and FCN-AlexNet have different weight formats. Moving these weights can be done using the Python command line

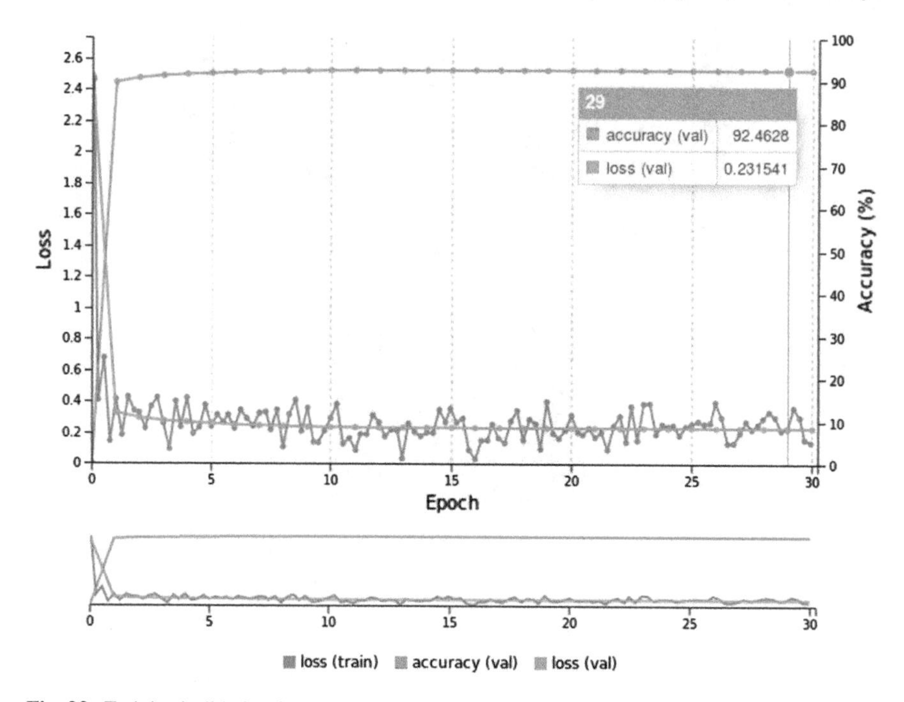

Fig. 22 Training/validation loss and validation accuracy when training FCN-AlexNet using pre-trained model

"net_surgery.py" in DIGITS repository in Github. The function of net_surgery.py is to transfer weights from fully connected layers to convolutional equivalents [63].

Also, another possible problem is how to start the upsampling layer added to create FCN-AlexNet since the upsampling layer is not part of the original AlexNet model. In [20], it is recommended that the corresponding weights are first randomly initiated and the network learns them. Later, however, the authors realized that it is easy to initialize these weights by doing bilinear interpolation, the way that the layer just acts like a magnifying glass [63].

As previously mentioned, training of FCN-AlexNet model was performed using the pre-trained model obtained by adapting the AlexNet model trained on the ImageNet dataset and the results in Fig. 22 were obtained.

Figure 22 shows that using the pre-trained FCN-AlexNet model, the validation accuracy quickly exceeded 90%, and the model achieved the highest accuracy at 92.4628% in 29th epoch. This means that 92.4628% of the pixels in the validation set of the model obtained in 29th epoch are labeled correctly. It has been shown to have fairly high accuracy compared to FCN-AlexNet initialized randomly weights.

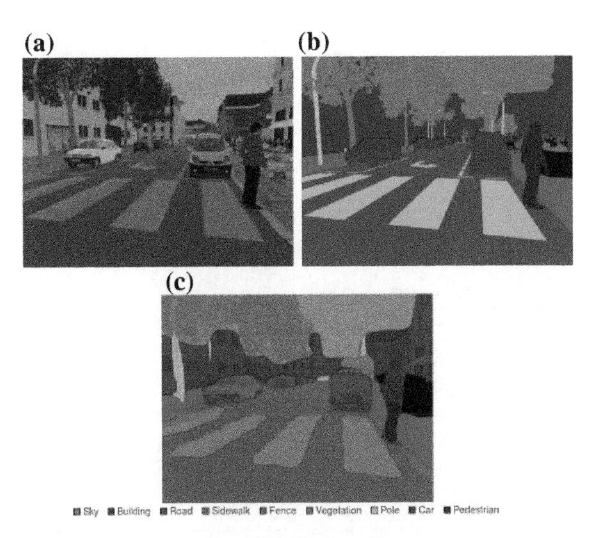

Fig. 23 A sample visualization of semantic image segmentation in DIGITS with FCN-AlexNet trained using pre-trained model-1: **a** Sample image, **b** Ground truth image, **c** Inference

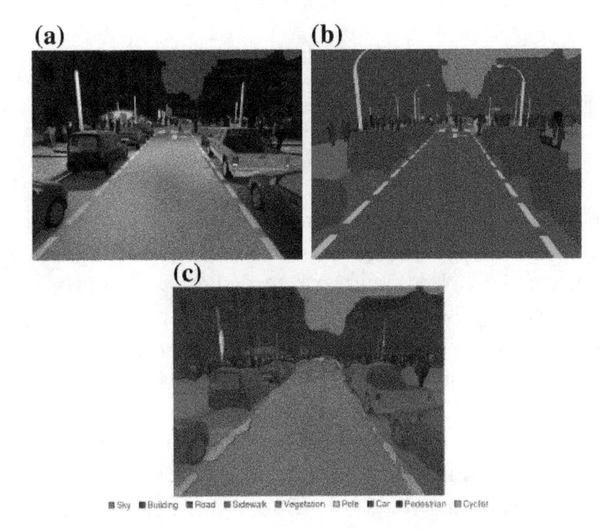

Fig. 24 A sample visualization of semantic image segmentation in DIGITS with FCN-AlexNet trained using pre-trained model-2: **a** Sample image, **b** Ground truth image, **c** Inference

When tested for sample images using the model obtained in 29th epoch, a semantic image segmentation was performed many times more satisfactorily by detecting different object classes as shown in Figs. 23 and 24. However, it can be clearly seen that the object contours are very rough.

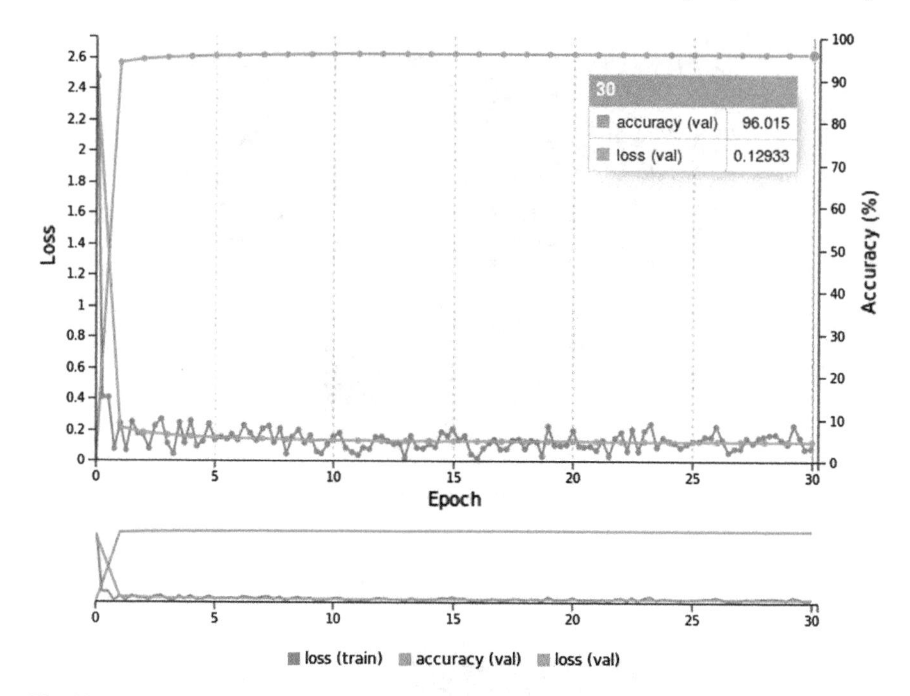

Fig. 25 Training/validation loss and validation accuracy when training FCN-8s using pre-trained model

FCN-8s network is used to further improve the precision and accuracy of the segmentation model. Using the pre-trained model in the PASCAL VOC dataset, validation accuracy of FCN-8s quickly exceeded 94% as shown in Fig. 25. The model reached to the highest accuracy with 96% in 30th epoch.

More importantly, when tested for sample images using the model obtained in 30th epoch, much sharper object contours are shown, as shown in Figs. 26 and 27.

FCN-8s architecture has been shown to provide segmentation with sharper object contours than FCN-AlexNet, which makes predictions in 32×32 pixel blocks, as it can make predictions at a fine-tuning down to 8×8 pixel blocks. Similarly, trainings of the models have been carried out by FCN-16s and FCN-32s architectures and it can be seen in Figs. 28 and 29 that the validation accuracy has exceeded 94% rapidly in a similar manner to FCN-8s. The highest validation accuracy was reached in 30th epoch as in FCN-8s with 95.4111% and 94.2595% respectively. Besides, Fig. 30 shows the comparison of segmentation inferences on the same images selected using FCN-AlexNet, FCN-8s, FCN-16s and FCN-32s architectures.

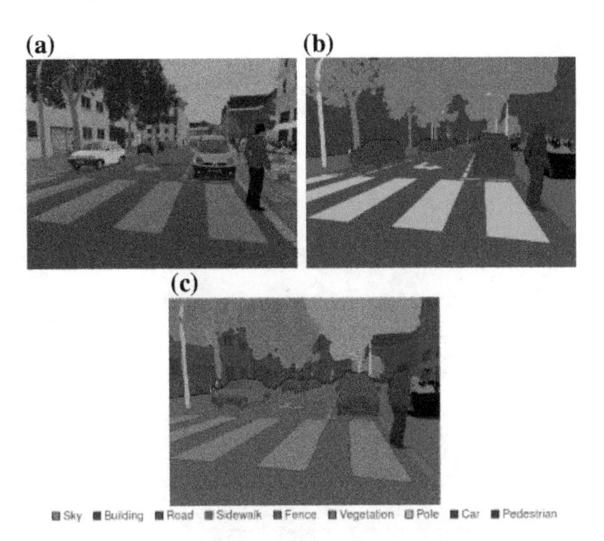

Fig. 26 A sample visualization of semantic image segmentation in DIGITS with FCN-8s trained using pre-trained model-1: **a** Sample image, **b** Ground truth image, **c** Inference

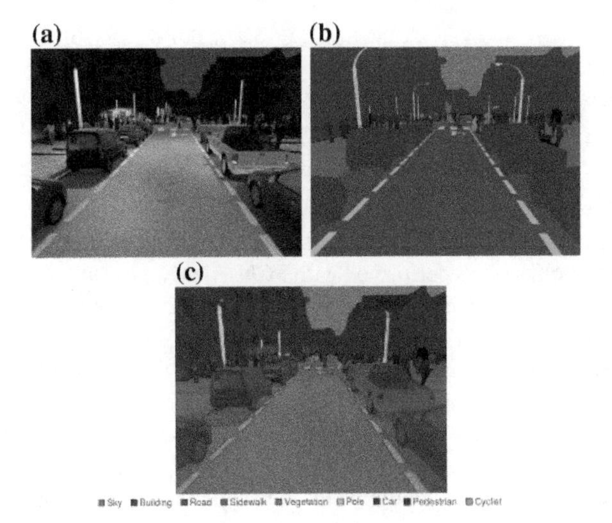

Fig. 27 A sample visualization of semantic image segmentation in DIGITS with FCN-8s trained using pre-trained model-2: **a** Sample image, **b** Ground truth image, **c** Inference

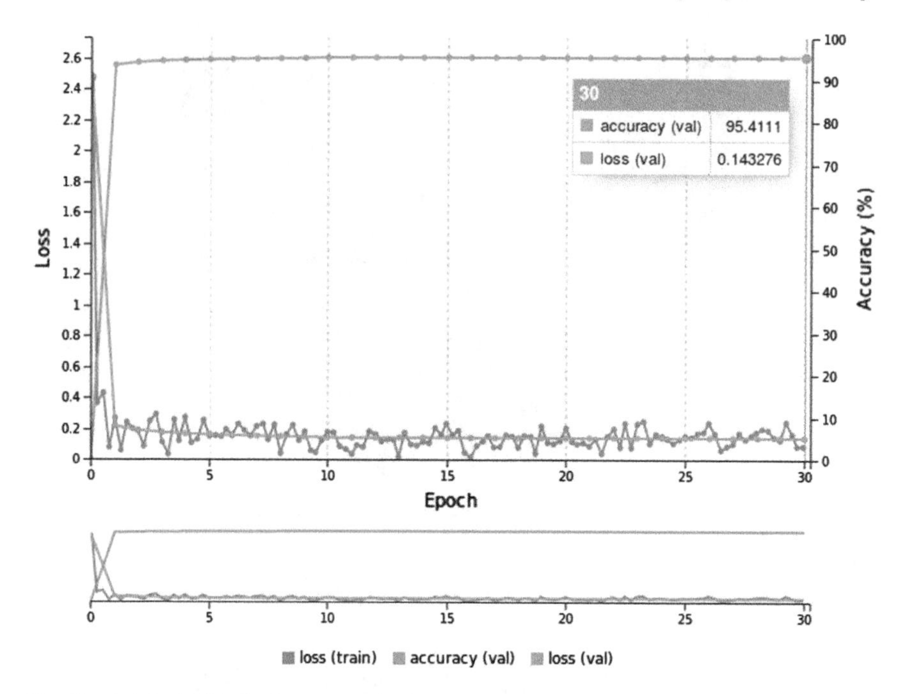

Fig. 28 Training/validation loss and validation accuracy when training FCN-16s using pre-trained model

When the segmentation process is analyzed considering the sample and the ground truth images in Fig. 30, it has been seen that the object contours are roughly segmented in the segmentation process performed with FCN-AlexNet model. Moreover, the fact that fine details such as pole could not be made out and segmented showed another limitation of this model for the application. With FCN-8s model, contrary to FCN-AlexNet, object contours are segmented sharply and the segmentation inferences are more similar to the ground truth images. Furthermore, the fact that the object classes can be detected completely indicates that FCN-8s is useful. Although FCN-16s model is not as sharp as FCN-8s, it can be seen that the object contours can be segmented successfully. Finally, when the segmentation inferences of FCN-32s model are analyzed, it can be said that the segmentations very close to FCN-AlexNet have been realized but may be a more useful model with small differences.

The training times of the trained models for semantic image segmentation in this paper are given in Fig. 31.

It was seen that the training time spent on FCN-AlexNet is considerably lower than the other FCN models with very close training times.

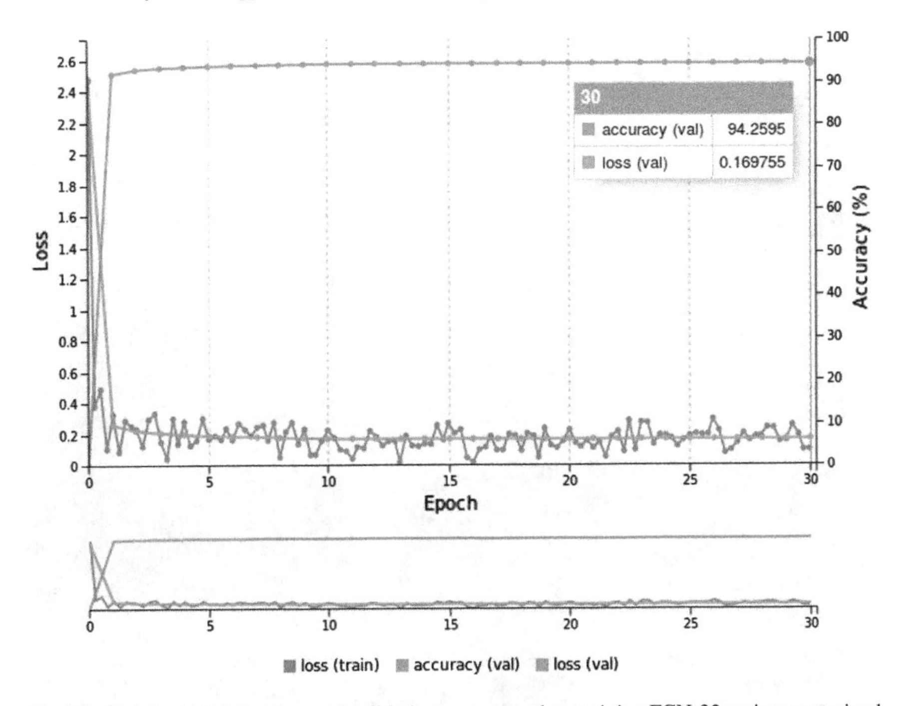

Fig. 29 Training/validation loss and validation accuracy when training FCN-32s using pre-trained model

6 Conclusions

For the application, firstly, FCNs are trained separately and the validation accuracy of these trained network models is compared on the SYNTHIA-Rand-CVPR16 dataset. Approximately 80% of the images in the dataset are used for the training phase and the rest are used during validation phase to validate the validity of the models. In addition, image segmentation inferences in this paper are visualized to see how precisely the used FCN architectures can segment objects.

Maximum validation accuracies of 92.4628%, 96.015%, 95.4111% and 94.2595% are achieved with FCN-AlexNet, FCN-8s, FCN-16 and FCN-32s models trained using weights in pre-trained models, respectively. Although these models can be regarded as successful at first sight when the accuracies are over 90% for the four models, it is seen that the object contours are roughly segmented in the segmentation process performed with FCN-AlexNet model. The impossibility of segmenting some object classes with small pixel areas is another limitation of FCN-AlexNet model. The segmentation inferences of FCN-32s model are also very close to FCN-AlexNet, but with this model it is seen that some better results can be obtained. However, with FCN-8s model, object contours are sharply segmented and the segmentation inferences are more similar to the ground truth images. Although FCN-16 models are

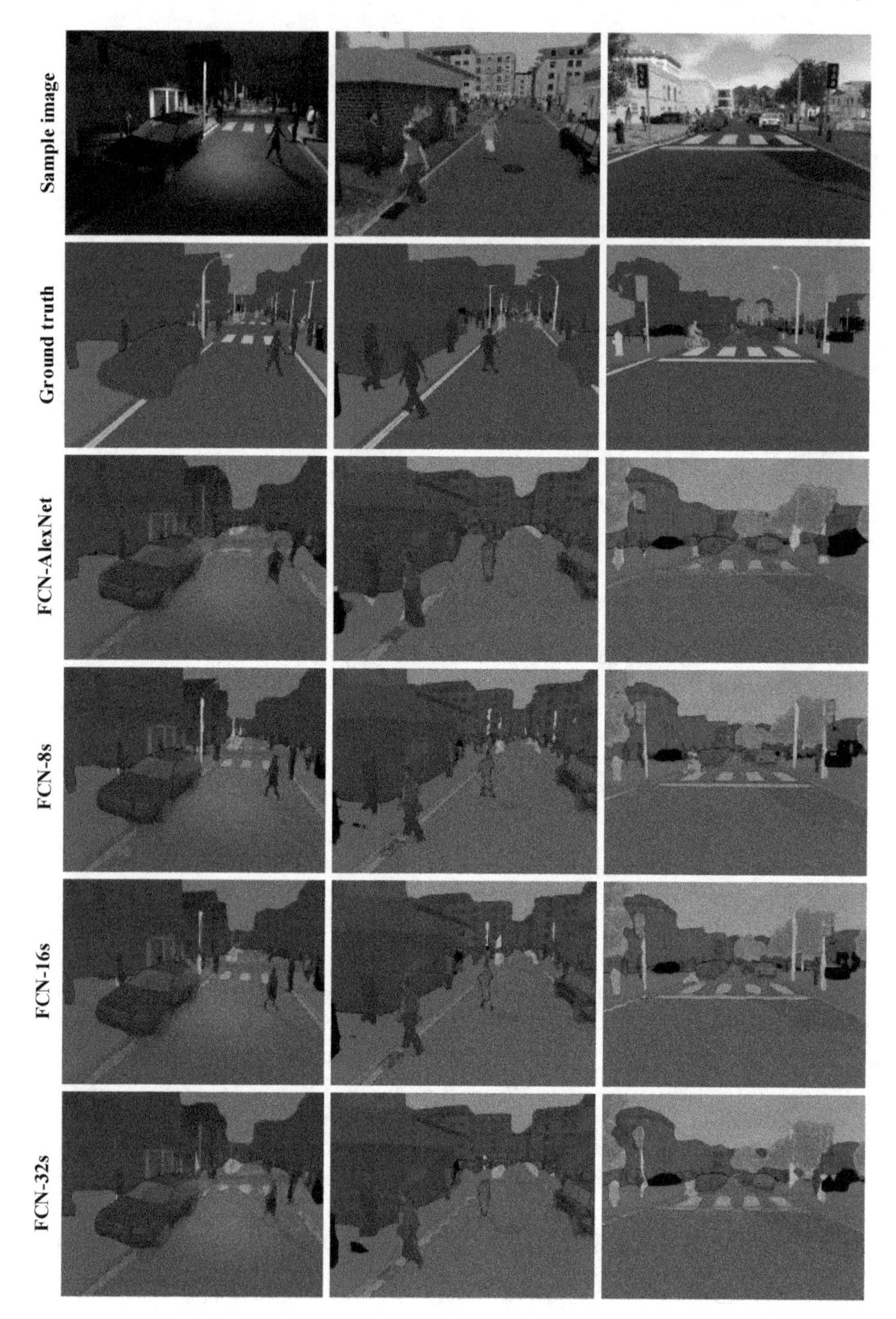

Fig. 30 Comparison of segmentation inferences according to the used FCN architectures for sample images

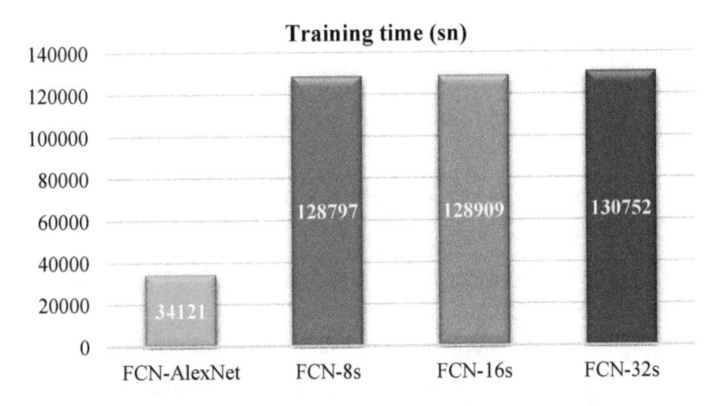

Fig. 31 Training times of the models

not as sharp as FCN-8s, it is seen that the object contours are successfully segmented according to the others.

When training times of FCN models are compared, it is seen that the training time spent on FCN-AlexNet is about one-fourth of the other FCN models with very close training times. However, considering that training of the model is carried out once for the application, it can be said that it does not have a very important place in the choice of the appropriate model. Therefore, it can be easily stated that the most suitable model for the application is FCN-8s.

The obtained experimental results show that the FCNs from deep learning approaches are suitable for semantic image segmentation applications. In addition, it has been understood that the FCNs are network structures in models that address many pixel-level applications, especially semantic image segmentation.

Acknowledgements We gratefully acknowledge the support of the NVIDIA Corporation, who donated the NVIDIA GTX Titan X Pascal GPU that was used for this research under the NVIDIA GPU Grant program.

References

1. R.C. Weisbin et al., Autonomous rover technology for mars sample return, in *Artificial Intelligence, Robotics and Automation in Space*, vol. 440 (1999), p. 1
2. R.N. Colwell, History and place of photographic interpretation, in *Manual of Photographic Interpretation*, vol. 2 (1997), pp. 33–48
3. I. Goodfellow, Y. Bengio, A. Courville, *Deep Learning* (Book in Preparation for MIT Press, 2016)
4. K. He, X. Zhang, S. Ren, J. Sun, Deep residual learning for image recognition, in *Proceedings of the IEEE Conference on Computer Vision and Pattern Recognition (CVPR)*, Seattle, WA, USA (2016), pp. 770–778
5. A. Karpathy, Convolutional neural networks for visual recognition. Course Notes. http://cs231n.github.io/convolutional-networks/. Accessed 5 Apr 2017

6. L. Van Woensel, G. Archer, L. Panades-Estruch, D. Vrscaj, Ten technologies which could change our lives. Technical report (European Parliamentary Research Service (EPRC), Brussels, Belgium, 2015)
7. D.G. Lowe, Distinctive image features from scale-invariant keypoints. Int. J. Comput. Vis. (IJCV) **60**(2), 91–110 (2004)
8. T. Ojala, M. Pietikainen, T. Maenpaa, Multiresolution gray scale and rotation invariant texture classification with local binary patterns. IEEE Trans. Pattern Anal. Mach. Intell. **24**(7), 971–987 (2002)
9. N. Dalal, B. Triggs, Histograms of oriented gradients for human detection, in *Proceedings of the IEEE Conference on Computer Vision and Pattern Recognition (CVPR)*, vol. 1, San Diego, CA, USA (2005), pp. 886–893
10. G.E. Hinton, S. Osindero, Y.W. Teh, A fast learning algorithm for deep belief nets. Neural Comput. **18**(7), 1527–1554 (2006)
11. C. Cortes, V.N. Vapnik, Support vector networks. Mach. Learn. **20**(3), 273–297 (1995)
12. Y. Freund, R.E. Schapire, A decision-theoretic generalization of on-line learning and an application to boosting. J. Comput. Syst. Sci. **55**, 119–139 (1997)
13. A. Uçar, Y. Demir, C. Güzeliş, A penalty function method for designing efficient robust classifiers with input space optimal separating surfaces. Turk. J. Electr. Eng. Comput. Sci. **22**(6), 1664–1685 (2014)
14. O. Russakovsky et al., ImageNet large scale visual recognition challenge. Int. J. Comput. Vis. **115**(3), 211–252 (2015)
15. Q.V. Le, Building high-level features using large scale unsupervised learning, in *Proceedings of the IEEE International Conference on Acoustics, Speech and Signal Processing (ICASSP)*, Vancouver, BC, Canada (2013), pp. 8595–8598
16. Y. Taigman, M. Yang, M.A. Ranzato, L. Wolf, Deepface: closing the gap to human-level performance in face verification, in *Proceedings of the IEEE Conference on Computer Vision and Pattern Recognition (CVPR)*, Colombus, Ohio, USA (2014), pp. 1701–1708
17. L. Deng, D. Yu, Deep learning: methods and applications, in *Foundations and Trends® in Signal Processing*, vol. 7 (2014), pp. 197–387
18. D. Amodei et al., Deep speech 2: end-to-end speech recognition in English and Mandarin, in *Proceedings of the International Conference on Machine Learning (ICML)*, New York, USA (2016), pp. 173–182
19. Trend Search of "Deep Learning" in Google, https://trends.google.com/trend/explore?q=deep%20learning. Accessed 12 Apr 2017
20. J. Long, E. Shelhamer,T. Darrell, Fully convolutional networks for semantic segmentation, *Proceedings of the IEEE Conference on Computer Vision and Pattern Recognition (CVPR)*, Boston, Massachusetts, USA (2015), pp. 3431–3440
21. L.C. Chen et al., Semantic image segmentation with deep convolutional nets and fully connected CRFs, in *Proceedings of the International Conference on Learning Representations (ICLR)*, San Diego, CA, USA (2015), pp. 1–14
22. H. Noh, S. Hong, B. Han, Learning deconvolution network for semantic segmentation, in *Proceedings of the IEEE International Conference on Computer Vision (ICCV)*, Los Alamitos, CA, USA (2015), pp. 1520–1528
23. S. Zheng et al., Conditional random fields as recurrent neural networks, in *Proceedings of the IEEE International Conference on Computer Vision (ICCV)*, Los Alamitos, CA, USA (2015), pp. 1529–1537
24. G. Papandreou, L.C. Chen, K. Murphy, A.L. Yuille, Weakly-and semi-supervised learning of a DCNN for semantic image segmentation, in *Proceedings of the IEEE International Conference on Computer Vision (ICCV)*, Los Alamitos, CA, USA (2015), pp. 1742–1750
25. F. Yu, V. Koltun, Multi-scale context aggregation by dilated convolutions, in *Proceedings of the International Conference on Learning Representations (ICLR)*, San Juan, Puerto Rico (2016), pp. 1–13
26. C. Farabet, C. Couprie, L. Najman, Y. LeCun, Learning hierarchical features for scene labeling. IEEE Trans. Pattern Anal. Mach. Intell. **35**(8), 1915–1929 (2013)

27. V. Badrinarayanan, A. Kendall, R. Cipolla, SegNet: a deep convolutional encoder-decoder architecture for image segmentation. arXiv:1511.00561 (2015)
28. A. Kendall, V. Badrinarayanan, R. Cipolla, Bayesian SegNet: model uncertainty in deep convolutional encoder-decoder architectures for scene understanding. arXiv:1511.02680 (2015)
29. D. Fourure et al., Semantic segmentation via multi-task, multi-domain learning, in *Joint IAPR International Workshop on Statistical Techniques in Pattern Recognition (SPR) and Structural and Syntactic Pattern Recognition (SSPR)*, Mérida, Mexico (2016), pp. 333–343
30. M. Treml et al., Speeding up semantic segmentation for autonomous driving, in *Proceedings of the Conference on Neural Information Processing Systems (NIPS)*, Barcelona, Spain (2016), pp. 1–7
31. J. Hoffman, D. Wang, F. Yu, T. Darrell, FCNs in the wild: pixel-level adversarial and constraint-based adaptation. arXiv:1612.02649 (2016)
32. D. Marmanis et al., Semantic segmentation of aerial images with an ensemble of CNSS. ISPRS Ann. Photogramm. Remote Sens. Spat. Inf. Sci. **3**, 473–480 (2016)
33. K. Simonyan, A. Zisserman, Very deep convolutional networks for large-scale image recognition, in *Proceedings of the International Conference on Learning Representations (ICLR)*, San Diego, CA, USA (2015), pp. 1–14
34. S. Song, S.P. Lichtenberg, J. Xiao, Sun RGB-D: a RGB-D scene understanding benchmark suite, in *Proceedings of the IEEE Conference on Computer Vision and Pattern Recognition (CVPR)*, Boston, Massachusetts, USA (2015), pp. 567–576
35. G.J. Brostow, J. Fauqueur, R. Cipolla, Semantic object classes in video: a high-definition ground truth database. Pattern Recognit. Lett. **30**(2), 88–97 (2009)
36. A. Geiger, P. Lenz, C. Stiller, R. Urtasun, Vision meets robotics: the KITTI dataset. Int. J. Robot. Res. **32**(11), 1231–1237 (2013)
37. F.N. Iandola et al., SqueezeNet: AlexNet-level accuracy with 50$\times$ fewer parameters and <0.5 MB model size. arXiv:1602.07360 (2016)
38. P.O. Pinheiro, T.Y. Lin, R. Collobert, P. Dollár, Learning to refine object segments, in *Proceedings of the European Conference on Computer Vision (ECCV)*, Amsterdam, Netherlands (2016), pp. 75–91
39. S.R. Richter, V. Vineet, S. Roth, V. Koltun, Playing for data: ground truth from computer games, in *Proceedings of the European Conference on Computer Vision (ECCV)*, Amsterdam, Netherlands (2016), pp. 102–118
40. G. Ros et al., The SYNTHIA dataset: a large collection of synthetic images for semantic segmentation of urban scenes, in *Proceedings of the IEEE Conference on Computer Vision and Pattern Recognition (CVPR)*, Seattle, WA, USA (2016), pp. 3234–3243
41. M. Cordts et al., The cityscapes dataset for semantic urban scene understanding, in *Proceedings of the IEEE Conference on Computer Vision and Pattern Recognition (CVPR)*, Seattle, WA, USA (2016), pp. 3213–3223
42. Machine Learning, http://www.nvidia.com/object/machine-learning.html. Accessed 25 Apr 2017
43. Deep Learning Architectures, https://qph.ec.quoracdn.net/main-qimg-4fbecaea0b4043d5450a1ca0ebe30623. Accessed 1 May 2017
44. S.S. Haykin, *Neural Networks: A Comprehensive Foundation* (Tsinghua University Press, 2001)
45. CUDA, http://www.nvidia.com/object/cuda_home_new.html. Accessed 5 May 2017
46. BVLC Caffe, http://caffe.berkeleyvision.org/. Accessed 7 May 2017
47. What is Torch?, http://torch.ch/. Accessed 7 May 2017
48. Introduction to the Python Deep Learning Library Theano, http://machinelearningmastery.com/introduction-python-deep-learning-library-theano/. Accessed 8 May 2017
49. About TensorFlow, https://www.tensorflow.org/. Accessed 9 May 2017
50. Keras: Deep Learning Library for Theano and TensorFlow, https://keras.io/. Accessed 9 May 2017
51. NVIDIA CuDNN, https://developer.nvidia.com/cudnn. Accessed 10 May 2017

52. Y. LeCun, Backpropagation applied to handwritten ZIP code recognition. Neural Comput. **1**(4), 541–551 (1989)
53. M. Shivaprakash, Semantic segmentation of satellite images using deep learning. Master's thesis (Czech Technical University in Prague & Luleå University of Technology, Institute of Science, Prague, Czech Republic, 2016)
54. V. Dumoulin, F. Visin, A guide to convolution arithmetic for deep learning. arXiv:1603.07285 (2016)
55. Convolution, https://leonardoaraujosantos.gitbooks.io/artificial-inteligence/content/convolution.html. Accessed 15 May 2017
56. A. Krizhevsky, I. Sutskever, G.E. Hinton, ImageNet classification with deep convolutional neural networks, in *Advances in Neural Information Processing Systems* (2012), pp. 1097–1105
57. An Intuitive Explanation of Convolutional Neural Networks, https://ujjwalkarn.me/2016/08/11/intuitive-explanation-convnets/. Accessed 16 May 2017
58. N. Srivastava et al., Dropout: a simple way to prevent neural networks from overfitting. J. Mach. Learn. Res. **15**(1), 1929–1958 (2014)
59. Y. Bengio, Practical recommendations for gradient-based training of deep architectures, in *Neural Networks: Tricks of the Trade* (Springer Berlin, Heidelberg, 2012), pp. 437–478
60. C. Szegedy et al., Going Deeper with convolutions, in *Proceedings of the IEEE Conference on Computer Vision and Pattern Recognition (CVPR)*, Boston, Massachusetts, USA (2015), pp. 1–9
61. Full-day CVPR 2013 Tutorial, http://mpawankumar.info/tutorials/cvpr2013/. Accessed 22 May 2017
62. Unity Development Platform, https://unity3d.com/
63. Image Segmentation using DIGITS 5, https://devblogs.nvidia.com/parallelforall/image-segmentation-using-digits-5/. Accessed 25 May 2017
64. A. Uçar, Y. Demir, C. Güzeliş, Object recognition and detection with deep learning for autonomous driving applications. Simulation **93**(9), 759–769 (2017)
65. J. Yosinski et al., How transferable are features in deep neural networks?. Adv. Neural Inf. Process. Syst. 3320–3328 (2014)

Phase Identification and Workflow Modeling in Laparoscopy Surgeries Using Temporal Connectionism of Deep Visual Residual Abstractions

Kaustuv Mishra, Rachana Sathish and Debdoot Sheet

Abstract The phase recognition task has been performed on different types of surgeries which ranges from cataract to neurological to laparoscopic intervention. The visual features of a surgical video can be used to identify the surgical phases in laparoscopic interventions. Owing to the significant improvement in performance exhibited by convolutional neural networks (CNN) on various challenging tasks like image classification, action recognition etc., they are widely used as feature extractors. In the proposed framework, features extracted by a CNN are used for phase recognition. The task of phase recognition in surgical videos is rendered challenging because of the presence of motion blur produced to the mobile nature of the recording device and the high variance in scenes observed during the course of the surgery. Also, blood stains on the camera lens and complete or partial occlusion of the scene captured by the laparoscopic camera poses additional challenges. These challenges can be overcome by using temporal features in addition to the spatial visual features. A long short-term memory (LSTM) network is used to learn the temporal information of the video. The m2cai16-workflow dataset consisting of videos of cholecystectomy is used for experimental validation of the performance. Surgical workflow, which refers to the statistical modelling of activities taking place in an operating room during a surgery be done in terms of the surgical phases.

K. Mishra · R. Sathish (✉) · D. Sheet
Indian Institute of Technology Kharagpur, Kharagpur, India
e-mail: rachana.sathish@iitkgp.ac.in

K. Mishra
e-mail: kaustuvmishra293@gmail.com

D. Sheet
e-mail: debdoot@ee.iitkgp.ernet.in

V. E. Balas et al. (eds.), *Handbook of Deep Learning Applications*,
Smart Innovation, Systems and Technologies 136,
https://doi.org/10.1007/978-3-030-11479-4_10

1 Introduction

Laparoscopic procedures are increasingly being practiced on account of various factors including reduced discomfort to patients and lesser post-surgical recovery time. However, these procedures are strenuous as the surgeons only have an indirect visual access to the area being operated. The surgical site is visualized using endoscopes and the surgical tools are inserted into the body through small incisions. As a result, the freedom of movement of the tools is limited. This calls for meticulous training of the surgeons to acquire the desired dexterity. Traditionally, surgical training is evaluated by experienced surgeons in person or by using pre-recorded videos of a trainee surgeon performing the procedure which is evaluated at a later time for workflow standards conformity. Both of the methods are time-consuming and tedious. Tracking of the workflow during the course of a surgery can also help in modeling the procedural commonalities across surgeons which in turn can be used to optimize the workflow. Automated techniques for surgical skill evaluation and workflow modeling can save much of the time and efforts wasted in traditional methods and can also aid in post-surgical quality monitoring. This can be achieved by modeling the surgical workflow in terms of the surgical phases using the video acquired while performing the procedure.

The modeling of the surgical procedures involved in laparoscopy surgery can aid in developing a graphical user interface (GUI) for visualization of the progress of an ongoing surgery. Typically, the various procedures involved in a laparoscopy surgery is segregated into certain phases, which enables their modeling in terms of these phases. Such a model helps in hassle free pre-operative planning of surgeries which helps in reducing much of the overhead incurred due to lack of proper preparation. In addition, it serves as a guide for intuitive visualization of the progress of a surgery during its course. This keeps the surgical staffs informed regarding the progress of a surgery and helps them prepare for the upcoming phases. Surgical workflow models also help in the automatic assessment of surgical skill and development of intelligent context-aware systems that needs information about the ongoing stage of surgery. Surgical models and their analysis using signals obtained from the operating room has several uses. Research in these area usually leverages signals which are obtained automatically, which can be videos, images, tool usage information or signal from robotic systems or from some additional sensor installed in the surgical instrument.

2 Prior Art

The task of recognition of surgical phases has been carried out for several procedures, including cataract surgery, neurological and laparoscopic surgeries for the purpose workflow modeling and monitoring. Several types of features like visual signals, surgical action triplets and signals regarding tool usage have been experimented with for this purpose. A Hidden Markov Model (HMM) based method that combines visual features from the frames of laparoscopy video and the tool usage signals was proposed by Padoy et al. [1]. In this method, surgical phases are recognized using tool usage signals. The major challenge here is the interrupted and delayed availability of these signals. On the other hand, tool usage signals were used by Blum et al. [2] to perform dimensionality reduction using canonical correlation analysis (CCA) on the visual features. However, this method requires handcrafted visual features. Lalys et al. [3] proposed a framework using only visual features like shape, color and texture for surgical phase recognition in cataracts surgery. Usage of such handcrafted features is inadequate for Cholecystectomy surgeries as they are performed for a longer duration in comparison with cataract surgeries. Also, additional visual challenges like camera motions and presence of multiple tools that are more articulated are present in laparoscopy procedures. Lea et al. [4] used skip-chain conditional random fields (CRF) combining both kinematics and image features to determine surgical activities but the shortcoming is that it is tested on a dataset of shorter duration (around minutes). Further, the visual features used in this method are hand crafted. Klank et al. [5] proposed automatic learning of visual features from videos to determine surgical phase. The method uses genetic programming, which is limited by predefined operators which results in worse performance in comparison with hand crafted. Twinanda et al. [6] proposed to use CNN for surgical phase identification.

3 Mathematical Formulation

The various actvities involved in a laparoscopy may be divided into several phases. In a typical scenario the phases in a laparoscopic intervention are trocar placement, preparation, calot's triangle dissection, clipping and cutting, gallbladder dissection, gallbladder packaging, cleaning and coagulation and gallbladder retraction. Sample video frames depicting these phases are shown in Fig. 1. Thus the frames in the video can be classified as a particular phase p_i of the surgery. The problem of phase identification in a surgical video, comprising of phases $\mathscr{P} = \{p_1, p_2, \ldots, p_n\}$, can be considered as a multi-class classification task, in which each frame f is to be classified into one of the phase p_i out of the set of phases $\mathscr{P}$ occurring in the procedure.

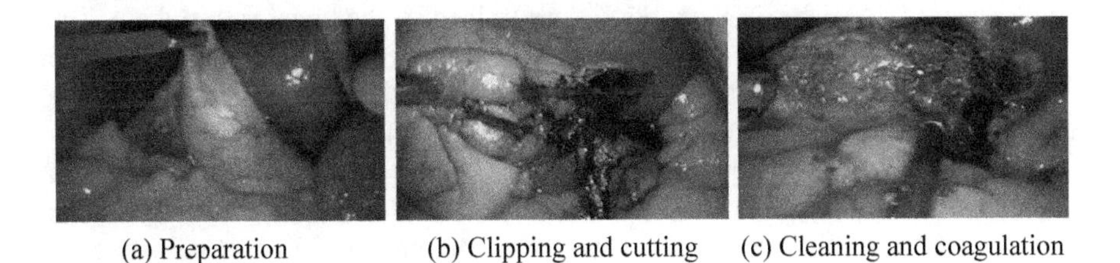

(a) Preparation (b) Clipping and cutting (c) Cleaning and coagulation

Fig. 1 Sample images at different phases of a laparoscopic surgery as retrieved from the m2cai16-workflow dataset

If each frame of the video is annotated with the corresponding phase, a statistical model can be developed by determining transition probability $P(p_i|p_j)$ which indicates the probability of the upcoming phase to be p_i given that the ongoing phase is p_j. Though the procedures involved in a particular type of surgery is pre-defined, the order at which they are executed may vary depending on patient's anatomy and the surgeon's skill. Therefore, once a workflow for a particular surgery is modeled using a dataset consisting of significant variations in the workflow, the progression of the workflow in a new unseen surgery can be determined with the knowledge of the current phase of the surgery.

4 Methodology

4.1 Phase Recognition in Surgical Videos

The appearance of surgical phases is characterized by the various tools used and the actions performed which are well represented by the visual features. These visual features can be used for the identification of phases. In the proposed framework, a CNN [7] is used to learn to extract visual features from the surgical video frames. These features contains only the spatial information and fails to incorporate temporal information. As the various surgical procedures are mostly executed methodically, the temporal information can aid in improving the performance of frameworks for automatic phase recognition. Therefore, an LSTM [8] network is trained on the visual features from the CNN to learn the temporal information and thereby improve the performance of the predictive framework. The problem of surgical phase recognition can be modeled as a single-label multi-class classification task. Typically, there are 8 phases in the cholecystectomy procedure which is recorded in the m2cai16-workflow dataset. Therefore, each frame can be classified into one out of the 8 phases.

Owing to the success of CNN in single label classification tasks, a similar approach is used. Inherently, CNN has a huge number of parameters and thus requires a large number of labelled training instances which are scarcely available in the domain of medical images. However, by the method of transfer learning [9], good performance on a target domain with fewer labelled data can be achieved by means of fine-tuning [10]. The pre-trained networks used are AlexNet [11], GoogLeNet [12] and ResNet-50 [13]. These networks are pre-trained on the Imagenet database. The final layer is then replaced with an 8-way log-sigmoid layer followed by finetuning on the m2cai-workflow[1] dataset. The proposed framework using ResNet-50 architecture for surgical phase identification in laparoscopic videos is shown in Fig. 2.

The spatial visual features learnt by the CNN does not encode any temporal information. However, temporal information is crucial in correctly identifying the phases in the surgical video. Long Short Term Memory (LSTM) [8] networks are

[1] http://camma.u-strasbg.fr/datasets.

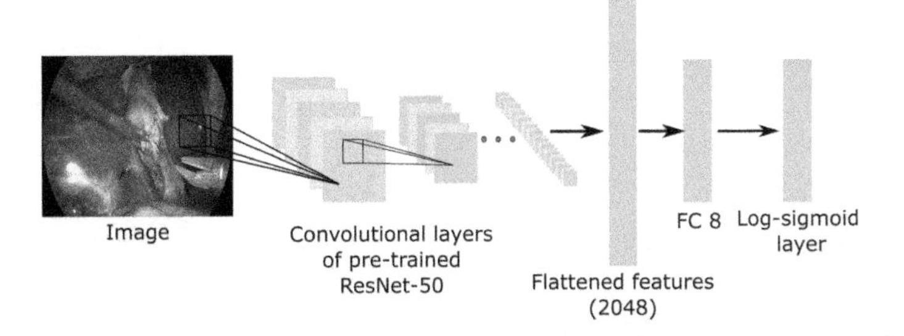

Fig. 2 Illustration of ResNet used for surgical phase identification

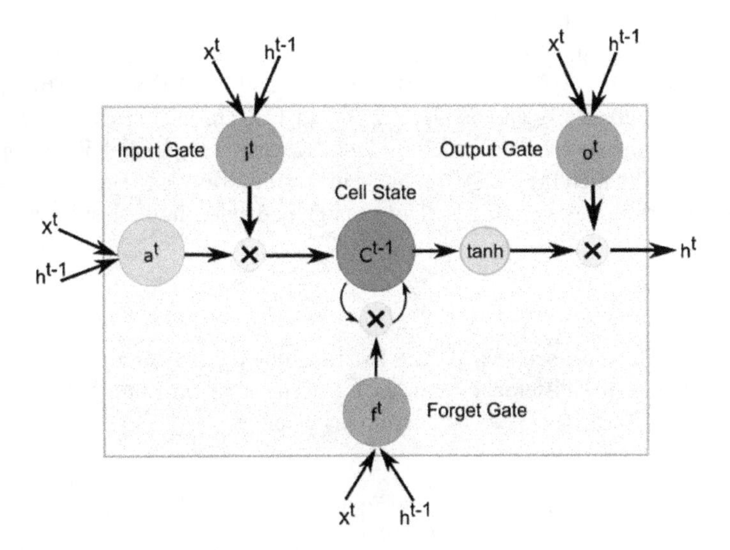

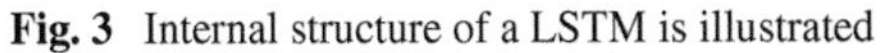
Fig. 3 Internal structure of a LSTM is illustrated

recurrent neural networks [14] capable of encoding temporal information in their internal states. Additionally, there are facilitated with three gates for regulating the information flow and the internal states. The structure of the LSTM module along with various gates are shown in Fig. 3.

The various operation at different gates of the LSTM are given below.

$$\mathbf{a}^t = \tanh(\mathbf{W}_c\mathbf{x}^t + \mathbf{U}_c\mathbf{h}^{t-1}) \tag{1}$$

$$\mathbf{i}^t = \sigma(\mathbf{W}_i\mathbf{x}^t + \mathbf{U}_i\mathbf{h}^{t-1}) \tag{2}$$

$$\mathbf{f}^t = \sigma(\mathbf{W}_f\mathbf{x}^t + \mathbf{U}_f\mathbf{h}^{t-1}) \tag{3}$$

$$\mathbf{o}^t = \sigma(\mathbf{W}_o\mathbf{x}^t + \mathbf{U}_o\mathbf{h}^{t-1}) \tag{4}$$

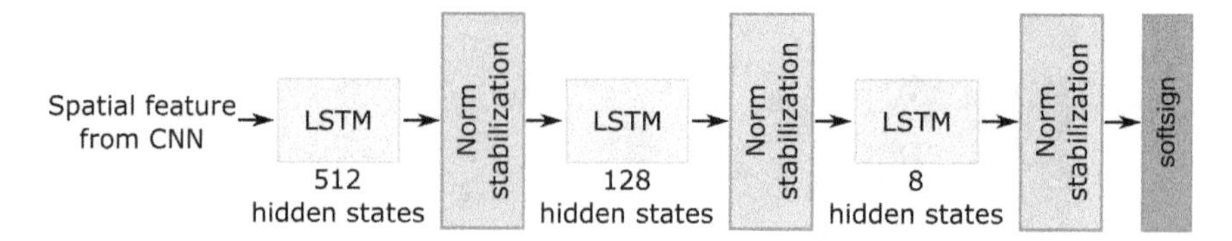

Fig. 4 Illustration of LSTM network with depth of 3

where $\mathbf{x}^t$ is the input to the LSTM at time t, $\mathbf{h}^{t-1}$ is the output vector of previous time step, $\mathbf{i}^t$, $\mathbf{f}^t$, $\mathbf{o}^t$ are the response of input, forget and output gates respectively, $\mathbf{W}$, $\mathbf{U}$ are the weights associated with the gates and $\sigma(.)$ represents the sigmoid non-linearity.

The features obtained from the fine-tuned AlexNet, GoogLeNet and ResNet-50 is used to train the LSTM. The depth of LSTM network can be increased by stacking one LSTM module above another. Here 2, 3, and 4 LSTM modules are used to learn the temporal information. The sequence length used is 3, 5, 10 and 50 which states the duration till which the LSTM retains temporal information. The architecture of LSTM network is shown in Fig. 4. The softsign activation in the last layer is given as,

$$f(x) = \frac{x}{(1 + |x|)} \tag{5}$$

Norm stabilization between the LSTM modules regularizes the hidden states. Minimization of the difference between L2 norms of each time step is done for regularization. The associated cost function is given as,

$$loss = \beta \times \frac{1}{T} \sum_{t=1}^{T} (\| \mathbf{h}^t \| - \| \mathbf{h}^{t-1} \|)^2 \tag{6}$$

where T is the total number of time steps, β is a constant and $\mathbf{h}$ represents the hidden state.

4.2 Statistical Modeling of Workflow

An intuitive visualization of the surgical workflow of cholecystectomy is presented using the a set of video recorded during the surgery with the per frame annotation of the phase. The m2ccai16-workflow dataset consisting of 27 annotated videos of cholecystectomy consisting of a maximum of eight different phases is used to validate the proposed modeling and visualization scheme. The total number of transitions from one phase to the other across 20 videos of the dataset is used to model the workflow. It is observed that the first five phases of the surgery are executed sequentially in almost all the videos with variation in the order of subsequent phases. The transition probabilities for the different phases is shown in Table 1.

The workflow scheme modelled from the dataset is shown in Fig. 5. The different workflow schemes observed in the dataset is graphically illustrated in Fig. 6.

Table 1 Transition probability of phases obtained on the m2cai-workflow dataset for laparoscopy procedure

Current phase	Next phase							
	P1	P2	P3	P4	P5	P6	P7	P8
P1	0	1	0	0	0	0	0	0
P2	0	0	1	0	0	0	0	0
P3	0	0	0	1	0	0	0	0
P4	0	0	0	0	1	0	0	0
P5	0	0	0	0	0	0.71	0.29	0
P6	0	0	0	0	0	0	0.67	0.33
P7	0	0	0	0	0	0.32	0	0.68
P8	0	0	0	0	0	0	1	0

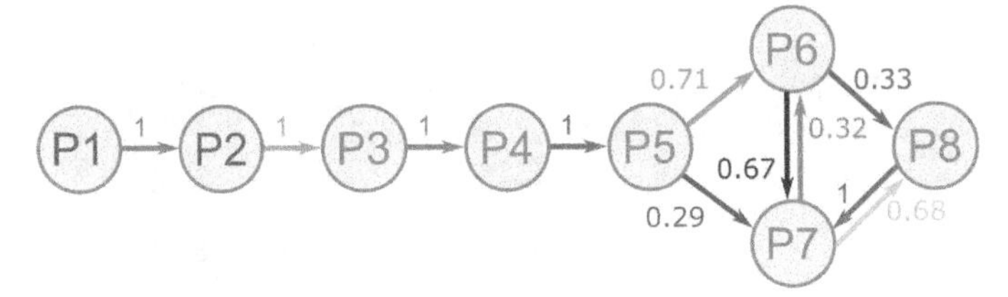

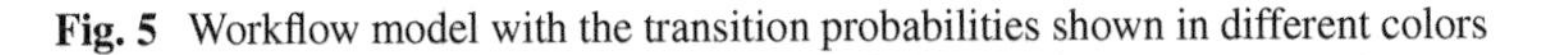
Fig. 5 Workflow model with the transition probabilities shown in different colors

The framework presented in Sect. 4.1 can be used to determine the current phase of the surgery. It makes use of the video stream recorded during the surgery. Knowledge of current phase of the surgery can inturn be leveraged to determine the next phase using the transition probabilities and the workflow model.

5 Experiments and Results

The approach to the problem is based on CNN and LSTM, which learns both the temporal and spatial features in the videos. CNNs have been proven to be very successful in object recognition and image classifications tasks where features from the CNN are used to make the LSTM learn sequence information. Thus spatio-temporal information is leveraged to identify the surgical phases in a laparoscopic intervention.

5.1 Dataset Description

The m2cai16-workflow dataset[2] is used for performance evaluation. The dataset is composed of 27 laparoscopic surgery videos which have their phase annotated at

[2]http://camma.u-strasbg.fr/datasets.

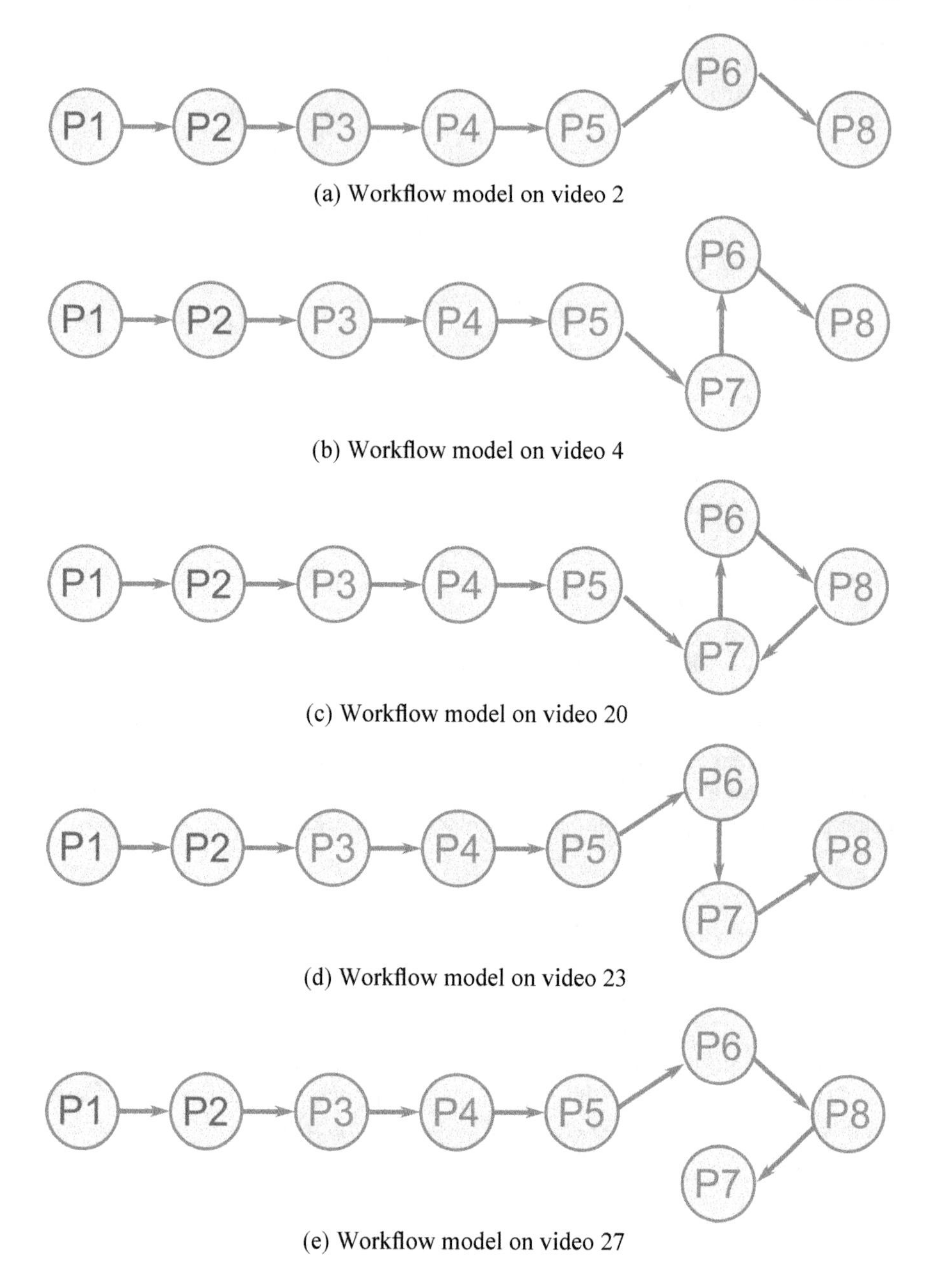

(a) Workflow model on video 2

(b) Workflow model on video 4

(c) Workflow model on video 20

(d) Workflow model on video 23

(e) Workflow model on video 27

Fig. 6 Workflow models for a set of selected videos from the m2cai-workflow dataset

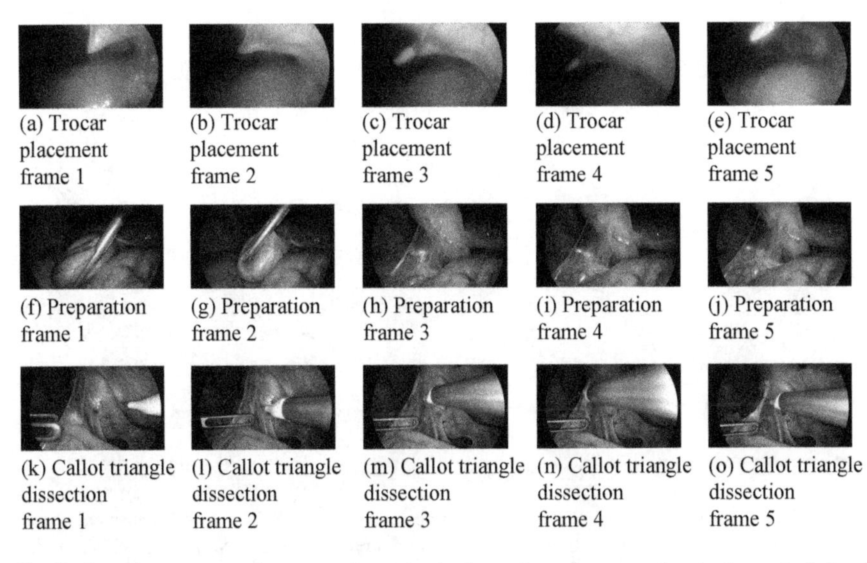

Fig. 7 Sample sequences for trocar placement is shown in **a–e**, preparation is shown in **f–j** and Callot triangle dissection is shown in **k–o**

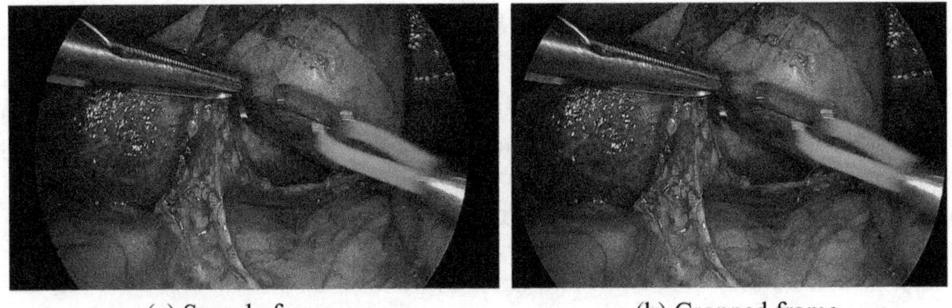

Fig. 8 Sample **a** frame from a video in the dataset and **b** the cropped region of interest

25 fps. 17 videos are used for training, 4 videos are used for validation and 6 videos are used for testing purpose. The different phases in laparoscopy are trocar placement, preparation, calot's triangle dissection, clipping and cutting, gallbladder dissection, gallbladder packaging, cleaning and coagulation and gallbladder retraction which is shown in Fig. 7.

In the videos it is observed that the region of interest (ROI) in the frames of the videos, where surgical activity takes place is generally confined to a smaller circular region as compared to the whole image as shown in Fig. 8a. The frames are cropped to extract the region of interest and reshaped to 224×224 as shown in the Fig. 8b before being fed forward through the CNN.

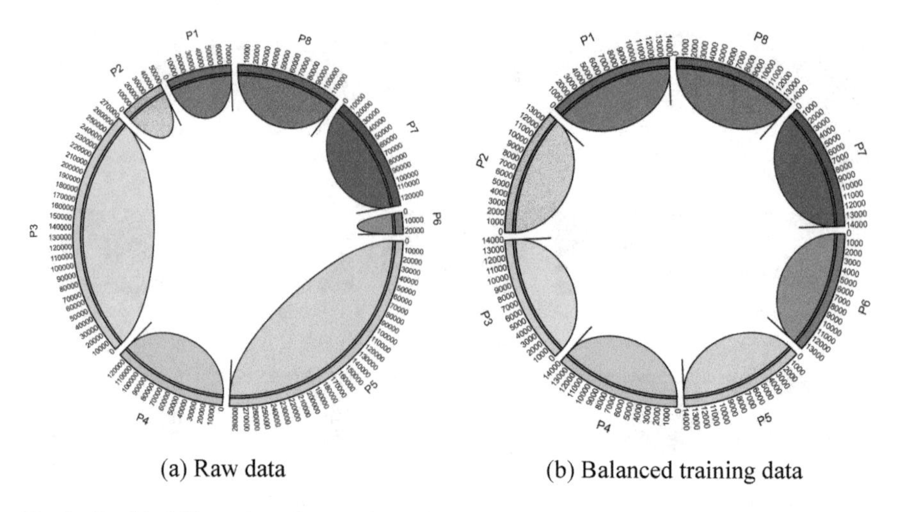

(a) Raw data (b) Balanced training data

Fig. 9 Graphical illustration of proportion of frames corresponding to each of the surgical phases in the **a** raw and **b** balanced training data. Each color represents a surgical phase. While the raw video has class imbalance, the trained data has been balanced

5.2 Compensating Class Imbalance in Training Dataset

The entire duration of the surgery can be divided into eight phases. The phases can be differentiated on the basis of different tools or tool combinations used during the surgery. The varied duration of these phase results in severe class imbalance in the dataset. Some phases occurs for longer duration and some phases occurs for very short duration. The phases trocar placement, preparation, calot's triangle dissection, clipping and cutting, gallbladder dissection, gallbladder packaging, cleaning and coagulation and gallbladder retraction are labeled as P1, P2, P3, P4, P5, P6, P7 and P8 respectively. The imbalance in class distribution in the dataset is graphically illustrated in Fig. 9a. Each sector corresponds to one of the phases and the size is proportional to the number of frames labelled as that phase.

Table 2 Number of frames corresponding to each phase in the training data before and after class balancing

Phase	Raw data	Balanced data
Trocar placement (P1)	2,907	14,008
Preparation (P2)	2,029	13,982
Calot's triangle dissection (P3)	10,805	14,008
Clipping and cutting (P4)	5,108	14,008
Gallbladder dissection (P5)	11,494	14,008
Gallbladder packaging (P6)	891	13,532
Cleaning and coagulation (P7)	5,064	14,008
Gallbladder retraction (P8)	4,693	14,008

In order to balance distribution of frames corresponding to each phase, the frames are extracted methodically from the videos. The annotation of the video is given at 25 fps and the training data is created by taking frames at 1 fps so the class balancing is done by taking into account the additional 24 frames per second for the class which has lower count. Consider the example case of phase P6 which has 891 frames in the training set. The data is balanced by extracting the intermediate frames. Table 2 shows the number of frames containing each phase before and after class balancing. The phase distribution balanced data is shown in Fig. 9b.

5.3 Baselines

Evaluation of the proposed method is done using the comparing its performance with the following baselines (BL).

BL1: Multi-class classification using AlexNet [11]. The five convolutional layers and the first two fully-connected layers of the AlexNet pre-trained on ImageNet dataset were retained. The final classification layer was replaced with a log-sigmoid layer preceded by a fully-connected layer of length eight. The network was then fine-tuned on the class balanced m2cai16-workflow dataset.

BL2: Multi-class classification using deep stacked LSTM network using the features extracted by AlexNet fine-tuned on m2cai-workflow dataset. The features from the fully-connected layer of length 4096 of the network trained in **BL1** was used to train the LSTM network detailed in Sect. 4.1.

BL3: Multi-class classification using GoogLeNet [12]. Similar to **BL1**, the fully-connected linear layer after the fifth inception module was replaced with a one of length eight followed by a log-sigmoid layer which is detailed in Sect. 4.1. The modified GoogLeNet with the inception modules pre-trained on ImageNet dataset was fine-tuned to detect surgical phase in laparoscopy videos.

BL4: Multi-class classification using deep stacked LSTM network using the features extracted by GoogLeNet fine-tuned on m2cai-workflow dataset. The features from the last inception module of length 1,024 from the network trained in **BL3** was used to train the LSTM network detailed in Sect. 4.1.

BL5: Multi-class classification using ResNet-50. The framework presented in Sect. 4.1 was trained on the m2cai16-workflow dataset.

Proposed method: Multi-class classification using deep stacked LSTM network presented in Sect. 4.1 using the features extracted by ResNet-50 fine-tuned on m2cai-workflow dataset. The baselines are summarized in Table 3.

Table 3 The training and testing times of the different baselines are tabulated

Baseline	Description	Train. time per epoch (min)	Test. time per frame (ms)
BL1	Modified (multi-class) AlexNet [11]	19.50	0.60
BL2	Modified (multi-class) AlexNet [11] (BL1) + LSTM	20.25	2.34
BL3	Modified (multi-class) GoogLeNet [12]	27.00	1.54
BL4	Modified (multi-class) GoogLeNet [12] (BL3) + LSTM	28.17	3.20
BL5	Modified (multi-class) ResNet-50 [13]	39.00	2.35
Proposed method	Modified (multi-class) ResNet-50 [13] (BL5) + LSTM	40.30	3.22

5.4 Performance Comparison

Experiments are carried out on the number of LSTM modules and the length of the input sequence to stacked LSTM network in baselines **BL2, BL4** and proposed framework. Performance evaluation for the baselines is done with 1, 2 and 3 stacked LSTM networks and sequence lengths of 3, 5, 10, and 50. Figure 10a shows the performance of **BL2** with the different configurations. Figure 10b shows that of **BL4**. Performance of proposed framework is shown in Fig. 10c. Comparison of performance of the best performing configuration of **BL2** and **BL4** and the proposed method with **BL1, BL3** and **BL5** is shown in Fig. 11. Figure 12 illustrates the prediction error per frame for one of the videos in the testing dataset.

5.5 Visualization of Surgical Workflow

The graphical visualization shows the progress of an ongoing surgery in terms of the modelled workflow for the procedure by augmenting the workflow model presented in Sect. 4.2 on the surgical video. Sample frames showing the transitions of the phases in a sample video from the dataset is shown in Fig. 13. This video corresponds to the model shown in Fig. 6a. The current phase of the surgery in highlighted in red in the workflow model. The phases that have executed already are in green and the upcoming phase is shown in white along with the probability.

Fig. 10 Performance of **BL2, BL4** and proposed method with different depth and sequence lengths. The best performing configuration is marked with a red box in each case

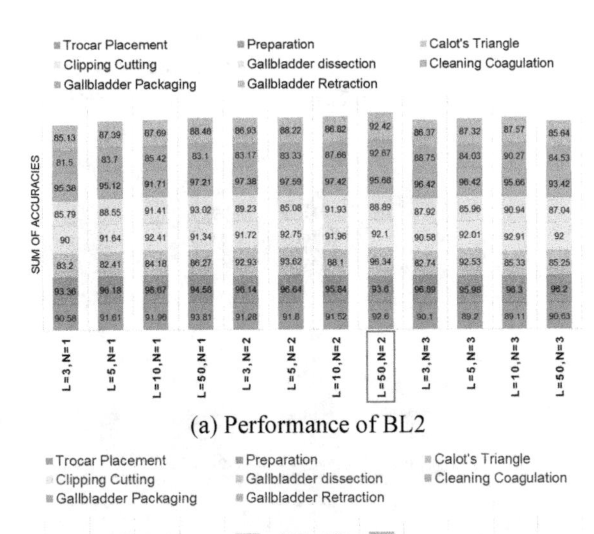

(a) Performance of BL2

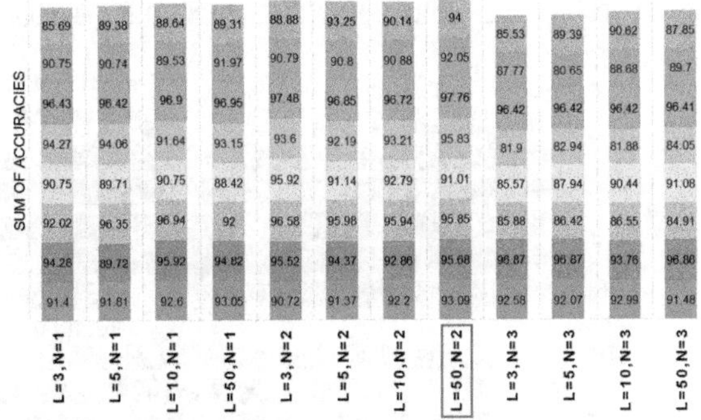

(b) Performance of BL4

(c) Performance of proposed method

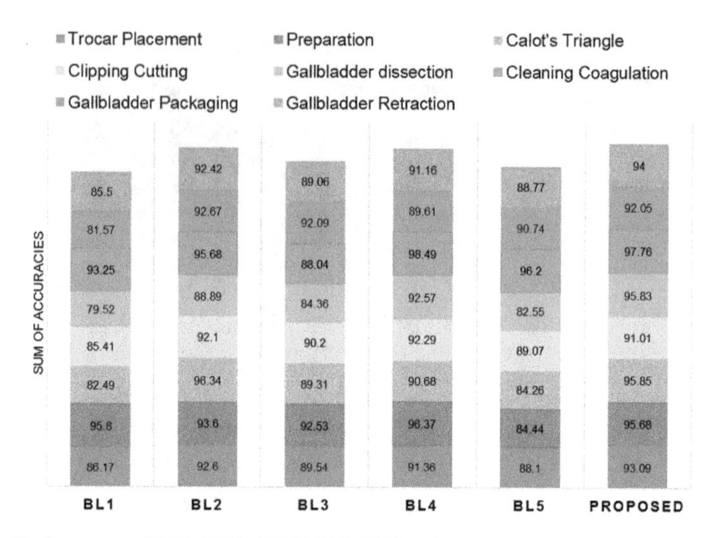

Fig. 11 Performance of **BL1, BL2, BL3, BL4, BL5** and proposed method

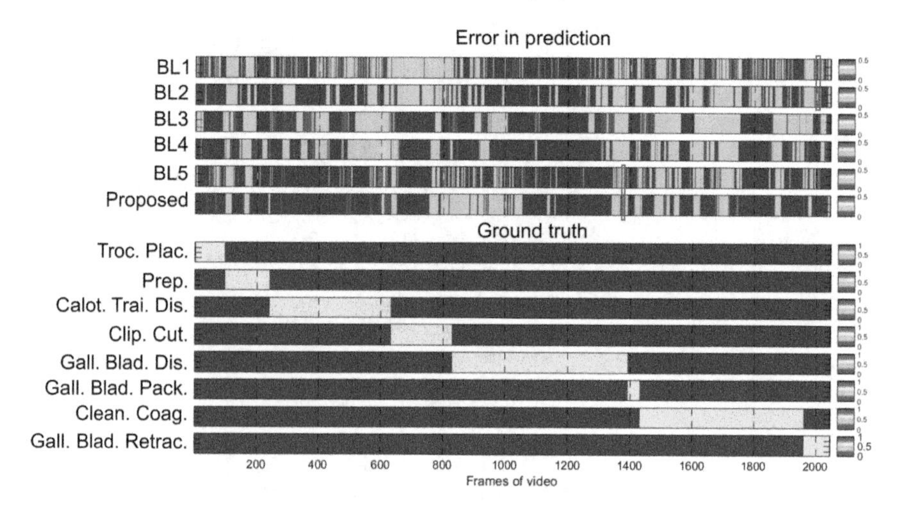

Fig. 12 Error in the prediction for each frame on one of the test videos no. 23. Along with the ground truth of the phase

6 Discussion

Figure 9a shows that the occurrence of surgical phases in a laparoscopy is heavily unbalanced. Among the eight phases P3 and P5 phases occur majority of the times whereas P2 and P6 has very less occurrence. Training any network with such imbalanced data biases the network to the phases which have high number of occurrence. So the class balancing is done as discussed in Sect. 5.2 and is graphically shown in Fig. 9b. As shown in Sect. 5.4 the network thus learns to identify the surgical phase

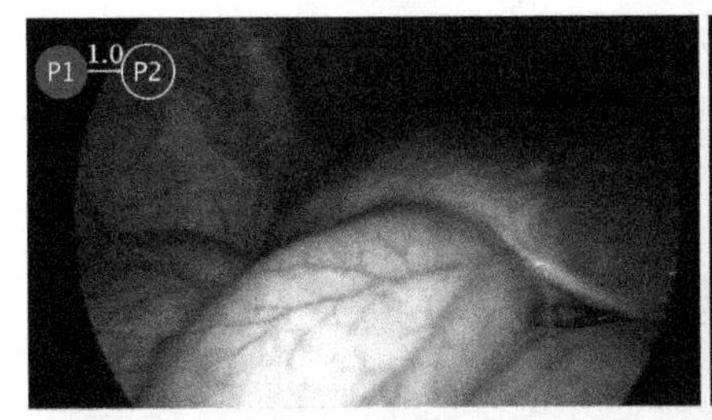

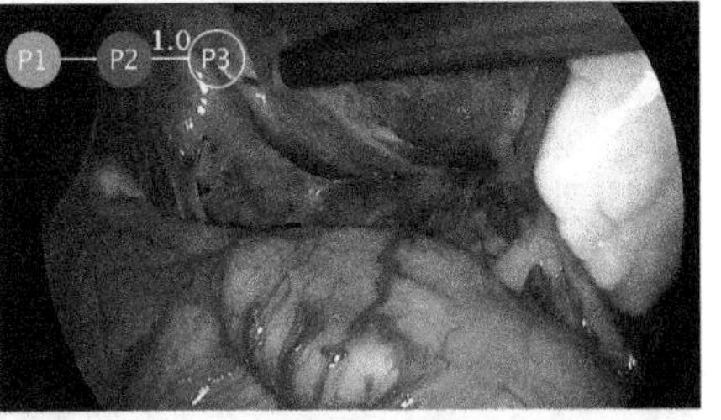

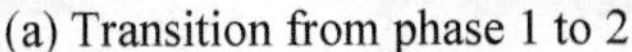

(a) Transition from phase 1 to 2 (b) Transition from phase 2 to 3

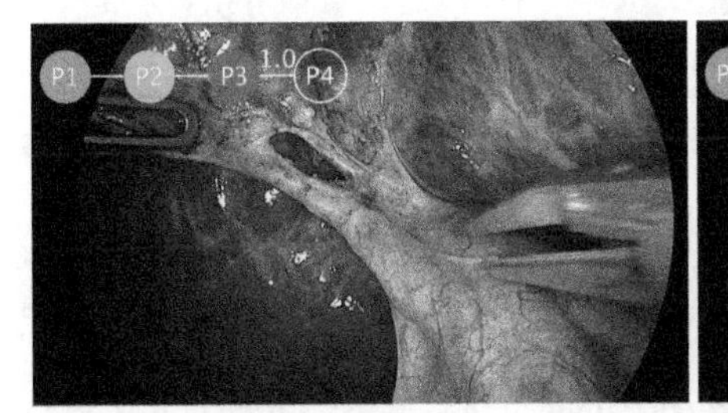

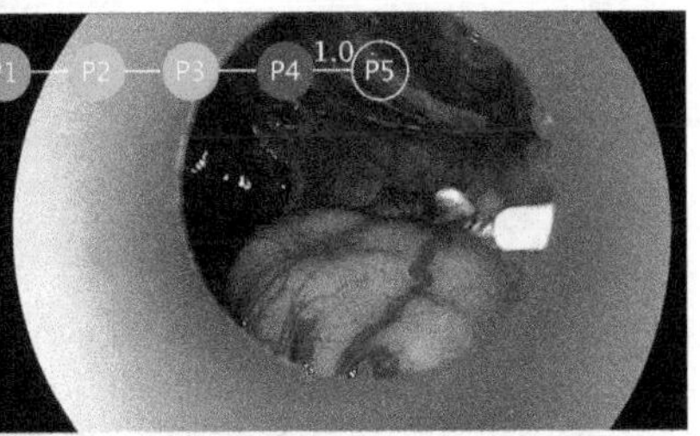

(c) Transition from phase 3 to 4 (d) Transition from phase 4 to 5

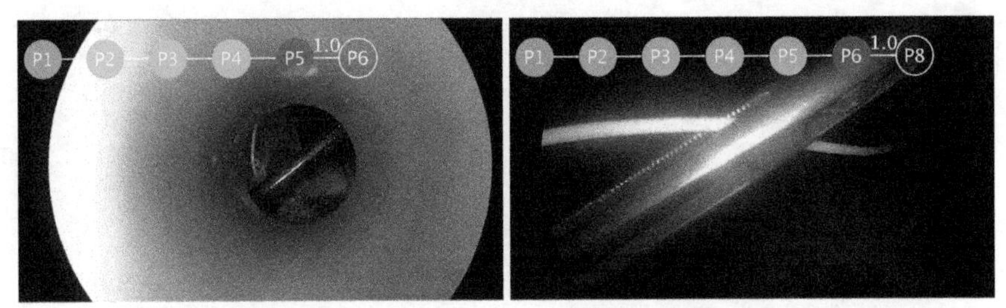

(e) Transition from phase 5 to 6 (f) Transition from phase 6 to 8

Fig. 13 Sample frames for the visualization of progress of the surgery in one of the videos of the dataset.eps

with better accuracy. The improvement in performance can also be attributed to the incorporation of temporal information in the prediction framework. In Fig. 12, the red boxes in BL1 and BL5 highlight the frames 1959 (Fig. 14c) and 1388 (Fig. 14h) respectively. In these frames, there is either a phase change or the video has some frames which could not be concluded as distinctly belonging to a particular phase. These transitions result in errors in detection. It has been observed that the error decreases considerably with the introduction of temporal learning. Figure 14 shows the sequence of frames succeeding and preceding region highlighted by the red box.

The proposed method consists two stages. Stage one comprised of training a CNN to detect the ongoing phase shown in the frames of the surgical video. In the next stage, a temporal model consisting of LSTM is trained using the features extracted by the CNN for identification of surgical phases. It is observed that the

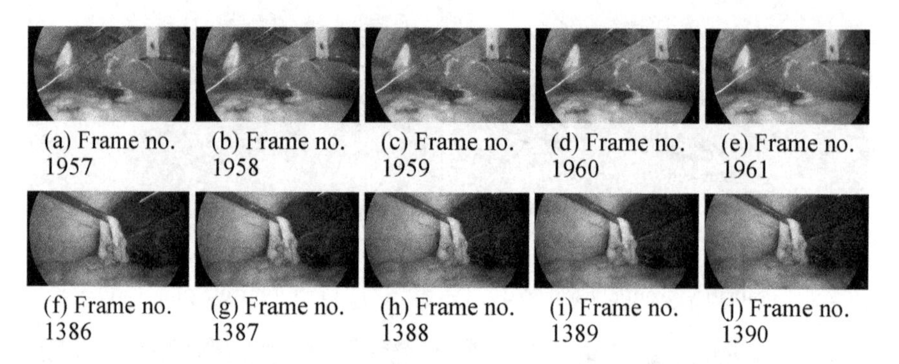

Fig. 14 Transitions in the video causing causes errors in phase detection as in observed in **a–e** corresponding to BL1 and BL2 in Fig. 11 and for **f–j** corresponding to BL5 and Proposed method in Fig. 11

accuracy of detection with the second stage (BL2, BL4 and proposed method) is better than with just stage one (BL1, BL3 and BL5). This could be because, in stage one, phase detection is done using individual frames of the video without considering the information relayed by the previous frames resulting in incorrect out of context detections. Whereas, the temporal learning in stage two considers sequence of frames for the detection of surgical phase in one frame of the video.

7 Conclusion

A framework that uses CNN as a feature extractor followed by temporal learning using a deep LSTM network is presented in Sect. 4.1. To train the framework to detect surgical phases in laparoscopy videos, a CNN trained on ILSVRC [15] is first fine-tuned on a publicly available laparoscopy video dataset. The performance of the proposed framework has been experimentally verified against a set of baselines. It is observed that the proposed framework comprising of using visual features of images extracted using deep residual networks followed by temporal learning with LSTM outperforms the different baselines in terms of detecting the surgical phase with high accuracy of {93.09, 95.68, 95.85, 91.01, 95.83, 97.76, 92.05 and 94.00%}for the surgical phases {Trocar placement, Preparation, Calot's triangle dissection, Clipping and curring, Gallbladder dissection, Gallbladder Packaging, Cleaning and coagulation, Gallbladder retraction} and average phase detection accuracy of 94.41%. The workflow modeling was done using a dataset consisting of surgical videos along with the per frame annotation of phase. The phase transitions in individual videos of the dataset was used to model the overall workflow model for the surgery. Visualization of the progress of the surgery was done by first recognizing the current surgical phase using a deep neural network framework trained to recognize the phase. The next probable next phase can then be determined using the workflow model for the surgery.

References

1. N. Padoy, T. Blum, H. Feussner, M.O. Berger, N. Navab, On-line recognition of surgical activity for monitoring in the operating room, in *Association for the Advancement of Artificial Intelligence*, pp. 1718–1724 (2008)
2. T. Blum, H. Feuner, N. Navab, Modeling and segmentation of surgical workflow from laparoscopic video, in *International Conference on Medical Image Computing and Computer-Assisted Intervention*, pp. 400–407 (2010)
3. F. Lalys, L. Riffaud, D. Bouget, P. Jannin, A framework for the recognition of high-level surgical tasks from video images for cataract surgeries. IEEE Trans. Biomed. Eng. **59**(4), 966–976 (2012)
4. C. Lea, G.D. Hager, R. Vidal, An improved model for segmentation and recognition of fine-grained activities with application to surgical training tasks, in *IEEE Winter Conference on Applications of Computer Vision (WACV)*, pp. 1123–1129 (2015)
5. U. Klank, N. Padoy, H. Feussner, N. Navab, Automatic feature generation in endoscopic images. Int. J. Comput. Assist. Radiol. Surg. **3**(3–4), 331–339 (2008)
6. A.P. Twinanda, S. Shehata, D. Mutter, J. Marescaux, M. de Mathelin, N. Padoy, Endonet: a deep architecture for recognition tasks on laparoscopic videos. IEEE Trans. Med. Imag. **36**(1), 86–97 (2017)
7. Y. LeCun, Y. Bengio, Convolutional networks for images, speech, and time series, *The Handbook of Brain Theory and Neural Networks*, vol. 3361, no. 10 (1995)
8. S. Hochreiter, J. Schmidhuber, Long short-term memory. Neural Comput. **9**(8), 1735–1780 (1997)
9. S.J. Pan, Q. Yang, A survey on transfer learning. IEEE Trans. Knowl. Data Eng. **22**(10), 1345–1359 (2010)
10. N. Tajbakhsh, J.Y. Shin, S.R. Gurudu, R.T. Hurst, C.B. Kendall, M.B. Gotway, J. Liang, Convolutional neural networks for medical image analysis: full training or fine tuning? IEEE Trans. Med. Imag. **35**(5), 1299–312 (2016)
11. A. Krizhevsky, I. Sutskever, G.E. Hinton, Imagenet classification with deep convolutional neural networks. Adv. Neural Inf. Process. Syst. 1097–1105 (2012)
12. C. Szegedy, W. Liu, Y. Jia, P. Sermanet, S. Reed, D. Anguelov, D. Erhan V. Vanhoucke, A. Rabinovich, Going deeper with convolutions, in *Proceedings of the IEEE Conference on Computer Vision and Pattern Recognition*, pp. 1–9 (2015)
13. K. He, X. Zhang, S. Ren, J. Sun, Deep residual learning for image recognition, in *Proceedings of the IEEE Conference on Computer Vision and Pattern Recognition*, pp. 770–778 (2016)
14. D.E. Rumelhart, G.E. Hinton, R.J. Williams, Learning representations by back-propagating errors. Cognit. Model. **5**(3), 1 (1988)
15. J. Deng, W. Dong, R. Socher, L.J. Li, K. Li, L. Fei-Fei, Imagenet: a large-scale hierarchical image database, in *Proceedings of the IEEE Conference on Conference Computer Vision and Pattern Recognition*, pp. 248–255 (2009)

Deep Learning Applications to Cytopathology: A Study on the Detection of Malaria and on the Classification of Leukaemia Cell-Lines

G. Gopakumar and Gorthi R. K. Sai Subrahmanyam

Abstract This chapter discusses a few applications of deep learning networks in cytopathology. Specifically, the detection of malaria from slide images of blood smear and classification of leukaemia cell-lines are addressed. The chapter starts with relevant theory for traditional (deep) multi-layer neural networks with back-propagation, followed by motivation, theory and training in Convolutional Neural Networks (CNN), the trending deep-learning based classifier. The detection of malaria from blood smear slide images using CNN is addressed followed by a discussion on the transfer learning capability of CNN by taking the classification of leukaemia cell-lines: K562, MOLT & HL60 as an example. The transfer learning capability of CNN is of particular interest especially when there are only very limited number of training samples to come up with a stand alone deep CNN classifier.

1 Introduction

Cytopathology is the cellular level study for the diagnosis of diseases. In cytopathology [1], often free cells are analysed unlike histopathology where the tissue as a whole get analysed. Each cell under investigation has a signature constituted by the morphology of the cells as well as their behavioural characteristics. The cytologists look for deviation from standard cell signature to report pathological state of the subject. The cells for analysis are either prepared as a smear on a glass slide or as

G. Gopakumar (✉)
Department of Computer Science & Engineering, Amrita Vishwa Vidyapeetham, Amritapuri, India
e-mail: gopakumarg@am.amrita.edu

G. R. K. Sai Subrahmanyam
Department of Electrical Engineering, IIT Tirupati, Tirupati, India
e-mail: rkg@iittp.ac.in

V. E. Balas et al. (eds.), *Handbook of Deep Learning Applications*, Smart Innovation, Systems and Technologies 136,
https://doi.org/10.1007/978-3-030-11479-4_11

a fluid suspension. Sometimes cyto-centrifugation is used to concentrate the cells under investigation without altering the morphology of the cells.

The microscopic examination even today remains as the gold standard for cell analysis for its low cost and well acceptance. However, the manual microscopic examination is a laborious task involving both slide preparation (fixation and staining) and analysis. It is a time consuming, repetitive and tedious job. Above all, the results may vary for the same sample among the clinicians depending on their level of expertise. In order to overcome these drawbacks, several efforts have been made in the recent past to automate the process of cytopathology. Research efforts in this direction have mostly been constrained to two approaches: instrumentation and data analysis.

In the instrumentation side, past few decades have seen developments in automated microscopy systems [2], flow cytometry [3] and imaging flow cytometry [4]. The automated microscopy systems such as the PathScope has considerably improved the throughput when compared to manual microscopy but is still well behind flow cytometry. Flow cytometry can analyse and identify cells of the order of a few thousands per second and it became an indispensable tool for clinicians. It uses a flow cell architecture where the cells are interrogated using lasers while they are in flow. A typical flow cytometry system measures the forward as well as side scatter profiles of the lasers. The forward scatter is a measure of the size of the cell and the side scatter is a measure of the complexity of the cell, and this knowledge is used to identify and count different cells under study. Though the acquisition speed of flow cytometry is very high, the per cell information it provides is usually low. The reason is that the flow cytometry will not capture specific morphological features other than the amount of scatter. On the other hand, traditional microscopic examination give detailed information with spatial localisation of components at sub-cellular level but has drawbacks both in terms of enumeration and speed. Imaging Flow Cytometry (IFC) [4, 5] is a relatively new technology which combines the speed of flow cytometry and the power of digital microscopy in providing the capability to analyse morphological features. However, the current commercially available imaging flow cytometry systems are bulky and are expensive [6] since they employ bulk fluid handling mechanisms for sample image acquisition and employ expensive and sophisticated image acquisition methods. Even the automated microscopy system uses automated slide preparation unit which employs extensive amount of robotic handling, rendering these systems bulky and expensive [2, 7, 8]. Thus these automated machines are not that affordable in resource limited settings. The new trend is to use microfluidic sample handling in combination with microscopy imaging modalities for high-throughput imaging of cells while they are in flow. These systems, which we call microfluidics microscopy (Mf-Ms), combine the power of flow cytometry (high throughput) and the power of digital microscopy (capability to provide spatial and quantitative morphology).

In the direction of developing automated system for point-of-care diagnosis, quite a large number of works have been done in developing algorithms necessary for processing the images and making the diagnostic decision. However, traditional classification systems are typically modelled to contain steps such as cell segmentation, feature extraction, followed by decision making using classifiers such as SVM [9].

It remained always a keen desire to build a decision making system that is as good as an expert clinician. It is an established fact that the power of human brain comes from massively interconnected large number of neurons capable of doing parallel computation [10]. A review on the use of artificial neural network in cytopathology can be found in [11, 12]. Fully connected normal artificial neural network is also not that feasible to learn such complex decision problems from big dataset. Such problems often requires large number of layers and neurons and hence large number of parameters need to be learned making the learning process slower. Also, the problem of vanishing of gradients [13] in lower layers of ANN during backpropagation based training makes the learning problem further difficult, if not impossible. Another difficulty with ANNs in supervised mode and SVMs is that the data for training need to have labels and the generalisation capability of the system is greatly dependant on the amount of labelled data used for training. Recently deep learning systems especially based on convolutional neural networks are emerging as reliable and default model for image analysis. Greenspan et al. [14] reviews recent deep learning related research in medical domain.

In this chapter, the application of convolutional neural network for the diagnosis of malaria [15] and in the classification of leukaemia cell lines are discussed. The datasets used in this study are generated, respectively, by cost-effective indigenously developed automated microscopy system [16] and microfluidic microscopy systems [17]. Section 2 provides the basic architecture of feed forward neural network and the backpropagation training algorithm which is instrumental in training convolutional neural network. Section 3 discusses the basic building blocks of convolutional neural networks and the mathematical details of associated training algorithm. Section 6 discusses the custom designed CNN operating on focus stack of blood smear slide images for malaria diagnosis followed by the transfer learning capability of CNN for the classification of leukaemia cell-lines in Sect. 7.

2 Multilayer Neural Network with Backpropagation

Multi layer feed forward neural networks and backpropagation [18] training algorithm was a milestone in developing a single classifier that automatically learn from examples just like our brain do. It has made the possibility that keeping the system architecture and algorithm the very same, we can learn classifiers addressing totally different classification tasks. Cybenko [19] and Hornick [20] have shown that any continuous function on compact subsets of R^n can be well approximated by a feed forward neural network (FFN) with a single hidden layer containing a finite number of neurons. Thus an FFN with single hissen layer can find any complex classification boundary on R^n. Though this universal approximation theorem is very exciting there is no guarantee for an algorithm which can automatically learn the very high number of parameters to address a complex classification task in reasonable amount of time using finite training set and with an architecture having only a single hidden layer. This constraint has motivated the engineers to design multi-layer neural network.

Also, human neural system (say, visual system [21]) often works at multiple levels in making inferences.

The backpropagation, is a widely used method in training artificial FFN. Training ANN has two steps: error backward propagation and weight update. An input vector when presented to the network, it is propagated forward through the network, layer by layer, until it reaches the output layer. Then the desired output corresponding to this input vector is compared to the generated output and the error value (using an appropriate loss function) is calculated for each neuron in the output layer. This error is then propagated backwards (hence the name backpropagation), starting from the output, until each neuron has an associated error value which is a measure of its contribution of the error at the output layer. Once the error at each neuron is determined, the weights are updated using an optimization method like gradient descent so as to minimize the loss function. The weight update at deeper layers is usually related to the final error by the use of chain rule and performed layer by layer. For this, the error backpropagation can be conveniently done in matrix notion in almost similar manner from layer to layer by observing the decoupling between the error back propagated and the corresponding weight update by gradient decent. As discusses here, this is feasible because the weight update term in gradient descent can be easily computed based on the back propagated error at the rear end of the connection (carrying the weight) and based on the input to the connection. This process, that we discuss here, gives a generalized procedure to interpret and derive the weight update in deep networks. Further, this understanding gives a nice base for the easy following of back propagation in deep convolutional networks.

Figure 1 shows an arbitrary 4 layer feed forward neural network. There are D neurons in the input layer, M neurons in 2nd layer, N neurons in 3rd layer and Z neurons in the output layer. There is a weighted connection between each neuron in a layer to every other neurons in the layer immediately following it. Also there is a bias to every neurons in all layers except at the input layer. During training these parameters (weights and biases) need to be updated so as to minimise the loss

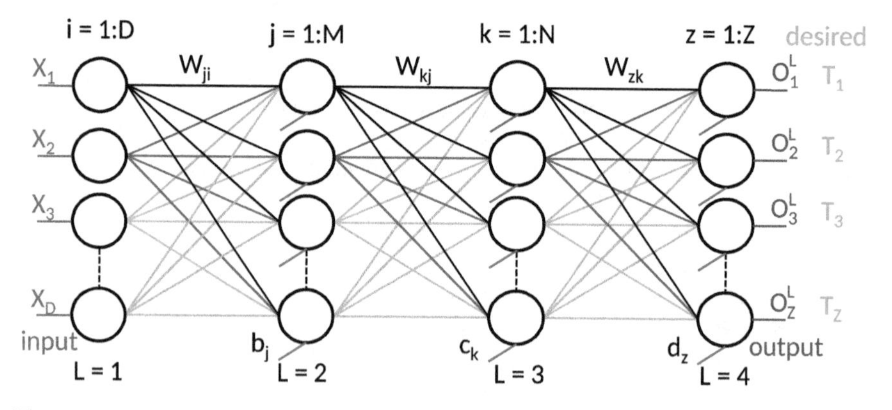

Fig. 1 Typical feed forward 4 layer neural network

function. A typical loss function could be the half of the sum of squared difference defined by

$$E = \sum_{z} \frac{1}{2}(T_z - O_z)^2 \tag{1}$$

Here T_z and O_z are the desired output and actual output produced at node z of the output layer for any specific input $X = \{X_i\}_{i=1}^{D}$. Note that the notion of input to the normal feed forward neural network is a $D \times 1$ column vector. If the input is an image, it has to be vectorised to form the input vector.

All nodes, except at the input layer, compute the weighted sum of inputs from the previous layer to produce an intermediate output *Net*. The activation function is applied to each of this *Net* to produce the final output at each neuron in an arbitrary layer *L*. We will consider the popular Sigmoid activation function defined by

$$O_z^L = f(Net_z^L) = \frac{1}{1 + \exp^{-Net_z^L}} \tag{2}$$

where Net_z^L is the intermediate output produced by node z in layer L and is defined by

$$Net_z^L = \sum_k W_{zk} O_k^{L-1} + d_z \tag{3}$$

Here O_k^{L-1} is the output at node k in $L-1$th layer and W_{zk} is the weights of the connections from node k in layer $L-1$ to node z in layer L.

The gradient descent is used to update the parameters. Each parameter is updated in the negative direction of the gradient of the loss function (Eq. 1) computed with respect to the parameter to be updated. To generalize this procedure, we first show that this weight update can be decoupled in two steps: back-propagating the error computed to lower layers and updating the weights based on the back-propagated error and input to the specific connection for which the weight update is done. Section 2.1 discusses the error back propagation procedure while Sect. 2.2 discusses the parameter updating procedure. In all these discussion, *L*, is an arbitrary layer, $L-1$ is one layer lower in the hierarchy, then $L-2$ and so on. The main theme of the presented formulation is that either error back propagation or the weight update based on it can be extended to any number of layers (on similar steps).

The simple update rule based on the gradient descent to update the weight connection W_{zk}^L between kth neuron in $L-1$th layer to zth neuron in Lth layer is

$$W_{zk}^{new} = W_{zk}^{old} - \eta \frac{\partial E^{L+1}}{W_{zk}^L} \tag{4}$$

Here η is the learning parameter and E^{L+1} becomes the error at the final layer (E) when updating the weights between the final and pre-final layers. The $\frac{\partial E^{L+1}}{W_{zk}^L}$ can be computed using chain rule

$$\frac{\partial E^{L+1}}{\partial W_{zk}^{L}} = \left[\frac{\partial E^{L+1}}{\partial O_{z}^{L}} \frac{\partial O_{z}^{L}}{\partial Net_{z}^{L}} \right] \frac{\partial Net_{z}^{L}}{\partial W_{zk}^{L}} = \Delta_{z}^{L} * O_{k}^{L-1} \tag{5}$$

Here in Eq. 5, $\frac{\partial Net_z^L}{\partial W_{zk}^L}$ turns out to be the input O_k^{L-1} to connection for which the weight update is seeking for and is due to the definition of Net_z^L in Eq. 3. The term in bracket Δ^L is the error back propagated at the rear end of the connection. The term $\frac{\partial O_z^L}{\partial Net_z^L}$ can be denoted as $f'\left(Net_z^L\right)$ and by definition (Eq. 2), this turns out to be $O_z^L * \left(1 - O_z^L\right)$ for Sigmoid. In short, the weight update for the branch connecting kth neuron in $L-1$th layer to zth neuron in Lth layer turns out to be $\Delta^L * O_k^{L-1}$, which is the product of the error back propagated at the rear end of the connection (Δ^L) and input to the connection from the lower layer neuron (O_k^{L-1}). This point will be clearly brought out mathematically in Sects. 2.1 and 2.2.

2.1 Backpropagating the Error Across the Layers

In this section, the error at each node in the output layer is computed and then back propagated to the layers lower in the hierarchy to determine the contribution of this error by each node in the network.

2.1.1 Finding Error at Output Layer L

Each node in the output layer contribute to the total error E. The contribution of Error by a node z at Layer L is due to its output O_z^L. Let ΔE_z^L denote this quantity and can be measured by computing $\frac{\partial E}{\partial O_z^L}$. By the definition of the error function (Eq. 1), this turns out to be

$$\Delta E_z^L = \frac{\partial E}{\partial O_z^L} = O_z^L - T_z \tag{6}$$

2.1.2 Propagating the Error from Layer L to Layer $L-1$

First, an intuitive idea behind the backpropagation of error is provided and is followed by deriving explicit expressions for it. In order to find the contribution of the final error E at layer L due to the output of a node k in layer $L-1$, the error found out at each node in L has to be backpropagated. Once the error ΔE_z^L at all nodes z in an arbitrary layer L is computed, it has to cross these neurons during backpropagation of the error to compute the error at all nodes in the lower layer $L-1$. This is done by weighing ΔE_z^L by the corresponding derivative of the sigmoid function $f'(Net_z^L)$, and

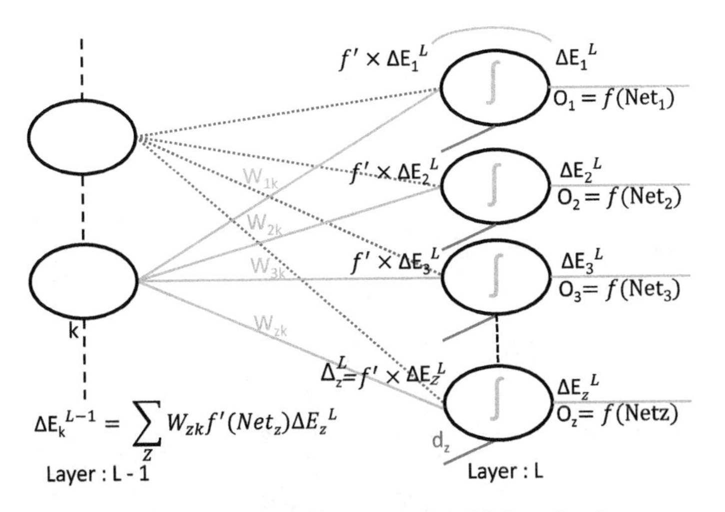

Fig. 2 Error backpropagation from nodes in layer L to node k in layer $L - 1$

then accumulating the shares (Δ^L) through the weighted connections at each node k in layer $L - 1$. The procedure can be explained better by referring to Fig. 2. It shows the relationship to each of ΔE_z^L from the node k in layer $L - 1$. It can be seen that node k in layer $L - 1$ influences the output at all nodes in L through the respective weight connections. Thus the error computed at each neuron in layer L contributes through the respective weight connection to each node k in layer $L - 1$ and it has to cross each neuron z in layer L. Thus the Error at a node k in layer $L - 1$ can be computed as

$$\Delta_k^{L-1} = f'\left(Net_k^{L-1}\right) \Delta E_k^{L-1} \tag{7}$$

where $f'\left(Net_k^{L-1}\right)$ is $O_k^{L-1}\left(1 - O_k^{L-1}\right)$ for Sigmoid and ΔE_k^{L-1} is defined by the recursive relation

$$\Delta E_k^{L-1} = \frac{\partial E_z^L}{\partial O_k^{L-1}} = \sum_z \frac{\partial E_z^L}{\partial O_z^L} f'(Net_z^L) W_{zk} \tag{8}$$

In matrix form, this can be written as

$$\Delta E^{L-1} = \mathbf{W}^{\mathbf{T}} * \left[\Delta E^L . * f'(Net^L)\right] \tag{9}$$

where '.*' represents the element by element multiplication, '*' represents the matrix multiplication, and **T** represents the matrix transpose. Equation 9 remains same for back propagation of error through any number of layers. The weight matrix **W** changes from layer to layer. The f' of the corresponding activation function (with associated outputs) and error at current layer are used to backpropagate the error to its previous layer.

The expression in Eq. 8 can also be explicitly computed using the normal chain rule for gradient computation and is shown below.

$$\frac{\partial E_z^L}{\partial O_k^{L-1}} = \sum_z \frac{\partial E_z^{L+1}}{\partial O_z^L} \frac{\partial O_z^L}{\partial Net_z^L} \frac{\partial Net_z^L}{\partial O_k^{L-1}} \tag{10}$$

where $\frac{\partial Net_z^L}{\partial O_k^{L-1}}$ can be computed as

$$\frac{\partial Net_z^L}{\partial O_k^{L-1}} = \frac{\partial}{\partial O_k^{L-1}} \sum_k W_{zk} O_k^{L-1} = W_{zk} \tag{11}$$

When substituted Eq. 11 in Eq. 10, it turns out that

$$\frac{\partial E_z^L}{\partial O_k^{L-1}} = \sum_z \frac{\partial E_z^{L+1}}{\partial O_z^L} \frac{\partial O_z^L}{\partial Net_z^L} W_{zk} \tag{12}$$

Being $O_z^L = f(Net_z^L), f'(Net_z^L)$ is another way of writing $\frac{\partial O_z^L}{\partial Net_z^L}$ for notational simplicity. Thus Eq. 12 is exactly the same equation provided in Eq. 8. The consequence of Eq. 9 is that the error backpropagation can be done independent of gradient update, and can be propagated back in each layers lower in the hierarchy one by one by simple matrix multiplication.

2.2 Updating the Parameters

As noted earlier, the optimisation method used to minimise the loss function defined in Eq. 1 could be gradient descent. The parameters are updated in the negative direction of the gradient of the error function computed with respect to the parameter to be updated.

2.2.1 Updating the Weights

The simple update rule based on the gradient descent is

$$W_{zk}^{new} = W_{zk}^{old} - \eta \frac{\partial E}{W_{zk}} \tag{13}$$

Here η is the learning parameter and $\frac{\partial E}{W_{zk}}$ is the gradient of the loss function E with respect to the parameter W_{zk}. The $\frac{\partial E}{W_{zk}}$ can be computed using chain rule

$$\frac{\partial E}{\partial W_{zk}} = \left[\frac{\partial E}{\partial O_z}\frac{\partial O_z}{\partial Net_z}\right]\frac{\partial Net_z}{\partial W_{zk}} \tag{14}$$

Here, the term in bracket is already computed by error backpropagation for all neurons in all layers as discussed in the last section. The second term turns out to be

$$\frac{\partial Net_z}{\partial W_{zk}} = \frac{\partial}{\partial W_{zk}}\sum_k W_{zk}O_k^{L-1} = O_k^{L-1} \tag{15}$$

W_{zk} is the weight connecting node k in layer $L-1$ to node z in layer L, $\frac{\partial Net_z}{\partial W_{zk}}$ turns out to be the output at the node k at layer $L-1$. Thus $\frac{\partial O_z^L}{\partial W_{zk}}$ turns out to be the output of the neuron k at layer $L-1$ weighted by the derivative of the sigmoid with respect to Net_z^L at node z in layer L.

2.2.2 Updating the Weights for the Network Shown in Fig. 1

Let $\mathbf{W}_{H_1}$, $\mathbf{W}_{H_2}$ and $\mathbf{W}_{H_3}$ be the weight matrices of the network holding respectively the weights between layer 1 & 2, 2 & 3 and 3 & 4.

$$W_{H_1} = \begin{bmatrix} W_{11} & W_{12} & W_{13} & . & .. & W_{1D} \\ . & . & . & . & .. & . \\ . & . & . & W_{ji} & .. & . \\ . & . & . & . & .. & . \\ W_{M1} & W_{M2} & W_{M3} & . & .. & W_{MD} \end{bmatrix}$$

$$W_{H_2} = \begin{bmatrix} W_{11} & W_{12} & W_{13} & . & .. & W_{1M} \\ . & . & . & . & .. & . \\ . & . & . & W_{kj} & .. & . \\ . & . & . & . & .. & . \\ W_{N1} & W_{N2} & W_{N3} & . & .. & W_{NM} \end{bmatrix}$$

$$W_{H_3} = \begin{bmatrix} W_{11} & W_{12} & W_{13} & . & .. & W_{1N} \\ . & . & . & . & .. & . \\ . & . & . & W_{zk} & .. & . \\ . & . & . & . & .. & . \\ W_{Z1} & W_{Z2} & W_{Z3} & . & .. & W_{ZN} \end{bmatrix}$$

Let $O^4 = \{O_z\}_{z=1}^{Z}$, $O^3 = \{O_k\}_{k=1}^{N}$ and $O^2 = \{O_j\}_{j=1}^{M}$ be the output produced at each node in layers 4th, 3rd and 2nd respectively when an input $X = \{X_i\}_{i=1}^{D}$ is fed to the network. The output at layer 1 is same as the input and hence $O^1 = \{O_i\}_{i=1}^{D} = \{X_i\}_{i=1}^{D}$. Also let $T = \{T_z\}_{z=1}^{Z}$ is the target or desired output at each node in the output layer. Now the error vector at layer L, Δ^L can be computed by error backpropagation.

$$\Delta^4 = O^4_{[z\times 1]} .*\left[1 - O^4\right]_{[z\times 1]} .*\left[O^4 - T\right]_{[z\times 1]} \quad (16)$$

$$\Delta^3 = \left[\mathbf{W}^{\mathbf{T}}_{H_3}\right]_{N\times Z} * \left[\Delta^4\right]_{Z\times 1} .* O^3_{N\times 1} .* (1 - O^3)_{N\times 1} \quad (17)$$

$$\Delta^2 = \left[\mathbf{W}^{\mathbf{T}}_{H_2}\right]_{M\times N} * \left[\Delta^3\right]_{N\times 1} .* O^2_{M\times 1} .* (1 - O^2)_{M\times 1} \quad (18)$$

Note that Eqs. 17 and 18 follow from the Eqs. 7 and 9. Now the gradient at each neuron with respect to the parameter to be updated can be found and the final update rule in matrix form will be

$$\mathbf{W}^{new}_{H_3} = \mathbf{W}^{old}_{H_3} - \eta\left[\Delta^4\right]_{Z\times 1} * \left[O^3\right]^{\mathbf{T}}_{1\times N} \quad (19)$$

$$\mathbf{W}^{new}_{H_2} = \mathbf{W}^{old}_{H_2} - \eta\left[\Delta^3\right]_{N\times 1} * \left[O^2\right]^{\mathbf{T}}_{1\times M} \quad (20)$$

$$\mathbf{W}^{new}_{H_1} = \mathbf{W}^{old}_{H_1} - \eta\left[\Delta^2\right]_{M\times 1} * \left[O^1\right]^{\mathbf{T}}_{1\times D} \quad (21)$$

Thus to update the weight (W_{kj}) between neuron j in layer $L-2$ and neuron k in layer $L-1$, the gradient of the error function at the final layer with respect to W_{kj} need to be computed. It can be seen from the above set of equations that this gradient can be interpreted as the product of two terms. The first term is the error computed at node k in layer $L-1$ multiplied by the derivative of the sigmoid activation function with respect to the output at node k. This can be treated as the error at the rear end of the connection holding the weight W_{kj}. The second term is the output at the node j in layer $L-2$, and can be considered as the input to the connection holding the weight W_{kj}. Thus, the change for updating the weight for a connection between any layers can be simply obtained by computing the product of error at rear end of the connection and input at front end of the connection. This is then weighted by the learning parameter η and the weight of the connection is updated by gradient descent.

3 Convolution Neural Network (CNN)

In traditional multi-layer feed forward neural network, image based classification introduces a few serious difficulties. Often feature extraction has to be pre-done and will be the input to neural network. These features are application dependent and hand-engineering them is an important but difficult job. Another option is vectorising the input image and feeding it to the network. However, the input images are typically bigger in size which leads to a large weight matrix and hence huge number of parameters to be learned. Convolution Neural Network (CNN) was introduced to address these constraints by sharing weights [22] across the neurons of the layer and was popularised by Lecun with his five layer CNN (LeNet-5 [23]) for digit recognition.

The CNN is a biologically inspired neural network designed to mimic our visual cortex system by its connectivity pattern. Basically there is a series of feature extraction layers one working on extracted features at its lower layer for producing certain inference (such as discriminating inter-class samples) at final layer. Sub-sampling

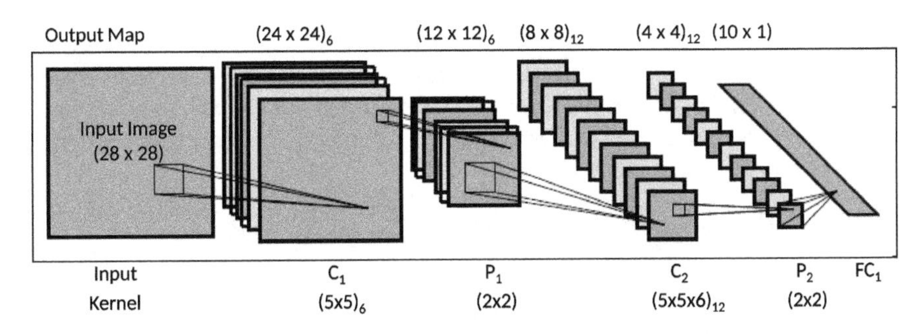

Fig. 3 LeNet architecture for digit recognition

of the feature map is done between subsequent feature extraction layers, in order to mitigate small difference due to shift, scale and noise which are often present in samples coming from same class (intra-class differences). Lecun has designed LeNet-5 [23] such that it has five layers as shown in Fig. 3. Feature extraction at first layer, followed by subsampling, again a feature extraction layer followed by subsampling layer, and a fully connected layer as the last layer. The same backpropagation algorithm discussed in last section forms the basis to train the convolutional neural network and it has produced great accuracy for the digit classification problem and for several other classification tasks.

The architecture was well received by the machine learning community and was applied to different problems by introducing more feature extraction and subsampling layers. However, extending the architecture by introducing more layers has a very serious drawback that the learning will not happen during backpropagation especially at the lower layers. The bottleneck is the activation function used in LeNet-5 which was the same sigmoid activation function provided in Eq. 2. Note that during backpropagation, while computing the error contribution E_j at any layer, ΔE_j computed need to be weighted by $f^{'}(Net_j)$. As the maximum value of $f(Net_j)$ is 1, the maximum possible value for $f^{'}(Net_j) = O_j(1 - O_j)$ turns out to be 0.25. Due to the chain rule of backpropagation, at a lower layer L, this means that E_L will be a very small quantity (since weighted by many $f^{'}(Net_j)$ at top layers), and weight update will not happen. In order to mitigate this problem due to vanishing of gradient, another activation function was introduced and is Rectified Linear Unit (ReLU). Thus the modern CNN architectures has three basic building blocks : feature extraction layer, subsampling layer and ReLU activation.

3.1 *Feature Extraction Layer*

In CNN, the feature extraction layers are designed as convolution layers, since the convolution operation on an input image with appropriate kernels always leads to different output feature-maps: just like Sobel kernel detecting edges and Laplacian kernel detecting blobs in an input image. Given a D dimensional image, the convolution block compute Y which can be thought of K instances of different feature maps. The operation performed a form of convolution and is done between the D dimensional image and K set of learned kernels f. Equation 22 depicts this operation where f^T represents the flipped kernel.

$$X \in \mathbb{R}^{H \times W \times D}, f \in \mathbb{R}^{H' \times W' \times D \times K}, Y \in \mathbb{R}^{H'' \times W'' \times K} \tag{22}$$

$$y_{i''j''k} = b_k + \sum_{i'=1}^{H'} \sum_{j'=1}^{W'} \sum_{d'=1}^{D} f^T_{i'j'd'k} \times x_{i''+i'-1, j''+j'-1, d'}$$

where $b_k \in \mathbb{R}$ are biases for the nodes in the layer.

Note that the weights learned in convolutional block of the CNN forms the kernel and biases. Intuitively, the learned kernel at convolution layer helps to extract very local features such as (though not exactly) low pass and high pass filters, edge features and corners. Thus during training a CNN, we are actually learning these feature detectors (not the features) like edge, line, corner detectors and contrast quantifiers that can be used to extract relevant features when operated on a test image.

Lecun has designed a five layer networ (LeNet-5) [23] with two convolution layers, two sub sampling (pooling) layers and a fully connected layer and is shown in Fig. 3. Note that the architecture has designed such that the input image is of dimension 28×28 and the output layer contain 10 nodes as it is addressing digit classification problem. There are 6 kernels of dimension 5×5 in first convolution layer and 12 kernels of dimension $5 \times 5 \times 6$ in second convolution layers. Thus, the number of parameters to be learned in the first convolutional layer is only 156: the $5 \times 5 = 25$ weights each for 6 kernels and their 6 biases. An equivalent layer in fully connected traditional feed forward neural network, will have 764 (28×28) input neurons and 3456 ($24 \times 24 \times 6$) output neurons which leads to 2,640,384 weights and 3456 biases. Compare the 156 parameters that we are trying to learn in the first convolutional layer with the 2,643,840 parameters in such a fully connected layer. However, note that there are almost same number of connections exists in convolutional layer as much as its equivalent fully connected counter part. The reduction in number of parameters is achieved by sharing the same parameters for multiple connection as defined by the convolution operation.

3.2 The Activation Function ReLU:

Given an input y_{ijk}, the Rectified Linear Unit (ReLU) suppresses an input, if it is negative else it retains the same value. Thus it is a non-linear activation function and usually follows the convolution operation. Without ReLU, the non-linear activation function, the whole network would have reduced to a simple linear transformation. Also, unlike other non-linear activation functions such as sigmoid, ReLU offers better resistance to slow learning (due to vanishing of gradients [13]) especially at lower layers of deep networks as its derivative is 0 or 1 depending on input being −ve or positive !.

$$y^1_{ijk} = \max(0, y_{ijk}) \tag{23}$$

3.3 *Subsampling:*

The sub-sampling typically follows the feature extraction layer and it introduces small shift invariance as well as scale invariance to the features, as the subsequent feature extraction is going to operate on the scale down version. In LeNet-5 architecture, the subsampling is defined by 2 × 2 average pooling. Another popular sub-sampling operation is by max-pooling as defined by

$$y^2_{ijk} = \underset{1 \leqslant i' \leqslant H', 1 \leqslant j' \leqslant W'}{max} \left(y^1_{i+i'-1, j+j'-1, k} \right) \tag{24}$$

4 Training CNN Using Backpropagation

The same backpropagation algorithm that discussed in Sect. 2 forms the basis o train convolutional neural networks as well. In this section, the training is discussed in detail by taking the architecture shown in Fig. 3 as an example. As noted earlier, it has two convolution layers (C_1, C_2), two average pooling layers (P_1, P_2) and one fully connected layer (FC_1). There are 6 kernels in C_1 each of size 5 × 5 and 12 kernels in C_2 each of size 5 × 5 × 6. The average pooling does a 2 × 2 pooling (thus the kernel weights are fixed as 0.25). The architecture was originally designed for digit recognition for input images of dimension 28 × 28. The dimensions of corresponding output maps generated at each layer is shown above the individual blocks. Note that the output dimension is 10 × 1 since each input has to be mapped to one of the 10 digits. Note that the parameters to be learned in this CNN are (1) the weights of the kernels and their biases at the convolutional layers ($5 \times 5 \times 6 + 6 = 156$ in C_1 and $5 \times 5 \times 6 \times 12 + 12 = 1812$ in C_2) and (2) the weights and biases at the fully connected layer ($4 \times 4 \times 12 \times 10 + 10 = 1930$).

4.1 *Parameter Initialization*

The parameters for each layer are initialised based on the number of input and output connections at that layer. Specifically these are initialised with random numbers selected from uniform distribution between $U[-\zeta, \zeta]$, where the bound ζ is deter-

Table 1 Parameters for weight initialisation for CNN in Fig. 3

	C_1	C_2	FC_1
Fan_{in}	$1 \times (5 \times 5) = 25$	$6 \times (5 \times 5) = 150$	$12 \times (4 \times 4) = 192$
Fan_{out}	$6 \times (5 \times 5) = 150$	$12 \times (5 \times 5) = 300$	10
ζ	0.0756	0.0471	0.0704

mined by the fan_{in} and fan_{out} of the layer.

$$\zeta = \sqrt{\frac{1}{fan_{in} + fan_{out}}} \tag{25}$$

Here, for convolutional layers fan_{in} is defined as the product of number of input maps and kernel size and fan_{out} is defined as the number of output maps and kernel size. For fully connected layer, fan_{in} is the product of number of input maps and size of the input map while fan_{out} is the number output nodes. Table 1 shows the parameters for weight initialisation for the CNN architecture shown in Fig. 3.

4.2 Forward Propagation

The respective operations are performed on the input maps at each layer. At convolutional layers, each kernel is used to convolve with the input map, then kernel bias is used to offset the result, and then the sigmoid activation function is applied to produce an output map. Note that only valid part of result of convolution is used to generate the output map. If the output map has to maintain the same dimension as input, the input need to be appropriately padded before feeding to the layers.

$$O\,[i] = \frac{1}{1 + \exp^{-\left(\sum_{d=1}^{D}[I(:,:,d)*K_i(:,:,d)+B(i)]\right)}} \tag{26}$$

In Eq. 26, $*$ represents the convolution operation (Eq. 22) and $B(i)$ represents the bias of ith kernel. $O[i]$ is the output map generated for ith kernel when applied on the dth input map. For convolutional layer 1, $D = 1$ and i varies from 1 to 6. For convolutional layer 2, $D = 6$ and i varies from 1 to 12. At the pooling layers, each 2×2 block is averaged to form a single pixel thereby reducing the output map size by 2 along each dimension. At the fully connected layer, the output is computed just like normal feed forward neural network as discussed in Sect. 2 (Fig. 4).

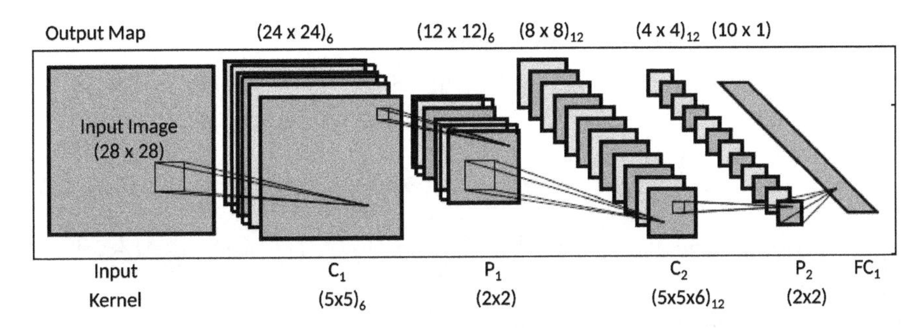

Fig. 4 LeNet architecture for digit recognition

4.3 *Backward Error Propagation*

The same loss function defined in Sect. 2 is used to compute the error. For any input image, we compute the result at the output layer which will be the 10×1 vector, and now we can compute $\frac{\partial E}{\partial O_k}$. Now this error has to be backpropagated across different layers.

Error backpropagation across fully connected layer FC_1 **:**

The derivative of the loss function with respect to the output at any node k in the final layer $\frac{\partial E}{\partial O_k}$ is $O_k - T_k$. Thus the error in fully connected layer FC_1 is a 10×1 vector holding $(O_k - T_k)_{k=1}^{10}$.

$$\Delta E_{FC_1} = O - T \tag{27}$$

The error at nodes in the FC_1 has to be backpropagated to the pooling layer 2 (P_2) and this turns out to be ΔE_{S_2}

$$\Delta E_{P_2} = W. * O. * (1 - O). * \Delta E_{FC_1} \tag{28}$$

Here in Eq. 28, '.*' represents the element by element multiplication. Here W is the weight matrix connecting 192×1 output map at P_2 to the 10×1 output vector. The 192×1 output map at P_2 is actually the vectorised representation of 12 maps each of size 4×4. Thus ΔE_{S_2} is of dimension ($[192, 10] \times [10, 1] \rightarrow [192, 1]$). This is reshaped into $(4 \times 4)_{12}$ for representational convenience in propagating the error further down the layers.

Error backpropagation across pooling layer P_2 :

Since the pooling operation does an average pooling in 2×2 neighbourhood, the error backpropagation across this layer is just an inverse operation of down-sampling i.e., up-sampling by 2.

$$\Delta E_{C_2} = US_2(\Delta E_{P_2}) \tag{29}$$

where US_2 represents the operation, upsampling by 2. This leads to ΔE_{C_2} of size $(8 \times 8)_{12}$.

Error backpropagation across convolutional layer C_2:

The error backpropagation across the convolution layer is also governed by the same error propagation equation used in FFN. But by the special construction of the convolution operation, the weights are related in a special way while propagating the error. It turns out to be the correlation of the kernel with the error to be backpropagated after weighted by the derivative of the activation function. First, the correspondence between the correlation operation and the weighted error propagation is shown by taking an example followed by explicit equation for error backpropgation across C_2. Consider the following convolution operation (only valid part of convolution is considered) in Eq. 30 and the correlation operation in Eq. 31 (Note that there is adequate padding by zero to reconstruct the dimension of the input).

In forward propagation of convolution,

$$\begin{bmatrix} x_1 & x_2 & x_3 & x_4 \\ x_5 & x_6 & x_7 & x_8 \\ x_9 & x_{10} & x_{11} & x_{12} \\ x_{13} & x_{14} & x_{15} & x_{16} \end{bmatrix} * \begin{bmatrix} w_1 & w_2 & w_3 \\ w_4 & w_5 & w_6 \\ w_7 & w_8 & w_9 \end{bmatrix} = \begin{bmatrix} y_1 & y_2 \\ y_3 & y_4 \end{bmatrix} \tag{30}$$

Here

$$\begin{aligned} y_1 &= w_9 x_1 + w_8 x_2 + w_7 x_3 + w_6 x_5 + w_5 x_6 + w_4 x_7 + w_3 x_9 + w_2 x_{10} + w_1 x_{11} \\ y_2 &= w_9 x_2 + w_8 x_3 + w_7 x_4 + w_6 x_6 + w_5 x_7 + w_4 x_8 + w_3 x_{10} + w_2 x_{11} + w_1 x_{12} \\ y_3 &= w_9 x_5 + w_8 x_6 + w_7 x_7 + w_6 x_9 + w_5 x_{10} + w_4 x_{11} + w_3 x_{13} + w_2 x_{14} + w_1 x_{15} \\ y_4 &= w_9 x_6 + w_8 x_7 + w_7 x_8 + w_6 x_{10} + w_5 x_{11} + w_4 x_{12} + w_3 x_{14} + w_2 x_{15} + w_1 x_{16} \end{aligned}$$

From the above set of equations/relations, note that x_1 only contribute to y_1 through the weight w_9, while x_2 contribute to y_2 through the weight w_9 and to y_1 through w_8. i.e., x_2 is connected to y_2 through w_9 and to y_1 through w_8. Thus error in x_2 is decided by the sum of back propagated errors in y_2 weighted by w_9 and error in y_1 weighted by w_8. i.e., $\Delta x_2 = w_9 \Delta y_2 + w_8 \Delta y_1$. Through similar rational relationship, the following set of equations can be obtained.

$$\begin{aligned} \Delta x_1 &= w_9 \Delta y_1; & \Delta x_2 &= w_9 \Delta y_2 + w_8 \Delta y_1; & \Delta x_3 &= w_8 \Delta y_2 + w_7 \Delta y_1; \\ \Delta x_4 &= w_7 \Delta y_2; & \Delta x_5 &= w_9 \Delta y_3 + w_6 \Delta y_1; & \Delta x_8 &= w_7 \Delta y_4 + w_4 \Delta y_2; \end{aligned}$$

$$\Delta x_{13} = w_3 \Delta y_3; \quad \Delta x_9 = w_6 \Delta y_3 + w_3 \Delta y_1; \quad \Delta x_{12} = w_4 \Delta y_4 + w_1 \Delta y_2;$$
$$\Delta x_{16} = w_1 \Delta y_4; \quad \Delta x_{14} = w_3 \Delta y_4 + w_2 \Delta y_3; \quad \Delta x_{15} = w_2 \Delta y_4 + w_1 \Delta y_3;$$

$$\Delta x_6 = w_9 \Delta y_4 + w_8 \Delta y_3 + w_6 \Delta y_2 + w_5 \Delta y_1; \quad \Delta x_7 = w_8 \Delta y_4 + w_7 \Delta y_3 + w_5 \Delta y_2 + w_4 \Delta y_1;$$
$$\Delta x_{10} = w_6 \Delta y_4 + w_3 \Delta y_2 + w_5 \Delta y_3 + w_2 \Delta y_1; \quad \Delta x_{11} = w_5 \Delta y_4 + w_2 \Delta y_2 + w_4 \Delta y_3 + w_1 \Delta y_1;$$

Very interestingly, the backpropagation can be compactly put as the following error and weight matrix correlation.

$$\begin{bmatrix} 0 & 0 & 0 & 0 & 0 & 0 \\ 0 & 0 & 0 & 0 & 0 & 0 \\ 0 & 0 & \Delta y_1 & \Delta y_2 & 0 & 0 \\ 0 & 0 & \Delta y_3 & \Delta y_4 & 0 & 0 \\ 0 & 0 & 0 & 0 & 0 & 0 \\ 0 & 0 & 0 & 0 & 0 & 0 \end{bmatrix} \odot \begin{bmatrix} w_1 & w_2 & w_3 \\ w_4 & w_5 & w_6 \\ w_7 & w_8 & w_9 \end{bmatrix} = \begin{bmatrix} \Delta x_1 & \Delta x_2 & \Delta x_3 & \Delta x_4 \\ \Delta x_5 & \Delta x_6 & \Delta x_7 & \Delta x_8 \\ \Delta x_9 & \Delta x_{10} & \Delta x_{11} & \Delta x_{12} \\ \Delta x_{13} & \Delta x_{14} & \Delta x_{15} & \Delta x_{16} \end{bmatrix} \tag{31}$$

Thus it can be seen that the relationship between the weights and the output during convolution is reproduced during correlation and hence can be used in back-propagating the error. For example, input x_6 at layer L influences all the output neurons (y_1, y_2, y_3, y_4) at layer $L + 1$ during convolution operation through weights w_5, w_6, w_8 and w_9 respectively. Therefore, when backpropagating the error Δy computed at layer $L + 1$, the error contribution at node corresponding to x_6 should be the aggregate sum of the error at Δy_1, Δy_2, Δy_3 and Δy_4 weighted exactly by the same weights w_5, w_6, w_8 and w_9. By analysing the expression for Δx_6 obtained after the correlation operation, it can be seen that this relationship is preserved. Thus back propagating the error across the convolution layers is equivalent to performing the correlation operation on the weighted error computed for the layer after weighting the derivative of the activation function. Thus ΔE_{P_1} can be computed as

$$\Delta EF_{C_2} = \Delta E_{C_2} .* O_{C_2} .* (1 - O_{C_2}) \tag{32}$$

$$\Delta E_{P_1}(:, :, i) = \sum_{l=1}^{L} \left[\Delta EF_{C_2}(:, :, l) \odot K_l(:, :, i) \right] \tag{33}$$

where $\odot$ represent the correlation operation, i varies from 1 to 6 and $L = 12$. This will result in backpropagated error dimension as $(12 \times 12)_6$.

Error backpropagation across pooling layer P_1:

As discussed earlier, the back propagated error is just an up-sampled version.

$$\Delta E_{C_1} = US_2(\Delta E_{P_1}) \tag{34}$$

4.4 Learning the Parameters by Gradient Descent

Once the error is propagated for all the neurons, the gradient is computed with respect to the parameter to be updated and the parameters are then updated in the negative direction of the gradient to minimise the loss function. As discussed before, the weight update for any connection is done based on the product of the backpropgated error found for the neuron at the rear end of the connection and input to the connection.

Derivative of gradient with respect to weights at fully connected layer

$$(\Delta W)_{FC_1} = \Delta E_{FC_1} .* O .* (1 - O) .* I_{FC_1} \tag{35}$$

In Eq. 35, I_{FC_1} represent the vectorized output map at the pooling layer 2 (P_2). Thus the dimension of the gradient is $[10, 1] \times [1, 192] \rightarrow [10, 192]$

Derivative of gradient with respect to kernel weights at convolution layers:

$$\Delta EF_{C_l} = \Delta E_{C_l} .* O_{C_l} .* (1 - O_{C_l}) \tag{36}$$

$$(\Delta K)_l = \Delta EF_{C_l} * I \tag{37}$$

In Eq. 37, '*' is the convolution operation. Note that here also weight update is based on the error back propagated at the neuron placed at the rear end of the connection for which we are updating the weights and the input to the connection. $l = 1$ for convolution layer 1 and $l = 2$ for convolution layer 2. I is the corresponding input map to the layer. Since only the valid part of convolution is taken, for C_1, the gradient dimension will be $[5, 5]_6$ ($[24, 24]_6 * [28, 28] \rightarrow [5, 5]_6$). Similarly for convolution layer 2, the gradient dimension will be $[5, 5, 6]_{12}$ ($[8, 8]_{12} * [12, 12]_6 \rightarrow [5, 5, 6]_{12}$)

Updating the parameters:

Once the gradient is determined with respect to the kernel K, it is updated using gradient descent.

$$K^{new} = K^{old} - \eta \Delta K \tag{38}$$

The gradient for the bias term for any node in the output layer, turns out to be the cumulative weight change computed for all the connections to that neuron. Similarly, the gradient for the bias of the kernels turns out to be the cumulative weight change computed for the weights in the respective kernel.

5 CNN in Cytopathology Applications

In last section, we have discussed the power of CNN over the traditional multi-layer feed forward neural network. CNNs have profound applications in health care and are widely used for disease screening/diagnosis applications [24]. Mainly they are used in two modes: stand alone classifier mode and transfer learning mode. In stand-alone classifier mode, the CNN is designed and learned from scratch for the intended application while in transfer learning based setting [25–27], the knowledge gained (the kernels learned) for a classification task is made use of in a totally different setting. As stand-alone classifier, CNNs have been used in pulmonary nodule detection from volumetric CT data [28], colorectal cancer detection [29], localizing the fetal abdominal standard plane in ultrasound [30], cervical cell analysis [31], prostate cancer detection [32].

In transfer learning based setting pre-trained CNN can be used just as a feature extractor and a different classifier such as SVM that needs less amount of training data to learn the classification problem is used for the intended task. Sometimes, the kernel weights from the pre-trained model is fine tuned by readjusting the weights learned by using the available training data. Thus, the transfer learning based setting is particularly useful, when the availability of training data is insufficient to come up with a stand-alone classifier. Today, transfer learning based setting has found applications in diverse domains of medical image analysis such as detecting colorectal cancer [33], interstitial lung diseases [34], localizing the fetal abdominal standard plane in ultrasound [35], mitosis detection in breast cancer [36].

In following sections we are going to discuss two specific applications of CNNs: the malaria detection and the leukemia cell-line classification. In Sect. 6, we discuss on the diagnosis of malaria using a stand-alone CNN classifier, while in Sect. 7, we discuss the use of CNN transfer learning to address the leukaemia cell-line classification.

6 Custom Designed CNN for Malaria Detection

In this section, a recent study [16] to assess the capability of the stack of slide images in different focal plane for the detection of *Plasmodium falciparum* infected malarial samples is discussed. Towards this, Leishman stained slide images of blood samples have been collected using custom-built focus stack collecting microscope [16]. The use of Leishman's stain is decided based on the study [37], which analysed the relative merit and demerit of using Giemsa and Leishman for malaria diagnosis and proposed that Leishman's stain is also a good choice particularly for analysing thin smears. The main reason is that the Leishman gives dark blue color to the chromatin structures in the nucleus there by allowing us to quickly discard the nucleated white blood cells from the non-nucleated Red Blood Cells (RBC) when the full blood sample is analysed.

In the direction of developing automated system for malaria diagnosis, large number of research works have been carried out both in thick [38–40] and thin [41–44] smear slide images. These works make use of data samples of different types, size and employs extensive use of computer vision techniques [45]. A large number of these techniques extract features at suspected parasite locations and use a classifier learned on these features to take the final decision. For example, [40] uses genetic programming to take the decision based on the statistical features including mean, variance, kurtosis of intensities at suspected parasite location while [39, 46, 47] use support vector machine (SVM [48]). Since these methods operate on a wide range of datasets, the performance measures reported largely varies. For examples, the reported sensitivity varies between 81.7 and 95% while specificity varies between 92.59 and 100%. More recently, both the transfer learning capability and the stand-alone classifier ability of CNN are analysed in detecting whether the sample smear is infected or not by inspecting the blood smear images [49]. The study was performed by analysing Giemsa stained slide image dataset containing 27578 RBCs and reported that the mean accuracy obtained by the stand alone CNN classifier was significantly better than that produced by the transfer learning based strategy (97.37 vs. 91.99%). However all the methods considered so far were using single image to address the parasite detection problem. One difficulty with this is that, parasites appear quite differently in slides based on which stage they are in their life cycle. It is also hard to get in a single image, all the parasites infected in focus. Also, if parasite infection is at its early stage, it may appear like a Gray spot and can be confused with the appearance of the Gray spot produced by dust particles on the lens/sensor of the camera. Thus if we are using single focus image, there is a chance that we may miss out a few true positives, if we want to avoid black spots due to dust or we will be having many false positives. In these situations it is better to use focus stack of images since the focus profile of a dust is different from the focus profile of a parasite infection. Following subsections detail experiments to evaluate the effectiveness of using focus stack of images for detecting malarial parasite from blood smear slide images. Because of the inherent capacity of CNN to deal with multi-channel images, a custom designed CNN is used for the experiments.

6.1 Dataset Collection

The dataset contains stack of images of different field of views, and for each stack it contains images captured at different focus by sweeping the same field of view across the depth-of-focus. The slide images are prepared to have thin smear and the stain used is Leishman. In order to mimic real experiments where full blood sample is used, the cultured *P. falciparum* malarial samples (O^+ red blood cells) are spiked with WBCs. The culture is maintained in 5% hematocrit. The prepared thin blood smear is then imaged using custom-built bright-field transmission microscope. The details of the microscope and the dataset preparation can be found in [16].

Unlike the methods there in the literature, we have used focus stack of images (Refer Fig. 7) for analysis. This provides us some additional information which can be useful while examining very fine details of the parasite that may not be directly observable if we are using only the best focused image from the focus stack. This in turn help us to differentiate parasite (especially at its early stage of infection) from image artefact producing similar features (due to dust on camera lens/sensor). Altogether the dataset contain 765 FoVs with 62,015 cells. The number

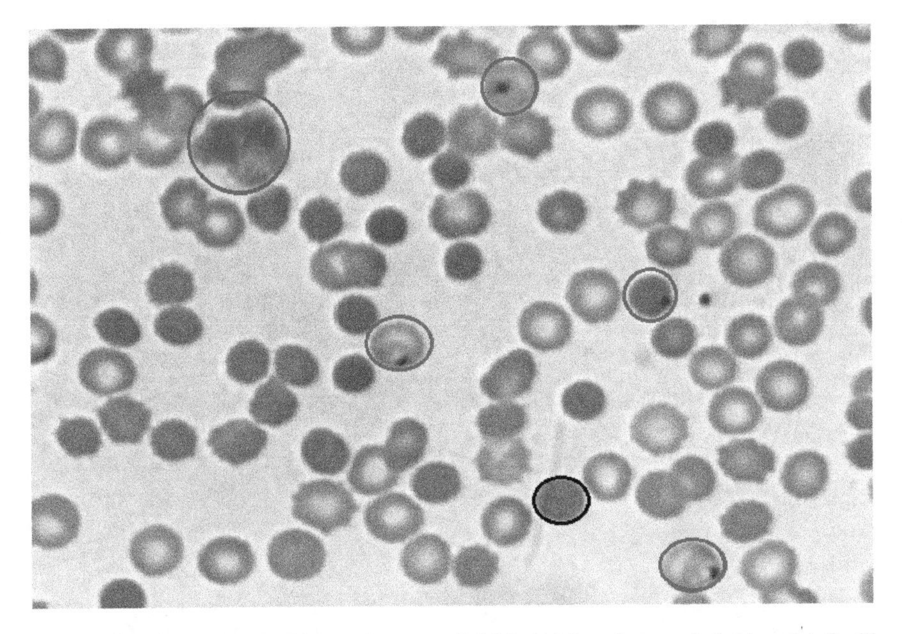

Fig. 5 One of the best focused image from a stack. Malaria infected are encircled in red, cell with artefact in black and a WBC in blue

of infected RBCs is 1191 and is decided by the experts working in the field after manually inspecting the focus stack of images for each FoV. Figure 5 shows one such slide image with marked infected cells. In order to identify these infected cells automatically, we use a classifier based system. However, rather than exhaustively inspecting each and every location on the slide image by the classifier, first we quickly identify all suspected parasite locations followed by closely examining these locations using a trained classifier. Following subsections explains the method used to identify suspected parasite locations and the features/patches to train the classifiers used to further analyse these locations.

6.2 *Identifying Suspected Parasite Locations*

It is the best focused image (from each stack) we analyse to identify the candidate locations for the parasite. The one with highest variance in each focus stack is identified as the best focused image. This image is then analysed in small neighbourhood sufficient to span the radius of a typical RBC, and the local minima points in each neighbourhood is identified as candidate locations for the parasite. This is from the observation that the parasites appear darker in a Leishman stained image especially when the parasite is at early stage of its life cycle. The candidate locations identified are then refined by excluding those candidate locations in the background region. As there is high contrast between the background and cell region, it is easy to get the

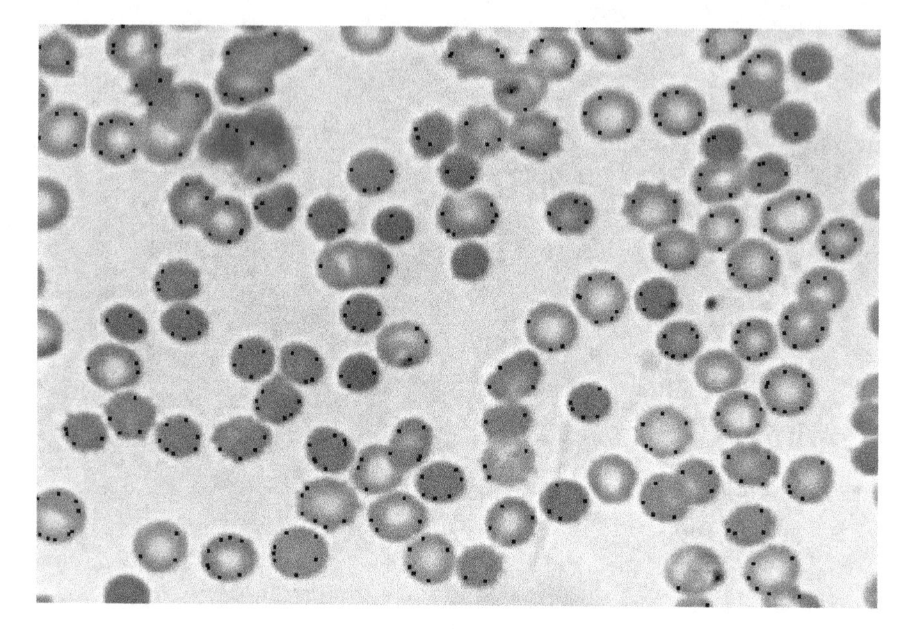

Fig. 6 Suspected parasite locations identified for the image shown in Fig. 5

masking image for excluding the candidate locations in the background by applying a threshold like Otsu [50]. The suspected parasite locations identified by this method for the slide image in Fig. 5 is shown in Fig. 6. Note that the noise such as dust on camera sensor can produce a candidate parasite location. However, the subsequent processing by classifier will identify it as an artefact since it is taking the decision based on the focus stack collected around suspected location and due to the fact that the focus stack profile of a dust is different from the focus profile of a parasite.

Out of 328,334 local regional minima locations identified from 765 slide images, 1400 locations are identified to be parasite locations (positive class) by experts and 326,934 are identified to be healthy (negative class). However the number of infected cells can be less since a cell can have more than one infected parasite locations and in our case there were 1191 infected cells. As the dataset is unbalanced (more negative class samples than positive samples), we have rotated the positive patches at degrees 90, 180 and 270 to take their total count to 5600.

6.3 Characterising Suspected Parasite Locations

Once the suspected parasite locations are identified, these locations has to be verified by a trained classifier. In order to facilitate this, RGB patch of size $32 \times 32 \times 3$ are extracted from the best focused image centred at the parasite location and are fed to a custom designed CNN. The effectiveness of this classifier is compared against

SVM classifier trained on hand engineered features extracted from the same set of patches. We have also assessed the performance of a CNN classifier working on the stack of image patches centred around the best focused image patches. The patches for the stack was set up by keeping the patch from the best focused image at centre and one at either end of the field of focus. It turned out that the CNN working on the focus stack has produced the best result as discussed in Sect. 6.4.

Statistical and textural features for SVM classifier

For the SVM classifier we have selectively picked fourteen features. In addition to the usage of regular statistical features [39, 46, 47], we have also used texture features in our experiment. As we know that there is gradient change and specific texture pattern in and around parasite locations, we use texture features such as 'Contrast', 'Correlation', 'Energy', and 'Homogeneity' calculated from the Gray level co-occurrence matrix (GLCM) [51] of the region. The mean and standard deviation (Std) of these features computed for infected and healthy patches are shown in Table 2. In order to characterise the statistical features at the parasite locations, we have taken the minimum, maximum, mean and variance of intensities of the patch and the minimum as well as maximum of gradient magnitude observed in the region. These features are listed in Table 3. These features reflect the statistical features for the whole patch. To get the localised feature, 3×3 non-overlapping regions are considered and minimum as well as maximum values of the mean intensity as well as the standard deviation observed in all subregions are measured. These features are reported in Table 4. Thus altogether there are 14 features and are used to train an SVM model.

Table 2 Mean (Std) of GLCM features for 32×32 Patches

	Contrast	Correlation	Energy	Homogeneity
Healthy	0.5698 (0.2335)	0.9825 (0.0094)	0.1020 (0.0445)	0.8189 (0.0455)
Infected	0.8850 (0.3439)	0.9754 (0.0119)	0.0584 (0.0357)	0.7577 (0.0498)

Table 3 Mean (Std) of global statistical features for 32×32 patches

	Min_Int	Max_Int	Mean_Int	Var_Int	Min_GMag	Max_GMag
Healthy	0.5023 (0.0615)	0.8891 (0.0345)	0.6797 (0.0534)	0.0168 (0.0069)	0.0007 (0.0019)	0.6426 (0.1303)
Infected	0.2720 (0.1057)	0.8748 (0.0453)	0.6109 (0.0641)	0.0195 (0.0090)	0.0028 (0.0034)	0.8545 (0.2147)

Table 4 Mean (Std) of local (3×3) mean and variance of patches

	Min_Mean	Max_Mean	Min_Var	Max_Var
Healthy	0.5095 (0.0620)	0.8817 (0.0337)	$3.4520e^{-6}$ $(2.7481e^{-6})$	0.0061 (0.0160)
Infected	0.2903 (0.1049)	0.8669 (0.0444)	$7.2392e^{-6}$ $(1.2817e^{-5})$	0.0092 (0.0137)

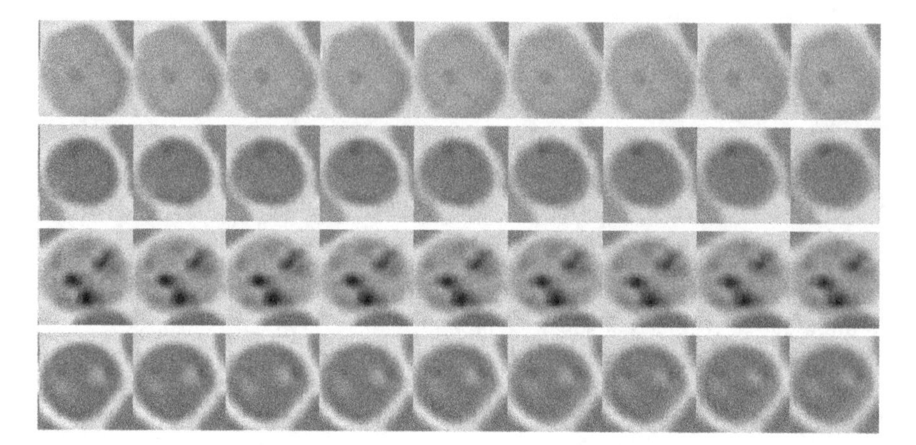

Fig. 7 First row shows a focus stack of healthy cell with an artefact due to sensor and the last three show focus stack of infected cell

RGB patches for CNN training

For CNN training we use two dataset: the RGB patches only from the best focused images, and focus stack of patches. The focus stack dataset was created based on the idea that the focus profile of an image artefact will be different from the profile created by a parasite. For example, consider the four focus stacks provided in Fig. 7. The cells except the one shown in first row are infected. If we closely observe, in the case parasite infection the parasites are coming into focus and then fades away. This is not the case with the cell having an artefact due to dust on camera sensor (Fig. 7, first row) where the artefact appears in constant intensity across the focus profile. In order to capture this difference in focus profile, we have chosen three images in the stack with the best focused patch at the centre and two patches far away from it one on both sides. Thus the dataset used to train the CNN operating on focus stack has size ($32 \times 32 \times 9$) (Fig. 8) and the one operating on the best focused patch contain RGB patches of dimension $32 \times 32 \times 3$.

Figure 9 shows the basic design of the CNN used in this study. Input is a D channel image patch of size 32×32. The C, R, and P are Convolution, ReLu and Pooling layers. The size of the kernel used for the convolution layers is provided directly under each block. The output map size at each block is provided just above the block. The subsampling is done at 2×2 scale by max-pooling. This model architecture was selected based on the following observations. The input image size was fixed at 32×32, since an RBC cell size as per our imaging setup is turned out to be 41×41. The max-pooling was used to avoid the average out behaviour of fine details if we had used average pooling at parasite locations and the standard size 5×5 is selected for kernels which is decent enough to work on a small image of size 32×32. Since we are addressing a binary classification problem, the number of output neurons is decided as 2. Note that once the input dimension and output dimension are fixed, a

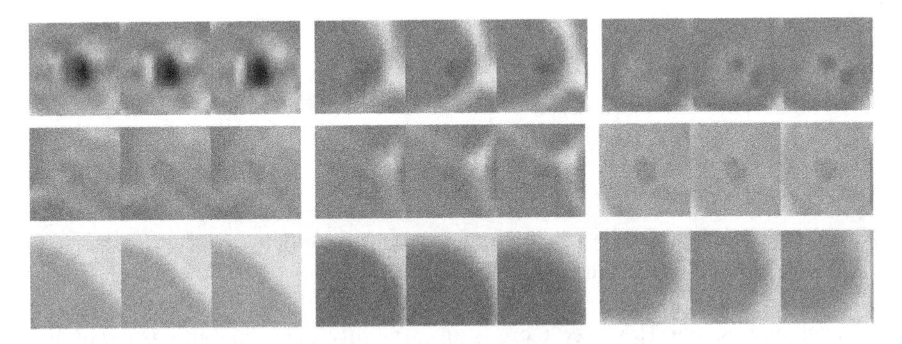

Fig. 8 Nine focus stack each containing 3 images (selectively picked from the full stack) used to train CNN. First row containg infected cells, second and third rows have healthy cells but with image artefact in second row

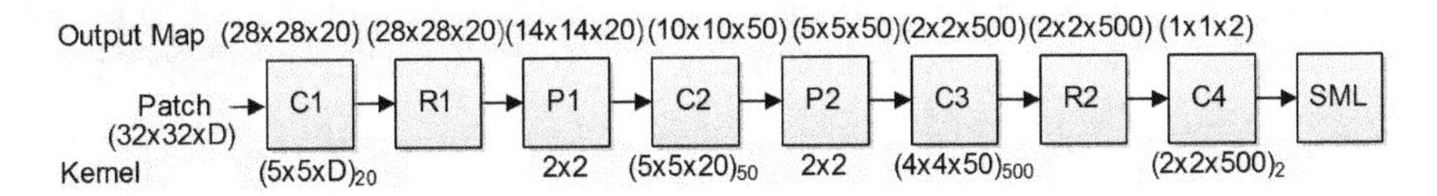

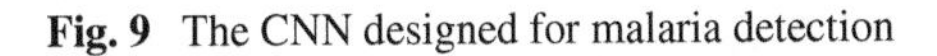
Fig. 9 The CNN designed for malaria detection

reasonably deep CNN should work for the problem, and we have selected neither a shallow nor deep architecture for classification.

As noted, for every input patch, the CNN is designed to produces a binary vector at the output (C4 in Fig. 9). Towards this goal, the CNN is trained using backpropagation algorithm, considering the log loss (Eq. 39) error function. The log loss J_i for an ith input patch is defined as

$$J_i = -\sum_{k=1}^{C} L_{ik} \log P_{ik} \tag{39}$$

In Eq. 39, C is the number of output classes (number of output neurons) which is 2 in this case and L_i is the boolean vector with value 1 only at the true class location for the input patch; i.e., output label vector is [1, 0] for an infected patch and [0, 1] for a healthy patch. P_{ik} is the model probability of assigning label k to the ith input instance. A perfect classifier should have a log loss of precisely zero.

Equation 39 can also be written as

$$J_i = -\log P_{iT} \tag{40}$$

where P_{iT} is the model probability of assigning label T (the true class label) to the ith input instance. Here T is 1 for infected patches and is 2 for healthy patches. The model probability P_{iT} can be calculated using soft-max function and the final loss function to be minimised turns out to be the soft-max log loss (Eq. 41).

$$J_i = -\log\left(\frac{\exp^{Z_{iT}}}{\sum_{t=1}^{C}\exp^{Z_{it}}}\right)$$
$$= -Z_{iT} + \log\sum_{t=1}^{C}\exp^{Z_{it}} \quad (41)$$

Note that, for each of the class there is one neuron in the output layer and Z_{iT} in Eq. 41 is the response produced at the final layer neuron corresponding to the true class T for the specific input (ith patch) at the input layer. The aim is to minimise this function as much as possible. However, there is an instability in evaluating this expression for any input. It is due to the difficulty in evaluating the second term, the log sum of exponentials. If any of the value Z_{it} becomes sufficiently large, its exponential becomes very large and the sum can over flow to positive infinity. Similar is the case, if any of the Z_{it} value becomes sufficiently smaller, the log sum of exponentials can underflow. However, we can get rid of this problem with little algebraic manipulation by taking $\Psi = \max\,(Z_{it})_{t=1}^{C}$.

$$\log\sum_{t=1}^{C}\exp^{Z_{it}} = \log\left(\sum_{t=1}^{C}\frac{\exp^{\Psi}}{\exp^{\Psi}}\exp^{Z_{it}}\right)$$
$$= \log\left(\exp^{\Psi}\sum_{t=1}^{C}\exp^{Z_{it}-\Psi}\right)$$
$$= \Psi + \log\sum_{t=1}^{C}\exp^{Z_{it}-\Psi} \quad (42)$$

Thus Eqs. 41 and 42 together defines soft-max log-loss (SML) function as

$$J_i = -Z_{iT} + \Psi + \log\sum_{t=1}^{C}\exp^{Z_{it}-\Psi} \quad (43)$$

Equation 43 ensures that the largest value passed to the exponential function is 0. If there are really tiny values after subtracting Ψ, it is forced to 0 and will be dropped (to deal limited precision arithmetic). This soft-max log loss has a simple derivative. From Eq. 41

$$\frac{dJ_i}{dZ_{it}} = -\left(\delta_{t=T} - \frac{\exp^{Z_{iT}}}{\sum_{t=1}^{C}\exp^{Z_{it}}}\right) \quad (44)$$

Here $\delta_{t=T}$ is a vector which has value 1 only at the true class location Z_{iT} and everywhere else it is 0. As noted the CNN is learned by backpropagation algorithm, in which the error at the final layer is computed for each input image considering its target class. Then this error is propagated down the layers. Whenever it crosses

ReLU layer, the derivative of ReLU (i.e., for positive values 1, else 0) is multiplied, and whenever it crosses the pooling layers, the error matrix gets up-sampled, and across the convolutional layers it is proportionately multiplied by the corresponding weight contributions from the kernel (correlation of the error with kernel). Then the parameters to be learned (the kernel weights and biases) are then updated by gradient descent, for which the gradient is computed by multiplying the error at the rear end of the layer and the input to the connection for which the parameters are to be computed. The explicit equations and the derivations for the weight update can be found in [52].

During testing we exclude SML layer, and assign for every test sample the label of the class for which the highest response was obtained during testing.

6.4 Results and Discussion

This section provides performance of the three classifiers that we have used to identify infected cells. The ground truth for the dataset is the marked infected locations (by experts) along with number of infected cells. The quantitative analysis of parasite detection is done by assessing the number of false positives and false negatives.

The effectiveness of the cnn working on the focus stack of patches is assessed by performing 10 fold cross validation experiment on FFN, SVM and on CNNs. Since we have used 14 hand-engineered features, the number of hidden layers is one and the number hidden neurons is set as 8. All the positive patches (5600) and 5600 negative patches selected at random are used in the cross validation experiment. The cross validation accuracy is then measured by considering the number of true positives (TP), true negatives (TN), false positives (FP) and false negatives (FN). The specific measures used are sensitivity (TP/(TP + FN)) and specificity (TN/(TN + FP)) and MCC. While sensitivity measures the ability of the system to correctly identify infected cells, specificity measures the ability to correctly classify healthy cells. Also we have measured a combined metric which is the Matthews correlation coefficient (MCC) as defined by

$$MCC = \frac{TP \times TN - FP \times FN}{\sqrt{(FP + TP)(TP + FN)(FN + TN)(TN + FP)}} \tag{45}$$

Table 5 shows the result of 10 fold cross validation experiments which report average sensitivity, specificity and MCC along with standard deviation and reveals that the CNN operating on focus stack has reduced false positives and false negatives. Note that this is reflected in MCC metric, which gives a measure of 1 for a perfect classifier, −1 for the worst classification and 0 for a random guess. The Receiver operator characteristics (ROC) curve drawn for the first fold revels that for CNN working on the focus stack the area under the curve is very close to the maximum possible value and is 0.9992. AUC for the one that operate on the best focused patch is 0.9987, 0.9910 for SVM and 0.9813 for FFN features. In subsequent results, we

Table 5 Results of 10 fold cross validation : average sensitivity, specificity, MCC and their standard deviation for 10 fold cross validation experiment on (O) FFN on features (A) SVM on features and CNN on (B) patches (C) focus stack

Metric	Method—O	Method—A	Method—B	Method—C
Sensitivity (%)	92.44 (0.84)	96.38 (0.88)	98.91 (0.36)	99.14 (0.37)
Specificity (%)	97.36 (0.73)	95.43 (0.85)	99.39 (0.31)	99.62 (0.18)
MCC	0.8991 (0.0140)	0.9181 (0.0150)	0.9831 (0.0039)	0.9877 (0.0032)

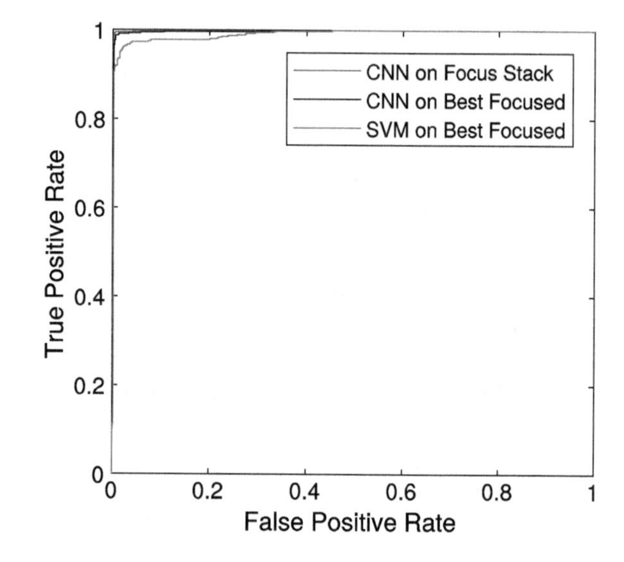

Fig. 10 ROC for CNNs on focus stack & best focused image and ROC for SVM on features

compare the results of CNNs (one operating on best focused patches and the other operating on focus stack) since CNN results produced the best results in 10 fold cross validation (Fig. 10).

Table 7 provides the result of classification experiment when the classifiers (SVM on features, CNNs on patches) are trained with relatively few samples: 60% positive samples for training and 20% samples for validation. All the classifiers are trained, validated and tested with the same input patches for fair comparison. The exact number of samples used for the training, validation and testing can be found in Table 6. The learning behaviour of CNN on the best focused patch and on the focus stack of patches is provided in Fig. 11. The blue plot shows how the network behaves on the validation set and the network that has produced minimum error was used for testing.

Table 6 Number of samples used for training, validation and testing

Patches	# Un-rotated	# Rotated	# Train	# Validation	# Test
+ve	1400	1400 × 4 (5600)	60% of 5600 (3360)	20% of 5600 (1120)	1400
−ve	326,934	–	3360	1120	326,934

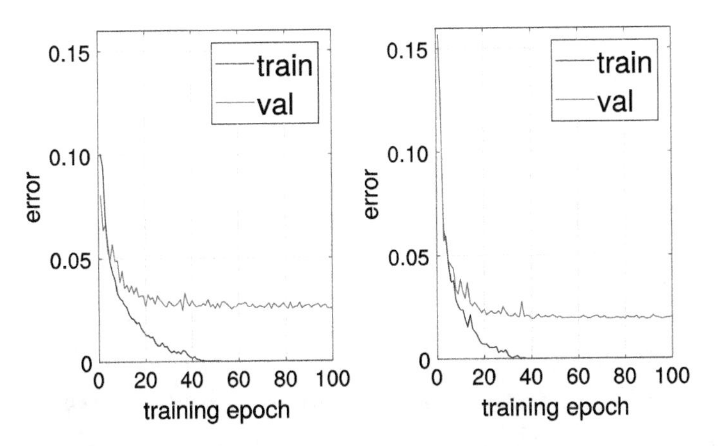

Fig. 11 **a** Training and validation error across epochs for **a** CNN on best focused image and **b** CNN of focus stack of patches

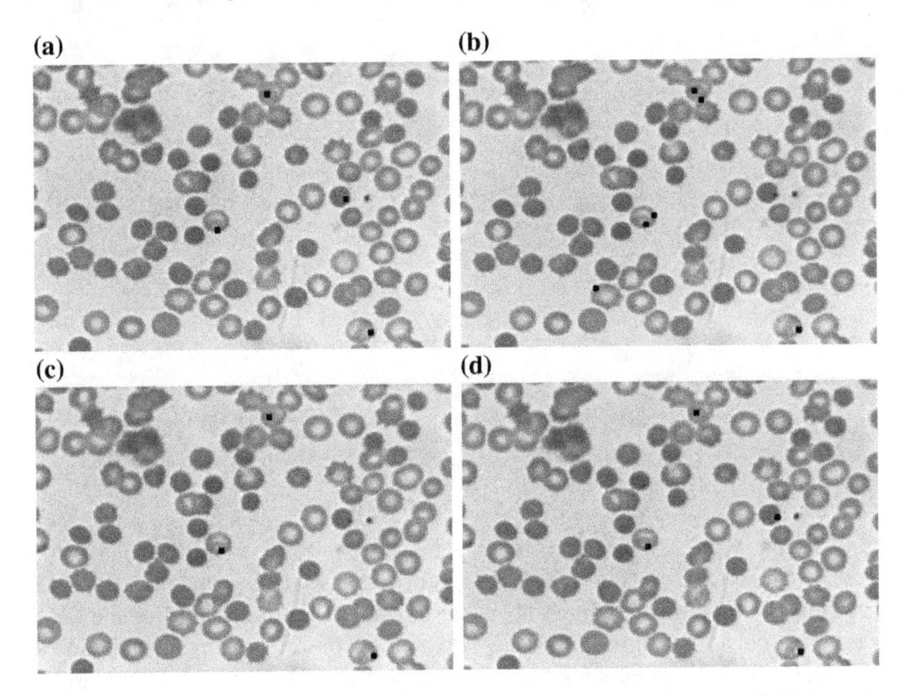

Fig. 12 **a** The ground-truth parasite locations in the slide image shown in Fig. 5. The parasite locations identified by **b** SVM trained on features **c** CNN trained on best focused image and **d** CNN trained on focus stack

For every test image the suspected locations are identified, patches are extracted and feature/patch/focus stack are fed directly to the trained classifiers. The classifier decides whether it is really a parasite location or a healthy location. The results for the test image shown in Fig. 5 is provided in Fig. 12. When we compare with the

Table 7 Confusion matrices by different classifiers (A) SVM on hand-engined features, CNNs on (B) best focused image and (C) on the focus stack of patches

	A		B		C	
	Infected	Healthy	Infected	Healthy	Infected	Healthy
Infected	1107	84	1151	40	1156	35
Healthy	3756	57068	1053	59771	912	59912

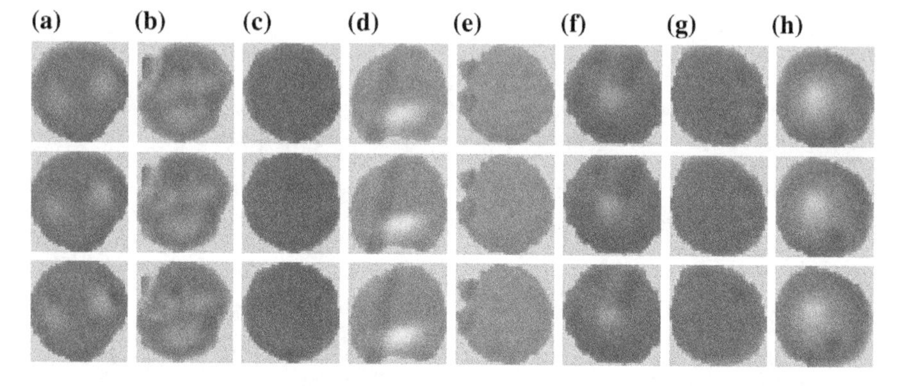

Fig. 13 **a–c** are the three true positives (focus stacks) only identified by CNN on focus stack **d–f** are three true negatives only identified by CNN on focus stack **g–h** are two infected cells missed out by all the classifiers

Table 8 Sensitivity, specificity and MCC computed from the matrix in Table 7

Metric	Method—A	Method—B	Method—C
Sensitivity (%)	92.95	96.64	97.06
Specificity (%)	93.82	98.27	98.50
MCC	0.4430	0.7036	0.7305

ground truth (Fig. 12a), it can be seen that the CNN operated on the focus stack has produced the best result (Fig. 12d) by producing no false positive and false negative. The confusion matrix shown in Table 7 also reveals the same result for the CNN trained on focus stack of patches. Figure 13 shows a few cases where the classifiers have failed. The first three stack are identified as infected only by the CNN working on the focus stack and is the case with the healthy cells shows in Fig. 13d–f. However, all the three classifiers missed in recognising the last two infected cells. It is clear from the shown images, that focus stack helps in producing a positive decision. This is also revealed by relatively high MCC measure shown in Table 8. With CNN operating on focus stack, we could identify 1156 infected cells correctly from 1191 infected cells producing a sensitivity 97.06% and has produced specificity of 98.50%. We have also compared the capability of CNN working on focus stack, with one of the state of the art method reported in [53]. Though the actual method was experimented on

Giemsa stained dataset, we have run the code provided by [54] to extract LBP/VAR [55] and SIFT features [56] and then used it for classification. The result is provided in Table CmpDetAccWithI which also reveals that the CNN working on the focus stack has produced the best result.

7 Pre-trained CNN for Classification of Leukaemia Cells

In this section we are going to discuss the transfer learning capability of CNN by addressing the classification of three important leukaemia cell-lines K562, MOLT and HL60. As discussed in Sect. 5, the transfer learning setting is particularly useful when there is insufficient training samples to come up with a stand-alone classifier. We are going to show that just by using the feature detectors from a pre-trained established model could extract discriminant features to produce classification accuracy on-par or much better than the classification result obtained from sophisticated features used along with traditional classifiers like SVM (Table 9).

7.1 Dataset Collection

Cancer is one of deadly diseases and take around 7–8 millions death worldwide [57]. The cancer treatment greatly depends on what stage it was detected. One important step in any cancer detection/treatment is the cytopatholoical study where samples are extracted from suspected tumour locations and are analysed by fine needle aspiration. The samples are then used to make smear and is inspected under microscope. Thus the cytopathological testing is very critical but is a skill demanding, time consuming job. The manual inspection also limit the throughput and is also prone to errors. Towards making automated system, microfluidics imaging flow cytomatry [17, 51] along with necessary image analysis turned out to be the need of the hour. For the present discussion, we use a dataset of leukemia cell-line captured using one such system. These are the leukaemia cell-lines made available by American Type Culture Collection (ATCC), and was of type K562, MOLT and HL60, which provided us the ground truth for the present study. Note that these are WBC lines, and in a real situation there can be RBCs as well. But separating out RBCs from WBCs in microfluidics system is not that difficult as can be seen in Fig. 14. There were

Table 9 The performance of SIFT, LBP/VAR feature based classification [53]

	Infected	Healthy		Method—[53]
Infected	984	208	Sensitivity	82.55
Healthy	4091	56628	Specificity	93.26

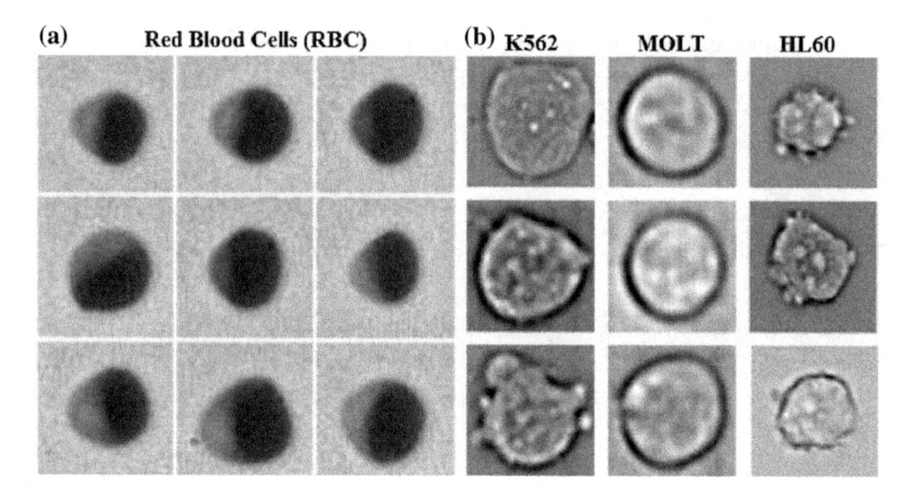

Fig. 14 **a** Red blood cells; **b** Leukaemia cells in Mf-Ms channels

altogether 618 cells in the dataset: 124 K562, 106 MOLT and 388 HL60. Though at first it seems as a small dataset, it is of significant size considering the fact that in unit microlitre of blood in healthy man, there are only $4 - 11 \times 10^3$ WBCs when compared to $4 - 5 \times 10^6$ RBCs. Hence for a screening tool for leukaemia, we should give more effort in classifying the abnormal WBCs.

The leukemia cell-lines are then imaged separately using the microfluid microscopy setup described in [17]. Inorder to enhance the features such as the contrast of the cell in the background simple background subtraction was done. The cells are then localised using bounding boxes to form the dataset. The detailed procedure can be found in [51].

7.2 *CNNs in Transfer Learning Mode*

In this section, the transfer learning capability of CNN is used in classifying Leukaemia cell-lines: K562, MOLT and HL60. In transfer learning setting, the knowledge that a CNN has learned for a relatively complex classification task using a large dataset is effectively transferred to a completely different setting. In such setting, CNN is used as a feature extractor and not as a classifier. These features are then used to train a classifier like SVM that needs only training data of moderate size. The transfer learning capability of CNN is studied in [58] and utilized in [59] to annotate the gene expression patterns in mouse brain. As noted in Sect. 3, the main building blocks of a CNN are (i) Convolution (ii) ReLU and (iii) Sub sampling. Note that, for every convolution layer, a number of Kernels are learned during training. Normal convolution is performed between the learned kernels and the input instance but select only valid part of the convolution. We know that the kernels can extract features by convolutions. For example, Sobel kernel find edges, Laplacian kernel detect blobs. Each of these kernel is applied only locally and extract features. Thus depending on the kernel, convolution can extract features like edges, blobs, corners, etc. and are valid features for any images. In CNN, the only learned parameters are the kernel weights and biases, and hence we are learning very local feature detectors rather than the actual features. Suppose that a CNN has to address a relatively complex classification task and there are sufficiently large number images to train the network. If the CNN is adequately trained using this large dataset, it is reasonable to believe that the learned kernels have the capability to extract an exhaustive set of features even to capture the small inter class variability on the original classification problem. Since the kernels learned act very locally, these exhaustive sets of feature detectors are most probably valid for any images irrespective of the classification problem. This fact is the corner stone of transfer learning capability of CNN.

In a nutshell, a heavily trained CNN for a complex classification task using a large image data set, must have learned an exhaustive set of feature detectors that can capture very local features valid for any images, and have the capability to introduce non-linearity in the detected features due to the architecture involving intermediate ReLU and sub-sampling layers. This enables us to use such a CNN as a feature extractor for the leukaemia cell-line classification dataset used in this experiment, and then use these features to build a suitable classifier using the small dataset available for training.

Trained CNN can outperform human in many visual recognition task [60]. However one bottleneck in using CNN as a classifier is the huge amount of labelled training samples required for training. In the present discussion, we show that cnn can be used in transfer learning mode in such situation. We make use of the deep cnn model [52, 61] trained on ImageNet [62] database to classify the leukemia cell-lines. It has to be noted that the model was not trained on any of these images but was extensively trained on very complex classification task containing 1000 classes of categories like animals, birds, etc. The architecture of the model that we have used is given in Fig. 15 and has 37 layers. In Fig. 15, C represnts convolution layer, R represents the ReLU activation function MP represents max-pooling based subsampling operation. We apply all the images in our dataset and generated deep features. Since the intention is to just show that the features are good enough to produce a fair classification, we have taken the features from 36th layer, which turns out to be vectors of dimension 4096. The dimensionality of these feature vectors are reduced to 20 (experimentally set) by PCA [63] and are then used to train number of classifiers including FFN, SVM, Naive Bayes and K Nearest neighbour classifiers. As noted earlier, it is reasonable to use these features for our classification task since it was generated by a deep CNN which was well trained for a complex classification task involving images from 1000 object categories. The kernels learned are therefore must be good feature detectors looking for extensive collection of very local features such as edges, blobs, corners etc which are relevant in any image classification task irrespective of the domain.

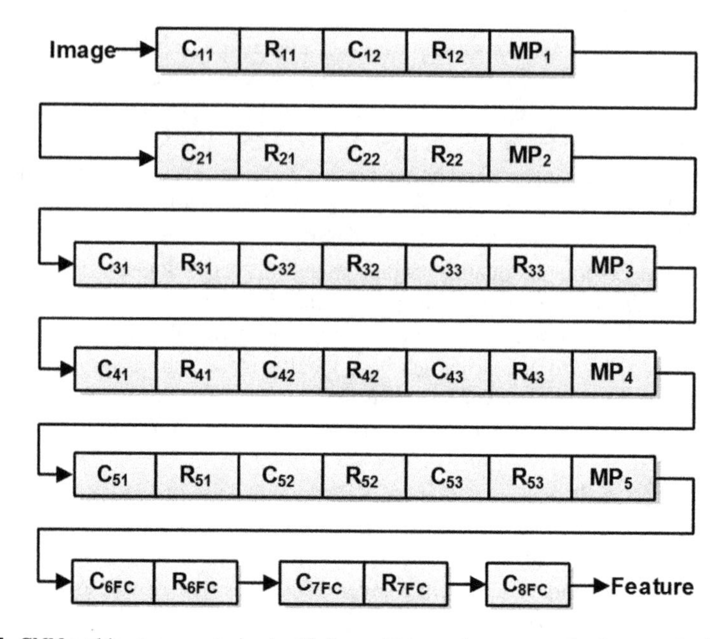

Fig. 15 CNN architecture pre-trained with ImageNet, used to extract the features for leukaemia cell-line classification

7.3 Results and Discussion

We compare the result of classification using the features generated by the pretrained cnn model by employing different classifiers. The results are compared against the result produced by the SVM model on the same dataset but using hand-engineered features reflecting size, shape, and texture features [51] generated from fine segmented cell images. The result for the three class leukemia classification problem that we are addressing when used the deep features from the pretrained model is shown in Table 10. We assess the accuracy by observing the average and standard deviation of the accuracy obtained for different cross validation experiments. The results show that the accuracy is better with less variance across different classifiers. It can also be seen that the accuracy produced on these features are high when compared to the classification accuracy reported in [51] as can be seen in Table 11.

Table 10 Cross validation accuracy in % (mean (std)) on CNN features

Kfold	10	5	4	3	2
SVM	97.80 (1.82)	97.69 (1.38)	97.62 (1.20)	97.47 (1.02)	97.19 (0.87)
FFN	98.26 (1.57)	98.18 (1.16)	98.16 (1.02)	98.05 (0.93)	97.96 (0.72)
NBS	98.40 (1.51)	98.37 (1.08)	98.36 (0.90)	98.40 (0.78)	98.38 (0.53)
KNN	98.42 (1.49)	98.40 (1.04)	98.37 (0.94)	98.35 (0.81)	98.36 (0.59)

Table 11 Cross validation accuracy in % (mean (std)) on morphological features [51]

Kfold	10	5	4	3	2
SVM	95.22 (2.83)	94.47 (1.85)	94.63 (1.52)	94.20 (1.40)	93.57 (1.16)
FFN	92.92 (3.48)	92.74 (2.63)	92.57 (2.35)	92.47 (2.23)	91.76 (3.00)
NBS	90.89 (3.60)	90.77 (2.34)	90.70 (2.18)	90.67 (1.74)	90.35 (1.44)
KNN	80.40 (4.81)	79.93 (3.13)	79.48 (2.90)	78.94 (2.63)	77.75 (0.59)

7.4 *Comparison with the Classification on Morphometric Features Discussed in [51]*

Table 12 reports the class specific accuracy for the cell-lines for 10 fold cross validation experiment. It can be seen that the classification is not biased to any particular cell-line. Table 12 A contain the confusion matrix for the SVM classifier on morphological features [51] and Table 12 B contain the confusion matrix for the SVM run on CNN features.The corresponding Precision and Recall measures are provided in Table 13. We have used one-vesus-all strategy in reporting these measures as they were primarily meant for binary classification. Thus the Case - I treat HL60 as positive class, Case II treat K562 as positive class while Case III treats MOLT as positive class. In each case, the cell-lines not used as positive class together contribute to negative class. Significant improvement can be found in both measures when compared to using the features reported in [51] as shown in Table 13. In short we can say that

Table 12 Comparison of class specific accuracy

	A. Morphology features [51]				B. CNN features		
	HL60	K562	MOLT		HL60	K562	MOLT
HL60	373	8	7	HL60	385	1	2
K562	4	116	4	K562	1	119	4
MOLT	2	5	99	MOLT	1	5	100

Table 13 Comparison of precision and recall in % (One against all)

	Case—1		Case—2		Case—3	
	Precision	Recall	Precision	Recall	Precision	Recall
SVM on morphology features [51]	98.42	96.13	89.92	93.55	90.00	93.40
SVM on CNN features	99.48	99.23	95.20	95.97	94.34	94.34

the features provided by the pretrained CNN model worked really well when compared to the carefully chosen hand engineered features [51] at least for the leukemia cell-line dataset that we have considered. This support the very well established field of transfer learning capability of CNN.

8 Summary

In this chapter, we have discussed the applications of convolutional neural network in cytopathology. We have provided basic understanding of the CNN training taking the popular LeNet architecture as an example. Both stand-alone capability of CNNs and transfer learning capability of CNNs for classification are addressed by specifically taking malaria detection from focus stack of blood smear images and leukaemia cell-line classification. The results shows that CNNs has produced better accuracy when compared to methods on similar problems using traditional classifiers like SVM. Using focus stack of blood smear images for automated malaria detection is first of its kind and the methods discussed are part of processing frameworks developed for custom-built low-cost microscopy and microfluidic microscopy devises meant for resource limited clinics.

References

1. R. Nayar, *Cytopathology in Oncology* (Springer, 2014), http://www.springer.com/medicine/oncology/book/978-3-642-38849-1
2. PathScope, Pathscope™ slide scanner; digipath inc. Pathology delivered digitally. http://www.digipath.biz/pr/PathScope.pdf. Accessed 7 Dec 2016
3. M. Rieseberg, C. Kasper, K.F. Reardon, T. Scheper, Flow cytometry in biotechnology. Appl. Microbiol. Biotechnol. **56**(3–4), 350–360 (2001)
4. D.A. Basiji, W.E. Ortyn, L. Liang, V. Venkatachalam, P. Morrissey, Cellular image analysis and imaging by flow cytometry. Clin. Lab. Med. **27**(3), 653–670 (2007), https://doi.org/10.1016/j.cll.2007.05.008
5. E. Schonbrun, S.S. Gorthi, D. Schaak, Microfabricated multiple field of view imaging flow cytometry. Lab Chip **12**, 268–273 (2012). https://doi.org/10.1039/C1LC20843H
6. Amnis Corporation® ISX - MKII Brochure (2016), https://www.amnis.com/documents/brochures/ISX-MKII20Brochure_Final_Web.pdf. Accessed 28 July 2016
7. L. Pantanowitz, P. Valenstein, A. Evans, K. Kaplan, J. Pfeifer, D. Wilbur, L. Collins, T. Colgan, Review of the current state of whole slide imaging in pathology. J. Pathol. Inform. **2**(1), 36–45 (2011). https://doi.org/10.4103/2153-3539.83746
8. M. Rojo, G. Garcia, C. Mateos, J. Garcia, M. Vicente, Critical comparison of 31 commercially available digital slide systems in pathology. Int. J. Surg. Pathol. **14**(4), 285–305 (2006). https://doi.org/10.1177/1066896906292274
9. H. Irshad, A. Veillard, L. Roux, D. Racoceanu, Methods for nuclei detection, segmentation, and classification in digital histopathology: a review - 2014; current status and future potential. IEEE Rev. Biomed. Eng. **7**, 97–114 (2014). https://doi.org/10.1109/RBME.2013.2295804
10. G. Deco, V.K. Jirsa, P.A. Robinson, M. Breakspear, K.J. Friston, The dynamic brain: from spiking neurons to neural masses and cortical fields. PLoS Comput. Biol. **4**(8) (2008)

11. A. Pouliakis, E. Karakitsou, N. Margari, P. Bountris, M. Haritou, J. Panayiotides, D. Koutsouris, P. Karakitsos, Artificial neural networks as decision support tools in cytopathology: past, present, and future. Biomed. Eng. Comput. Biol. **7**, 1–18 (2016). https://doi.org/10.4137/BECB.S31601
12. Z. Shi, L. He, Current status and future potential of neural networks used for medical image processing. J. Multimed. **6**(3) (2011)
13. K. Rohan, Vanishing of gradients (2016), https://ayearofai.com/rohan-4-the-vanishing-gradient-problem-ec68f76ffb9b. accessed: 2017-04-10
14. H. Greenspan, B. van Ginneken, R.M. Summers, Guest editorial deep learning in medical imaging: overview and future promise of an exciting new technique. IEEE Tran. Med. Imaging **35**(5), 1153–1159 (2016). https://doi.org/10.1109/TMI.2016.2553401
15. WHO, Basic malaria microscopy—Part I: Learner's guide. World Health Organization (2010)
16. G. Gopakumar, M. Swetha, G.S. Siva, G.R.K.S. Subrahmanyam, Convolutional neural network-based malaria diagnosis from focus-stack of blood smear images acquired using custom-built slide scanner. J. Biophoton. (2017). https://doi.org/10.1002/jbio.201700003
17. V.K. Jagannadh, G. Gopakumar, G.R.K.S. Subrahmanyam, S.S. Gorthi, Microfluidic microscopy-assisted label-free approach for cancer screening: automated microfluidic cytology for cancer screening. Med. Biol. Eng. Comput. 1–8 (2016). https://doi.org/10.1007/s11517-016-1549-y
18. D.E. Rumelhart, G.E. Hinton, R.J. Williams, Learning representations by back-propagating errors, *Neurocomputing: Foundations of Research*. MIT Press, Cambridge, MA, USA, pp. 696–699, http://dl.acm.org/citation.cfm?id=65669.104451
19. G. Cybenko, Approximation by superpositions of a sigmoidal function. Math. Control Signals Syst. **2**(4), 303–314 (1989). https://doi.org/10.1007/BF02551274
20. K. Hornik, Approximation capabilities of multilayer feedforward networks. Neural Netw. **4**(2), 251–257 (1991). https://doi.org/10.1016/0893-6080(91)90009-T
21. E.A. Buffalo, P. Fries, R. Landman, H. Liang, R. Desimone, A backward progression of attentional effects in the ventral stream. Proc. Natl. Acad. Sci. **107**(1), 361–365 (2010). https://doi.org/10.1073/pnas.0907658106
22. W. Zhang, K. Itoh, J. Tanida, Y. Ichioka, Parallel distributed processing model with local space-invariant interconnections and its optical architecture. Appl. Opt. **29**(32), 4790–4797 (1990). https://doi.org/10.1364/AO.29.004790
23. Y. LeCun, L. Bottou, Y. Bengio, P. Haffner, Gradient-based learning applied to document recognition. Proc. IEEE **86**, 2278–2324 (1998)
24. L. Lu, Y. Zheng, G. Carneiro, L. Yang (eds.), *Deep Learning and Convolutional Neural Networks for Medical Image Computing* (Springer International Publishing, 2017)
25. P. Nguyen, T. Tran, N. Wickramasinghe, S. Venkatesh, *mathttDeepr*: a convolutional net for medical records. IEEE J. Biomed. Health Inform. **21**(1), 22–30 (2017). https://doi.org/10.1109/JBHI.2016.2633963
26. H.C. Shin, H.R. Roth, M. Gao, L. Lu, Z. Xu, I. Nogues, J. Yao, D. Mollura, R.M. Summers, Deep convolutional neural networks for computer-aided detection: CNN architectures, dataset characteristics and transfer learning. IEEE Trans. Med. Imaging **35**(5), 1285–1298 (2016). https://doi.org/10.1109/TMI.2016.2528162
27. N. Tajbakhsh, J.Y. Shin, S.R. Gurudu, R.T. Hurst, C.B. Kendall, M.B. Gotway, J. Liang, Convolutional neural networks for medical image analysis: full training or fine tuning? IEEE Trans. Med. Imaging **35**(5), 1299–1312 (2016). https://doi.org/10.1109/TMI.2016.2535302
28. Q. Dou, H. Chen, L. Yu, J. Qin, P.A. Heng, Multilevel contextual 3-D CNNs for false positive reduction in pulmonary nodule detection. IEEE Trans. Biomed. Eng. **64**(7), 1558–1567 (2017). https://doi.org/10.1109/TBME.2016.2613502
29. L. Yu, H. Chen, Q. Dou, J. Qin, P.A. Heng, Integrating online and offline three-dimensional deep learning for automated polyp detection in colonoscopy videos. IEEE J. Biomed. Health Inform. **21**(1), 65–75 (2017). https://doi.org/10.1109/JBHI.2016.2637004
30. H. Chen, L. Wu, Q. Dou, J. Qin, S. Li, J.Z. Cheng, D. Ni, P.A. Heng, Ultrasound standard plane detection using a composite neural network framework. IEEE Trans. Cybern. **47**(6), 1576–1586 (2017). https://doi.org/10.1109/TCYB.2017.2685080

31. L. Zhang, L. Lu, I. Nogues, R.M. Summers, S. Liu, J. Yao, Deeppap: deep convolutional networks for cervical cell classification. IEEE J. Biomed. Health Inform. **21**(6), 1633–1643 (2017a). https://doi.org/10.1109/JBHI.2017.2705583
32. J.T. Kwak, S.M. Hewitt, Nuclear architecture analysis of prostate cancer via convolutional neural networks. IEEE Access **5**, 18,526–18,533 (2017). https://doi.org/10.1109/ACCESS.2017.2747838
33. R. Zhang, Y. Zheng, T.W.C. Mak, R. Yu, S.H. Wong, J.Y.W. Lau, C.C.Y. Poon, Automatic detection and classification of colorectal polyps by transferring low-level CNN features from nonmedical domain. IEEE J. Biomed. Health Inform. **21**(1), 41–47 (2017b). https://doi.org/10.1109/JBHI.2016.2635662
34. S. Christodoulidis, M. Anthimopoulos, L. Ebner, A. Christe, S. Mougiakakou, Multisource transfer learning with convolutional neural networks for lung pattern analysis. IEEE J. Biomed. Health Inform. **21**(1), 76–84 (2017)
35. H. Chen, D. Ni, J. Qin, S. Li, X. Yang, T. Wang, P.A. Heng, Standard plane localization in fetal ultrasound via domain transferred deep neural networks. IEEE J. Biomed. Health Inform. **19**(5), 1627–1636 (2015). https://doi.org/10.1109/JBHI.2015.2425041
36. S. Albarqouni, C. Baur, F. Achilles, V. Belagiannis, S. Demirci, N. Navab, Aggnet: deep learning from crowds for mitosis detection in breast cancer histology images. IEEE Trans. Med. Imaging **35**(5), 1313–1321 (2016). https://doi.org/10.1109/TMI.2016.2528120
37. S. Sathpathi, A.K. Mohanty, P. Satpathi, S.K. Mishra, P.K. Behera, G. Patel, A.M. Dondorp, Comparing Leishman and Giemsa staining for the assessment of peripheral blood smear preparations in a malaria-endemic region in india. Malar. J. **13**(1), 1–5 (2014). https://doi.org/10.1186/1475-2875-13-512
38. M. Elter, E. HaBlmeyer, T. ZerfaB, Detection of malaria parasites in thick blood films, in *2011 Annual International Conference of the IEEE Engineering in Medicine and Biology Society*, pp. 5140–5144 (2011). https://doi.org/10.1109/IEMBS.2011.6091273
39. A. Pinkaew, T. Limpiti, A. Trirat, Automated classification of malaria parasite species on thick blood film using support vector machine, in *2015 8th Biomedical Engineering International Conference (BMEiCON)*, pp. 1–5 (2015). https://doi.org/10.1109/BMEiCON.2015.7399524
40. I.K.E. Purnama, F.Z. Rahmanti, M.H. Purnomo, Malaria parasite identification on thick blood film using genetic programming, in *2013 3rd International Conference on Instrumentation, Communications, Information Technology, and Biomedical Engineering (ICICI-BME)*, pp. 194–198 (2013). https://doi.org/10.1109/ICICI-BME.2013.6698491
41. V.V. Makkapati, R.M. Rao, Segmentation of malaria parasites in peripheral blood smear images, in *2009 IEEE International Conference on Acoustics, Speech and Signal Processing*, pp. 1361–1364 (2009). https://doi.org/10.1109/ICASSP.2009.4959845
42. A. Mehrjou, T. Abbasian, M. Izadi, Automatic malaria diagnosis system, in *2013 First RSI/ISM International Conference on Robotics and Mechatronics (ICRoM)*, pp. 205–211 (2013). https://doi.org/10.1109/ICRoM.2013.6510106
43. Y. Purwar, S.L. Shah, G. Clarke, A. Almugairi, A. Muehlenbachs, Automated and unsupervised detection of malarial parasites in microscopic images. Malar. J. **10**(1), 364 (2011). https://doi.org/10.1186/1475-2875-10-364
44. A. Ravendran, K.W.T.R.T. de Silva, R. Senanayake, Moment invariant features for automatic identification of critical malaria parasites, in *2015 IEEE 10th International Conference on Industrial and Information Systems (ICIIS)*, pp. 474–479 (2015). https://doi.org/10.1109/ICIINFS.2015.7399058
45. F.B. Tek, A.G. Dempster, I. Kale, Computer vision for microscopy diagnosis of malaria. Malar. J. **8**(1), 153 (2009). https://doi.org/10.1186/1475-2875-8-153
46. W. Preedanan, M. Phothisonothai, W. Senavongse, S. Tantisatirapong, Automated detection of plasmodium falciparum from Giemsa-stained thin blood films, in *2016 8th International Conference on Knowledge and Smart Technology (KST)*, pp. 215–218 (2016). https://doi.org/10.1109/KST.2016.7440501
47. S.S. Savkare, S.P. Narote, Automated system for malaria parasite identification, in *2015 International Conference on Communication, Information Computing Technology (ICCICT)*, pp. 1–4 (2015). https://doi.org/10.1109/ICCICT.2015.7045660

48. B.E. Boser, I.M. Guyon , V.N. Vapnik, A training algorithm for optimal margin classifiers, in *Proceedings of the Fifth Annual Workshop on Computational Learning Theory*, ACM, New York, NY, USA, COLT '92, pp. 144–152 (1992), https://doi.org/10.1145/130385.130401
49. Z. Liang, A. Powell, I. Ersoy, M. Poostchi, K. Silamut, K. Palaniappan, P. Guo, M.A. Hossain, A. Sameer, R.J. Maude, J.X. Huang, S. Jaeger, G. Thoma, CNN-based image analysis for malaria diagnosis, in *2016 IEEE International Conference on Bioinformatics and Biomedicine (BIBM)*, pp. 493–496. https://doi.org/10.1109/BIBM.2016.7822567
50. N. Otsu, A threshold selection method from gray-level histograms. IEEE Trans. Syst. Man Cybern. **9**(1), 62–66 (1979)
51. G. Gopakumar, V.K. Jagannadh, S.S. Gorthi, G.R.K.S. Subrahmanyam, Framework for morphometric classification of cells in imaging flow cytometry. J. Microsc. **261**(3), 307–319 (2016). https://doi.org/10.1111/jmi.12335
52. A. Vedaldi, K. Lenc, Matconvnet—convolutional neural networks for MATLAB. CoRR abs/1412.4564. http://arxiv.org/abs/1412.4564
53. N. Linder, R. Turkki, M. Walliander, A. Mårtensson, V. Diwan, E. Rahtu, M. Pietikäinen, M. Lundin, A malaria diagnostic tool based on computer vision screening and visualization of plasmodium falciparum candidate areas in digitized blood smears. PLoS ONE **9**(8), e104,855 (2014)
54. LBP/VAR implementation; centre for machine vision and signal analysis. University of Oulu (2016), http://www.cse.oulu.fi/CMV/Downloads/LBPMatlab. Accessed 15 Oct 2016
55. T. Ojala, M. Pietikainen, T. Maenpaa, Multiresolution gray-scale and rotation invariant texture classification with local binary patterns. IEEE Trans. Pattern Anal. Mach. Intell. **24**(7), 971–987 (2002). https://doi.org/10.1109/TPAMI.2002.1017623
56. D.G. Lowe, Distinctive image features from scale-invariant keypoints. Int. J. Comput. Vis. **60**(2), 91–110 (2004). https://doi.org/10.1023/B:VISI.0000029664.99615.94
57. B.W. Stewart, C. Wild, *World Cancer Report 2014* (World Health Organization, 2014)
58. W. Zhang, R. Li, T. Zeng, Q. Sun, S. Kumar, J. Ye, S. Ji, Deep model based transfer and multi-task learning for biological image analysis, in *Proceedings of the 21th ACM SIGKDD International Conference on Knowledge Discovery and Data Mining*, ACM, New York, NY, USA, KDD '15, pp. 1475–1484 (2015). https://doi.org/10.1145/2783258.2783304
59. T. Zeng, R. Li, R. Mukkamala, J. Ye, S. Ji, Deep convolutional neural networks for annotating gene expression patterns in the mouse brain. BMC Bioinform. **16**(1), 1–10 (2015). https://doi.org/10.1186/s12859-015-0553-9
60. K. He, X. Zhang, S. Ren, J. Sun, Delving deep into rectifiers: surpassing human-level performance on imagenet classification. ArXiv e-prints **1502**, 01852 (2015)
61. K. Chatfield, K. Simonyan, A. Vedaldi, A. Zisserman, Return of the devil in the details: delving deep into convolutional nets, in *British Machine Vision Conference* (2014)
62. J. Deng, W. Dong, R. Socher, L.J. Li, K. Li, L. Fei-Fei, ImageNet: a large-scale hierarchical image database. In: *CVPR09* (2009)
63. I. Jolliffe, *Principal Component Analysis*. Springer Series in Statistics (Springer, 2002)
64. Y. Bar , I. Diamant , L. Wolf , S. Lieberman, E. Konen, H. Greenspan, Chest pathology detection using deep learning with non-medical training, in *2015 IEEE 12th International Symposium on Biomedical Imaging (ISBI)*, pp. 294–297, https://doi.org/10.1109/ISBI.2015.7163871
65. E.J. Breen, R. Jones, Attribute openings, thinnings, and granulometries. Comput. Vis. Image Underst. **64**(3), 377–389 (1996). https://doi.org/10.1006/cviu.1996.0066

Application of Deep Neural Networks for Disease Diagnosis Through Medical Data Sets

Alper Baştürk, Hasan Badem, Abdullah Caliskan and Mehmet Emin Yüksel

Abstract In this chapter, a novel classification methodology for medical disease diagnosis is proposed. The proposed classification operator comprises a stacked autoencoder network cascaded with a softmax layer. The classifier is trained by applying a special training approach, where each layer of the proposed classifier is trained individually and sequentially. The performance of the proposed classifier is compared with a number of representative classification methods from the literature. The experimental results on medical data sets show that the proposed classifier performs better than or at least competitive with classifiers used in this chapter. It is also seen that the proposed classifier can efficiently be used for the diagnosis of medical diseases provided that it is trained with a suitable data set with a sufficient number of medical features obtained from a sufficient number of patients.

1 Introduction

A great majority of human deaths are due to a number of major diseases such as cardiovascular diseases and various forms of cancer. The early diagnosis of these diseases is vitally important regarding the success as well as the cost of the treatment and survival of the patients. Therefore, it is of high importance to develop and employ methods that allow accurate diagnosis of diseases as at an early stage as possible.

A. Baştürk
Department of Computer Engineering, Erciyes University, 38039 Kayseri, Turkey

H. Badem
Department of Computer Engineering, Kahramanmaras Sutcu Imam University, 46100 Kahramanmaras, Turkey

A. Caliskan
Department of Biomedical Engineering, Iskenderun Technical University, 31200 Hatay, Turkey

M. E. Yüksel (✉)
Department of Biomedical Engineering, Erciyes University, 38039 Kayseri, Turkey
e-mail: yuksel@erciyes.edu.tr

V. E. Balas et al. (eds.), *Handbook of Deep Learning Applications*, Smart Innovation, Systems and Technologies 136,
https://doi.org/10.1007/978-3-030-11479-4_12

The most straightforward way to diagnose a particular disease is to get the patient to the hospital and perform appropriate medical tests and procedures to acquire information that enables physicians to reach a decision regarding the presence of that particular diseases on the patient. This approach of disease diagnosis is very accurate and results obtained in this way are usually very reliable, which together make this approach highly desirable.

On the other hand, some of the above mentioned medical tests and procedures are usually very sophisticated, and in most cases, they can only be conducted in well established hospitals and health institutions. Sometimes, these procedures involve invasive operations that can only be practiced by highly trained medical experts. Furthermore, these procedures are frequently very costly and, in some cases, very risky regarding patient health. Finally, only very few number of patients living in the developed countries and regions of the world have access to this facility. Therefore, it is highly desirable to develop novel, cost effective and easily accessible methods for disease diagnosis.

In the last few years, the applications of deep neural networks (DNN) in classification problems have gained high popularity in various areas of science and technology [1–7]. One good example of such applications is disease diagnosis by classification, which is accomplished by acquiring and processing diagnostic information from medical data sets through intelligent classifiers based on DNNs [8]. In fact, a considerable number of different and well established medical data sets are available in the literature and DNN based classifiers can easily be utilized to operate on these data sets to generate diagnostic decisions regarding the presence of a certain disease related with the database. This approach can greatly eliminate the need for sophisticated medical diagnosis procedures discussed above and replace them with a much simpler, cheaper, non-invasive and risk-free method as long as suitable network structures and learning strategies are employed.

Many different algorithms are available in the literature for the solution of classification problems. Among these algorithms; *k-nearest neighbor classifiers (KNN)* [9, 10], *decision trees (DT)* [9, 11, 12], *Naive Bayes classifiers (NB)* [10–12], *neural and fuzzy networks* [9, 13] and *support vector machines (SVM)* [10, 14, 15] are the most commonly used ones. Each of these methods actually represent a group of similar classification methods with certain advantages and disadvantages.

The KNN is the most basic and well known non-parametric classification algorithm [16]. The algorithm utilizes the information from a training set, which comprises a certain number of feature vectors in the feature space with known class labels. The class label of a new feature vector, which is not a member of the training set, is made equal to the class label of the majority of the k nearest neighboring feature vectors in the training set. The main advantage of this approach is its computational efficiency since there is no need for exclusively training the classifier. Despite its simplicity, the KNN yields satisfying results for classification problems that are not too complex. However, it has usually a relatively low accuracy rate due to its sensitivity to noise [16].

DT is another class of simple classifiers that are expressed as a recursive partition of the feature space [17]. There are many different kinds of successful examples of decision trees in the literature which generate classifiers such as ID3 [18], C4.5 [19], and CART [20]. DTs have also been successfully used for the classification of medical data sets [9, 11, 12]. They are capable of handling data sets that may have errors and missing values. However, this method has a replication problem and over-sensitivity to the training set. Moreover, its performance is degraded by irrelevant attributes and noise in the data set [19, 21].

Bayesian classifiers are statistical classifiers which may predict class membership probabilities. The NB, from the family of the Bayesian statistical classifiers, are widely studied in machine learning and the classification of data sets from different fields of science.They have also been applied in the classification of medical data sets [10–12]. The structure of the NB mechanism is simple in implementation but it exhibits good performance. However, the NB classifier assumes that the classes are conditionally independent, which causes a loss of accuracy. In addition, dependencies existing among variables cannot be modeled by the NB classifier [22].

The SVM is a supervised learning model that is used as a binary classifier. Given a set of training feature vectors where each vector is labeled in advance as belonging to one of the two available classes, an SVM classifier is optimized to become a model that maps the training vectors with points in the feature space in such a manner that the training vectors of the two classes are divided by a clear gap as wide as possible. The SVM constructs a hyperplane or a set of hyperplanes in a high-dimensional space which functions as a decision boundary in the feature space. The classification process is performed by determining the appropriate side of this decision boundary that the input feature vector belongs to [23]. The SVM does not usually gets trapped in local minima during training [24] and generally yields high accuracy rates in most classification problems including the classification of medical data sets provided that a suitable kernel is chosen [10, 14, 15]. However, it has a number of disadvantages including the difficulty in choosing the most appropriate kernel for a given data set, high computational complexity, and demand for memory in large scale classification problems [25, 26].

In addition to the KNN, DT, NB and SVM methods discussed above, a number of classification methods based on soft computing methodologies have also been presented [9, 13]. These methods generally offer relatively better classification results with higher accuracy rates than the KNN, DT, NB and SVM methods. However, the structures of these classifiers are much more complicated and their implementation as well as training is more difficult.

Great majority of the classification methods have some drawbacks as mention above. In addition, their performances also depend on one or more user-supplied external parameters such as, the choice of the kernel, inducers and structure; values of some tuning parameters, and so on. These choices are heuristically made, externally supplied and experimentally validated by the user for each individual data set since there is no analytical method to choose best value of these parameters for an effective classification.

Recently, there has been a huge research in the applications of deep learning based tools such as DNNs comprising Stacked Autoencoders (sAE) and Softmax Classifiers to the problems in classification [1–4, 27, 28]. Indeed, DNNs offer the ability of sAE to learn feature hierarchies, and the capability of softmax classifiers to predict the correct class label for efficient disease diagnosis. Therefore, DNNs may be utilized to classify medical data sets, and diagnose diseases.

In this chapter, we present an efficient general-purpose classification tool and evaluate its performance on medical data sets. The presented tool is based on a DNN constructed by appropriately combining a two cascaded autoencoder (AE) layers and a softmax classifier layer. The performance of the presented DNN classifier is tested on various medical data sets. In addition the DNN classifiers are compared with representative conventional and state-of-the-art classification methods, which is available in the literature. Results of the classification experiments show that the presented DNN classifier has superior performance over the other classifiers used in the experiments and is capable of efficiently classifying 7 different medical data sets.

The rest of this chapter is organized as follows. Section 2 defines the presented DNN and its training procedure. Section 3 reports the results of the classification experiments and their discussions. Section 4 presents the conclusions.

2 Method

The proposed method for medical data set classification is based on a DNN operator having a sAE network and a softmax classifier.

2.1 The Autoencoder

An AE is a neural network containing three layers, including input, hidden and output. Although an AE is structurally similar to a feed forward neural network, its puspose of use is very different. The AE is trained to form its own input at the output. For this reason, the number of neurons in the output layer is always the same as the size of the input vector. The main aim of AE is to generate a different representation of the input in the hidden layer. The generated new feature vector at the hidden layer is called a *code* [29]. The number of the neurons in the hidden layer determines the dimension of the code, which generally is fewer than the dimension of the input. The AE maps the input space to the code space in order to define the input more clearly than before [30–32].

Figure 1 demonstrates an AE with three layers. The AE consists basically of two parts: *encoder* and *decoder*. The encoder generates the code space, which represents the input space in a different manner. The decoder reproduces the input from the code space. The dimensions of the input and code space are M and N, respectively.

The encoder is given by:

$$\mathbf{c} = f(\mathbf{b} + \mathbf{W}^{\mathrm{T}}\mathbf{x}) \tag{1}$$

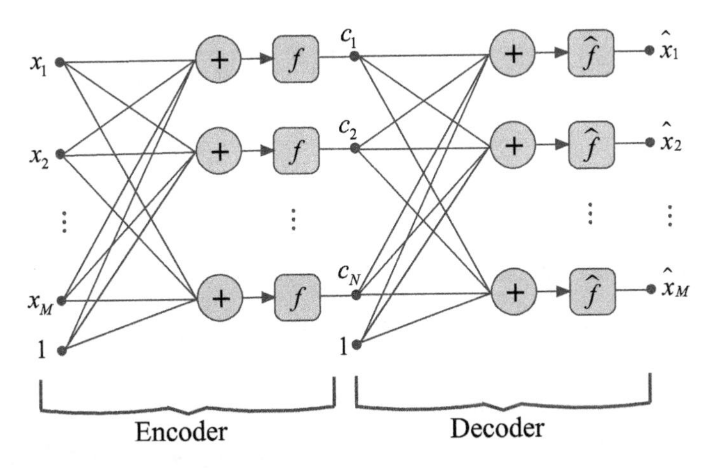

Fig. 1 An autoencoder network

where

$$\begin{aligned}\mathbf{c} &= [c_1\ c_2\ \ldots\ c_N]^{\mathrm{T}}\\ \mathbf{x} &= [x_1\ x_2\ \ldots\ x_M]^{\mathrm{T}}\\ \mathbf{b} &= [b_1\ b_2\ \ldots\ b_N]^{\mathrm{T}}\\ \mathbf{W} &= [\mathbf{w}_1\ \mathbf{w}_2\ \ldots\ \mathbf{w}_N]\end{aligned}$$

with f being a sigmoid activation function.

Here the **b** vector and **W** matrix contain the biases and weights of the encoder, respectively. The columns of the **W** can be defined as follows:

$$\mathbf{w}_i = [w_{i1}\ w_{i2}\ \ldots\ w_{iM}]^{\mathrm{T}} \tag{2}$$

with $i = 1, 2 \ldots N$.

The encoder part can shortly be given by

$$\mathbf{c} = g_E(\mathbf{W}, \mathbf{b};\ \mathbf{x}) \tag{3}$$

where g_E is the *encoding function.*

The decoder part is very similar to encoder which can be defined as follows

$$\hat{\mathbf{x}} = \hat{f}(\hat{\mathbf{b}} + \hat{\mathbf{W}}^{\mathrm{T}}\mathbf{c}) \tag{4}$$

where

$$\begin{aligned}\hat{\mathbf{x}} &= [\hat{x}_1\ \hat{x}_2\ \ldots\ \hat{x}_M]^{\mathrm{T}}\\ \hat{\mathbf{b}} &= [\hat{b}_1\ \hat{b}_2\ \ldots\ \hat{b}_M]^{\mathrm{T}}\\ \hat{\mathbf{W}} &= [\hat{\mathbf{w}}_1\ \hat{\mathbf{w}}_2\ \ldots\ \hat{\mathbf{w}}_N]\end{aligned}$$

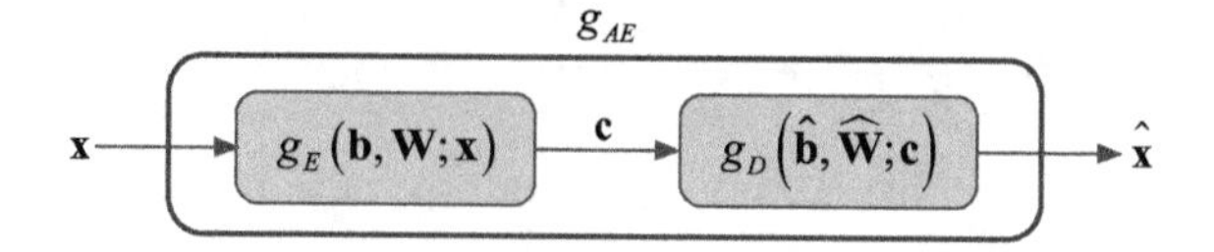

Fig. 2 The diagram of the autoencoder network

and $\hat{f}$ is similar to f. The $\hat{\mathbf{W}}$ consist of the weights, which can be defined as follows:

$$\hat{\mathbf{w}}_i = [\hat{w}_{i1}\ \hat{w}_{i2}\ \dots\ \hat{w}_{iN}]^{\mathrm{T}} \tag{5}$$

with $i = 1, 2 \dots M$.

The decoder part can shortly be given as

$$\hat{\mathbf{x}} = g_D(\hat{\mathbf{W}}, \hat{\mathbf{b}};\ \mathbf{c}) \tag{6}$$

where g_D is the *decoding function.*

The AE complete with its encoder and decoder sections may be illustrated as in Fig. 2.

2.2 *Training Procedure of the Autoencoder*

Let $\{\mathbf{x}^{(1)}, \mathbf{x}^{(2)} \dots \mathbf{x}^{(m)}\}$ denote the m input vectors, which are recruited for training the AE mentioned in the above subsection. Objective function of the AE training process can be divided into two parts. The first part is the objective function of the network that corresponds to the error term and it can be defined as follows [32]:

$$E_N = \frac{1}{m}\sum_{k=1}^{m} e_k^2 + \frac{\lambda}{2}\left(\sum_{i=1}^{N} \|\mathbf{w}_\mathbf{i}\| + \sum_{i=1}^{M} \|\hat{\mathbf{w}}_i\|\right) \tag{7}$$

Here, error vector e_k is the difference between actual $\hat{\mathbf{x}}$ and the desired $\mathbf{x}$ outputs. e_k is defined as follows:

$$e_k = \left\|\mathbf{x}^{(k)} - \hat{\mathbf{x}}^{(k)}\right\| \qquad k = 1, 2, \ \dots, m. \tag{8}$$

λ is a regularization term (also known as a *weight decay term*) which is utilized to prevent overfitting [32].

It should be noted that the E_N is a function of the weights of the AE.

$$E_N = g_{AE}(\mathbf{W}, \mathbf{b}, \hat{\mathbf{W}}, \hat{\mathbf{b}}) \tag{9}$$

The second part of the objective function is E_S, which is employed for imposing sparsity constraint to AE network. E_S defined as follows:

$$E_S = \beta \sum_{j=1}^{N} KL(\rho \parallel \hat{\rho}_j) \tag{10}$$

where β is the weight of the sparsity penalty term. In this expression, the KL is the Kullback–Leibler divergence [30, 32] given as

$$KL(\rho \parallel \hat{\rho}_j) = \rho log\frac{\rho}{\hat{\rho}_j} + (1-\rho)log\frac{1-\rho}{1-\hat{\rho}_j} \tag{11}$$

where ρ is a constant named *sparsity parameter* and $\hat{\rho}_j$ is the mean activation value of jth neuron for all training set, which may be defined as follows

$$\hat{\rho}_j = \frac{1}{m}\sum_{i=1}^{m} f_j(\mathbf{x}^{(i)}) \tag{12}$$

Finally, the total cost function of the AE is given as follows:

$$E = \frac{1}{S}\sum_{k=1}^{S} e_k^2 + \frac{\lambda}{2}\left(\sum_{i=1}^{N} \|\mathbf{w_i}\| + \sum_{i=1}^{M} \|\hat{\mathbf{w}}_i\|\right) + \beta \sum_{j=1}^{N} KL(\rho \parallel \hat{\rho}_j) \tag{13}$$

2.3 *The Stacked Autoencoder*

A sAE network, as the name implies, is a network of stacked autoencoder layers. All layers are actually encoder sections of previously trained autoencoders. The number of cascaded encoder layers are determined depending on the particular application. The input-output mapping equation of the sAE with L layers can be conceptually defined as in the following equation and may be demonstrated as in Fig. 3.

$$g_{SAE} = g_E^1 \circ g_E^2 \circ \cdots \circ g_E^L \tag{14}$$

It should be observed that the decoder parts of the trained autoencoders are not included in the structure when constructing the sAE as they are only needed during the training.

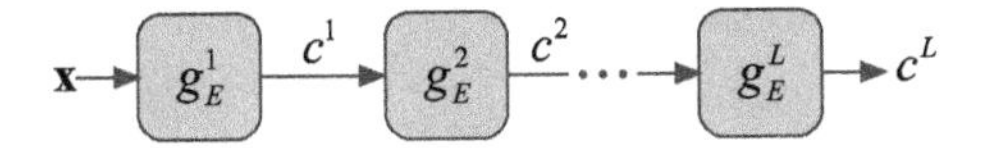

Fig. 3 A stacked autoencoder with L autoencoder layers

2.4 The Softmax Classifier

The softmax classifier is based on the *softmax function*, which is defined as follows:

$$v_j = \frac{\exp(u_j)}{\sum_{k=1}^{K} \exp(u_k)} \quad (j = 1, 2, \ldots K) \tag{15}$$

It is easily observed from the above equation that the softmax function maps a K-dimensional vector of arbitrary real values (u_j) to another K-dimensional vector of real values (v_j) that are all within the interval [0, 1] and add to 1.

Inspired by the softmax function, the *softmax classifier* is a classification operator for general multiclass classification problems [33]. It is constructed by cascading a neural layer and a normalization layer, as illustrated in Fig. 4a.

The neural layer is the input layer of the softmax classifier and is structurally very similar to the encoding section of an AE. The only difference is that the neuron activation function here is the exponential function.

The input-output relationship of the neural layer of a softmax classifier that maps input code vectors from an N-dimensional space to K classes is as follows:

$$\mathbf{e} = \exp(\mathbf{d} + \mathbf{S}^{\mathrm{T}}\mathbf{c}) \tag{16}$$

where

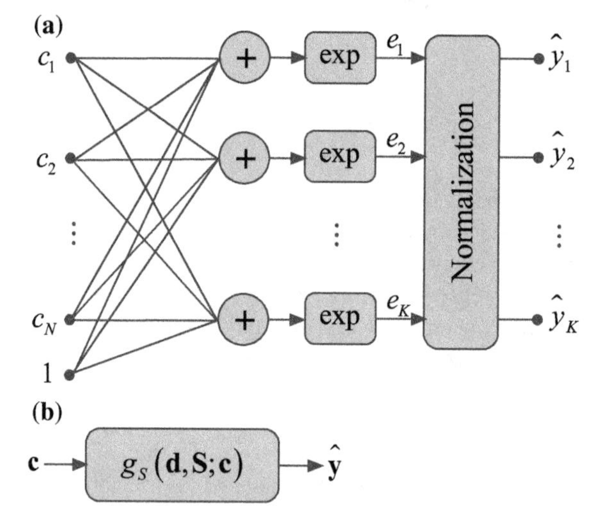

Fig. 4 **a** A softmax classifier classifying N-dimensional input vectors into K output classes and **b** its short representation

$$\mathbf{e} = [e_1\ e_2\ \ldots\ e_K]^{\mathrm{T}}$$
$$\mathbf{d} = [d_1\ d_2\ \ldots\ d_K]^{\mathrm{T}}$$
$$\mathbf{S} = [\mathbf{s}_1\ \mathbf{s}_2\ \ldots\ \mathbf{s}_K]$$
$$\mathbf{c} = [c_1\ c_2\ \ldots\ c_N]^{\mathrm{T}}$$

Here, the elements of the **d** vector are the neuron biases and the columns of the **S** matrix are the weights connecting the input nodes to the kth neuron:

$$\mathbf{s}_k = [s_{k1}\ s_{k2}\ \ldots\ s_{kN}]^{\mathrm{T}} \tag{17}$$

with $k = 1, 2 \ldots K$.

The normalization layer is the output layer and it is simply used for normalizing the output values of the neural layer of the softmax classifier:

$$y_j = \frac{e_j}{\sum_{k=1}^{K} e_k} \quad (j = 1, 2, \ldots K) \tag{18}$$

which may alternatively be expressed in terms of the input vector **c** as

$$y_j = \frac{\exp(\mathbf{s}_j^{\mathrm{T}}\mathbf{c})}{\sum_{k=1}^{K} \exp(\mathbf{s}_k^{\mathrm{T}}\mathbf{c})} \quad (j = 1, 2, \ldots K) \tag{19}$$

This normalization process ensures that the values obtained at outputs y_j of the softmax classifier are always contained in the interval [0, 1] and their sum is always equal to 1. Hence, the values obtained at the individual outputs of the softmax classifier for a given input vector may be regarded as the probabilities of the input vector being in the classes associated with individual outputs of the softmax classifier. Based on this interpretation, a given input vector is classified by placing it to the class represented by the individual softmax classifier output that yields the highest probability.

The input-output relationship of the softmax classifier may shortly be written as

$$\mathbf{y} = g_S(\mathbf{d}, \mathbf{S};\ \mathbf{c}) \tag{20}$$

where

$$\mathbf{y} = [y_1\ y_2\ \ldots\ y_K]^{\mathrm{T}} \tag{21}$$

and shortly be depicted as in Fig. 4b.

The training of the softmax classifier is achieved by using following objective function with respect to training set $\{\mathbf{c}^{(i)},\ y^{(i)}\}_{i=1}^{m}$.

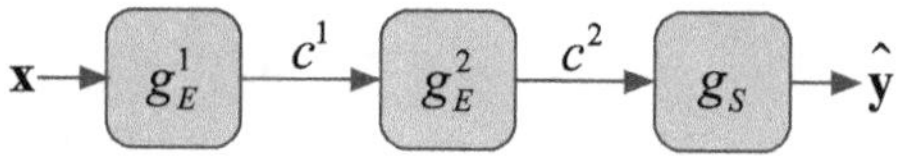

Fig. 5 The deep neural network

$$J = -\frac{1}{m}\sum_{i=1}^{m}\sum_{j=1}^{K}\mathbf{1}\left\{y^{(i)} = j\right\}\log\frac{\exp(\mathbf{s}_j^{\mathrm{T}}\mathbf{c}^{(i)})}{\sum_{l=1}^{K}\exp(\mathbf{s}_l^{\mathrm{T}}\mathbf{c}^{(i)})} + \frac{\lambda}{2}\sum_{i=1}^{K}\sum_{j=1}^{N}s_{ij} \tag{22}$$

In the above equation, $\mathbf{1}\{\cdot\}$ is known as *indicator function*, so that $\mathbf{1}\{\textit{a true statement}\} = 1$, and $\mathbf{1}\{\textit{a false statement}\} = 0$.

2.5 The Deep Neural Network Classifier

The proposed classification operator for processing medical data sets is a DNN constructed by cascading a sAE network with a softmax classifier layer. The sAE network contains two AE layers. The structure diagram of the DNN is illustrated in Fig. 5.

2.6 Training of the DNN Classifier

The training of the DNN is achieved by using a suitable data set (i.e medical data set) which is $\{\mathbf{x}^{(1)}, \mathbf{x}^{(2)}, \ldots, \mathbf{x}^{(m)}\}$ with the m input vectors and labels $\{y^{(1)}, y^{(2)}, \ldots, y^{(m)}\}$. Each of the labels is equal to either 0 or 1, where the value of 0 represents a negative decision and 1 represents a positive decision regarding disease diagnosis, respectively. The main purpose of the training is to tune the internal parameters of the DNN with respect to training data set.

The L-BFGS training algorithm [34] is used to fit the internal parameters of the proposed classifier to their optimal values. This algorithm is a powerful training algorithm especially suited to the kind of optimization problems where there are a large number of model parameters. The algorithm is especially popular because of its speed and relatively less memory requirements.

A special training procedure is adopted here to obtain the best performance. For this purpose, each layer of the proposed classifier is trained individually and sequentially. The training procedure comprises the following steps:

1. At the beginning, the first AE is trained by using the original input vectors $\mathbf{x}^{(i)}$ $(i = 1, 2, \ldots, m)$, as shown in Fig. 6a.
2. Then, the second AE is trained with $\mathbf{c}^{1,(i)}$ $(i = 1, 2, \ldots, m)$ code vectors, which is generated by first AE, as shown in Fig. 6b.
3. Following this, the softmax layer is trained with $\mathbf{c}^{2,(i)}$ $(i = 1, 2, \ldots, m)$ which is generated by first AE and targets $y^{(i)}$ $(i = 1, 2, \ldots, m)$, shown in Fig. 6c.

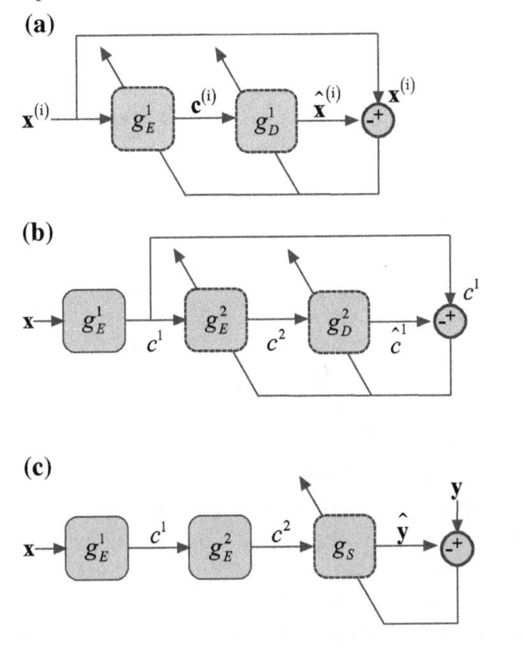

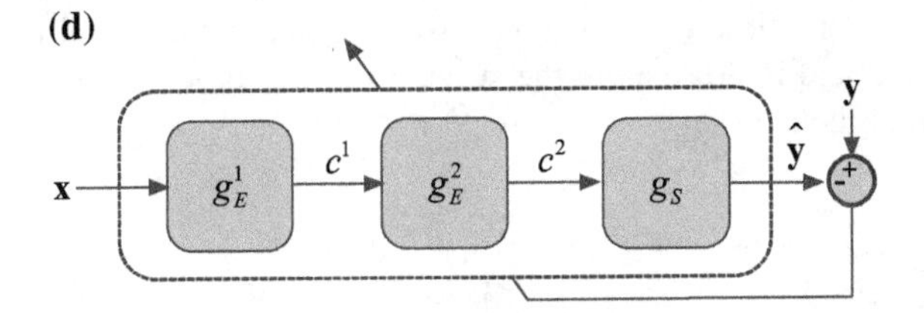

Fig. 6 Layer by layer training of the deep neural network based classification operator. **a** Training of the first autoencoder layer. **b** Training of the second autoencoder layer. **c** Training of the softmax layer. **d** Fine-tuning the whole network

4. Finally, an additional fine-tuning procedure is applied to DNN training task in order to improve the ability of the classification. This is shown in Fig. 6d.

3 Experimental Results and Discussion

In this chapter, the performance of the DNN classifier is tested on 7 medical data sets, which are taken from UCI repository [35, 36] with a variety of domains, binary classes, attributes and instances. Details about used data sets are given in following section. For sake of fair comparison, 10-fold cross-validation is applied to data sets. The DNN classifier is compared with state-of-the-art methods, which are SVM, NB, KNN, DT and some soft computing based classifiers for each data set. The DNN is implemented over a system with Intel i7 2600 3.4 GHz CPU and 12 GB DDR3 RAM.

3.1 Experimental Setup of the Designed DNN

In order to demonstrate the effectiveness of the proposed classifier, seven experimental setups are designed for each data set. The DNN classifier with two AE and a softmax layer is designed for the classification of medical data sets. The parameter settings of the DNN are given in Tables 1, 2 and 3 for each data set. Experimental setup of the designed DNN is shown in Fig. 7.

3.2 Principal Components Analysis

Dimensionality reduction is an important procedure in the storage, classification, communication and visualization of high dimensional data. An easy and commonly used procedure is the principal components analysis (PCA), which detects the ways of highest variance in the data set and depicts each data point by its coordinates along each of these directions [29].

For visualization, the first and second principle components are plotted against each other to acquire a two-dimensional graph of the data that captures the highest variance that contains relevant information [37]. Third principle component with the first and second principle components is used for tree-dimensional representation. Thus analyzing and interpretation the structure of a data set is easy thanks to this visualization of the data set.

3.3 Data Sets

Seven data sets are chosen to show the effectiveness of the DNN classifier. Medical data sets used in this chapter, which are binary class, have different number of features and instances. Detailed information about these data sets are as follows.

3.3.1 Heart Disease Data Sets

Heart diseases or cardiovascular diseases relate to a number of complications which afflict blood vessels in the heart. In particular, coronary artery disease (CAD) is a main reason of disorder and death in the contemporary community. When plaque such as a waxy substance builds up inside the coronary arteries, a serious disease named CAD emerges. These arteries carry oxygen-rich blood to the heart muscle [9, 38]. When the narrowing of at least one of the blood vessels in the heart is more than 50%, the CAD emerges, which is normally detected by using coronary angiogram or cardiac catheterization. This approach has high acceptance. However both methods are invasive, costly and not available for large populations. A few

Table 1 Setting parameters of DNN for heart diseases data sets

Data set	1st Autoencoder		2nd Autoencoder		Softmax layer		Fine-tuning	
Cleveland	Number of neuron	4	Number of neuron	4	Lambda (λ)	0.003	Lambda (λ)	0.003
	Sparsity (ρ)	0.4	Sparsity (ρ)	0.1	Training algorithm	L-FBGS	Training algorithm	L-FBGS
	Lambda (λ)	0.003	Lambda (λ)	0.003	Iteration	400	Iteration	400
	Beta (β)	4	Beta (β)	1	Input normalized?	Yes		
	Training algorithm	L-FBGS	Training algorithm	L-FBGS				
	Iteration	400	Iteration	400				
	Input normalized?	Yes	Input normalized?	Yes				
Hungarian	Number of neuron	4	Number of neuron	4	Lambda (λ)	0.003	Lambda (λ)	0.003
	Sparsity (ρ)	0.5	Sparsity (ρ)	0.1	Training algorithm	L-FBGS	Training algorithm	L-FBGS
	Lambda (λ)	0.003	Lambda (λ)	0.003	Iteration	400	Iteration	400
	Beta (β)	2	Beta (β)	4	Input normalized?	Yes		
	Training algorithm	L-FBGS	Training algorithm	L-FBGS				
	Iteration	400	Iteration	400				
	Input normalized?	Yes	Input normalized?	Yes				
Switzerland	Number of neuron	4	Number of Neuron	4	Lambda (λ)	0.003	Lambda (λ)	0.003
	Sparsity (ρ)	0.4	Sparsity (ρ)	0.4	Training algorithm	L-FBGS	Training algorithm	L-FBGS
	Lambda (λ)	0.003	Lambda (λ)	0.003	Iteration	400	Iteration	400
	Beta (β)	2	Beta (β)	1	Input normalized?	Yes		
	Training algorithm	L-FBGS	Training algorithm	L-FBGS				
	Iteration	400	Iteration	400				
	Input normalized?	Yes	Input normalized?	Yes				
VA Long Beach	Number of neuron	4	Number of neuron	4	Lambda (λ)	0.003	Lambda (λ)	0.003
	Sparsity (ρ)	0.2	Sparsity (ρ)	0.2	Training algorithm	L-FBGS	Training algorithm	L-FBGS
	Lambda (λ)	0.003	Lambda (λ)	0.003	Iteration	400	Iteration	400
	Beta (β)	3	Beta (β)	3	Input normalized?	Yes		
	Training algorithm	L-FBGS	Training algorithm	L-FBGS				
	Iteration	400	Iteration	400				
	Input normalized?	Yes	Input normalized?	Yes				

Table 2 Setting parameters of DNN for breast cancer data sets

Data set	1st Autoencoder		2nd Autoencoder		Softmax layer		Fine-tuning	
Breast cancer	Number of neuron	4	Number of neuron	4	Lambda (λ)	0.003	Lambda (λ)	0.003
	Sparsity (ρ)	0.3	Sparsity (ρ)	0.2	Training algorithm	L-FBGS	Training algorithm	L-FBGS
	Lambda (λ)	0.003	Lambda (λ)	0.003	Iteration	400	Iteration	400
	Beta (β)	4	Beta (β)	2	Input normalized?	Yes		
	Training algorithm	L-FBGS	Training algorithm	L-FBGS				
	Iteration	400	Iteration	400				
	Input normalized?	Yes	Input normalized?	Yes				
Wisconsin breast cancer	Number of neuron	8	Number of neuron	4	Lambda (λ)	0.003	Lambda (λ)	0.003
	Sparsity (ρ)	0.5	Sparsity (ρ)	0.2	Training algorithm	L-FBGS	Training algorithm	L-FBGS
	Lambda (λ)	0.003	Lambda (λ)	0.003	Iteration	400	Iteration	400
	Beta (β)	2	Beta (β)	5	Input normalized?	Yes		
	Training algorithm	L-FBGS	Training algorithm	L-FBGS				
	Iteration	400	Iteration	400				
	Input normalized?	Yes	Input normalized?	Yes				

methods including, electrocardiogram, image and heart sound analysis [9] are applied to diagnose the CAD in less expensive and non-invasive ways [9].

The main objective is to forecast the existence of CAD from a number of demographic, detected, and measured patient features by using four different CAD data sets (i.e Cleveland, Hungarian, Long Beach, Switzerland) received from Data Mining Repository of University of California, Irvine (UCI) [35].

Table 3 Setting parameters of DNN for Pima Indian diabetes data set

Data set	1st Autoencoder		2nd Autoencoder		Softmax layer		Fine-tuning	
Diabetes	Number of neuron	4	Number of Neuron	2	Lambda (λ)	0.001	Lambda (λ)	0.001
	Sparsity (ρ)	0.15	Sparsity (ρ)	0.15	Training algorithm	L-FBGS	Training algorithm	L-FBGS
	Lambda (λ)	0.003	Lambda (λ)	0.003	Iteration	1000	Iteration	1000
	Beta (β)	2	Beta (β)	2	Input normalized?	Yes		
	Training algorithm	L-FBGS	Training algorithm	L-FBGS				
	Iteration	1000	Iteration	1000				
	Input normalized?	Yes	Input normalized?	Yes				

The first data set is Cleveland heart disease data set consisting of 14 attributes with 303 instances [39–42]. Six of the them are removed due to missing values.

The second data set about CAD is Hungarian heart disease data set. Three of the attributes have been removed because of a large percentage of missing values. However the form of the data is completely the same as that of the first data set. 34 instances of the database is removed due to missing values and 261 examples were present [43].

The third data set is VA Long Beach heart disease data set that consists of 14 attributes and also 200 instances are used in this research. Four of the attributes and 67 instances have been removed because of a large percentage of missing values in the VA Long Beach data sets.

The last data set about CAD is Switzerland heart disease data set which contains of 14 attributes and also 123 instances are used in this research. Four of the attributes and 18 instances have been removed because of a large percentage of missing values in the Switzerland data sets.

3.3.2 Breast Cancer Data Sets

Breast cancer emerges with abnormal growth of breast cells which are either benign (non-cancerous) or malignant (cancerous). Early detection is very important the patients suffered form breast cancer [44]. Surgical biopsy is most reliable method for detection of breast cancer. However, biopsy is a costly and time consuming procedure. However, Fine Needle Aspiration biopsy (FNA) is one of the most accepted

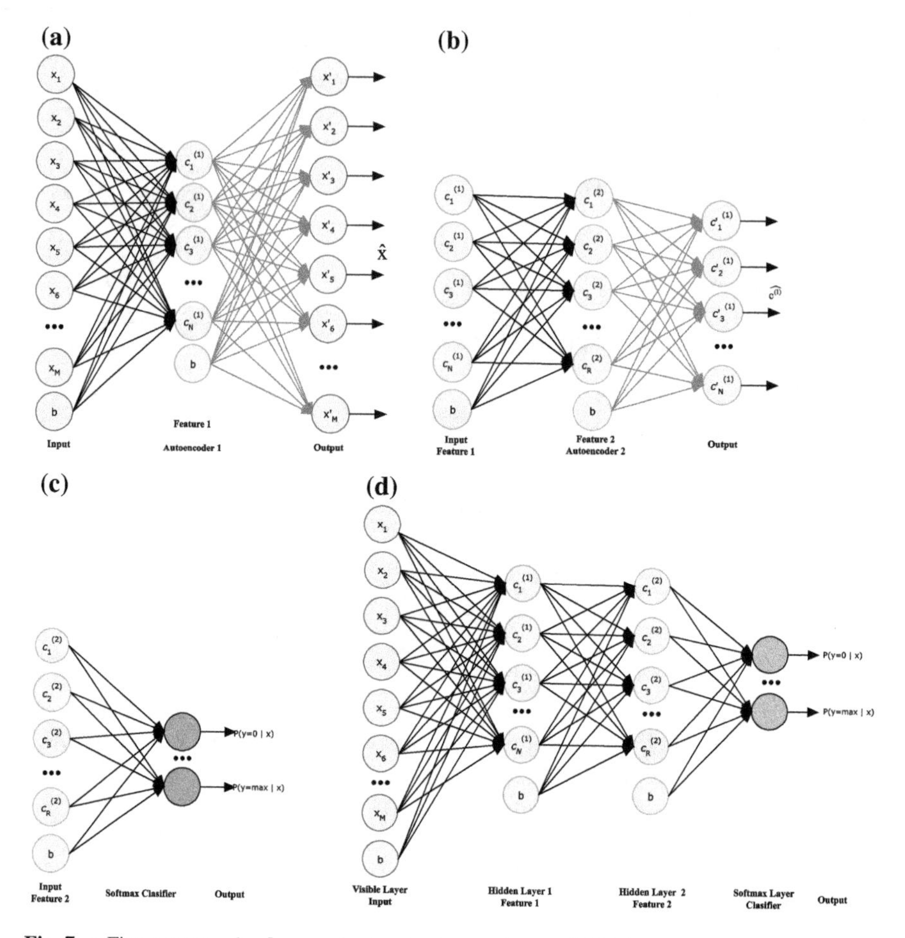

Fig. 7 **a** First autoencoder, **b** second autoencoder, **c** softmax, **d** DNN

diagnostic procedure for breast cancer without the adverse effects of surgical biopsy [45]. A breast FNA is a quick and easy method to make that a fine needle and syringe is used to take a sample like some fluids or cells from the breast lesion or cyst [45, 46]. The sample is examined under a microscope to extract the feature of the tumor. These features can be utilized to detect the breast cancer by using soft computing methods [47].

On the other hand, at present about 40% of all patients who suffer breast cancer face a recurrence and nearly all of them die due to this fact. Therefore, cancer-related loss in women is most commonly caused by the breast cancer. The risk of recurrence is highest in the first 2 or 3 years and then decline steadily, but this risk never disappears [48]. The prediction of recurrence with machine learning techniques is also possible like whether tumor is benign or malignant.

Wisconsin breast cancer data set and breast cancer data set are utilized to apply the proposed technique. Breast cancer data set is utilized in order to forecast the existence of recurrence from measured patient features. Th other data set is collected to decide whether breast cancer tumor is benign or malignant. This data set is received from Data Mining Repository of University of California, Irvine (UCI) [35].

First data set is breast cancer data set and contains patients with known diagnostic conditions 5 years after the operation were performed [49, 50]. Data sets with 286 instances and 9 attributions investigate recurrence or non-recurrence of breast cancer. Nine instances are not used because of missing attributes.

Another data set is Wisconsin breast cancer data set that contains 699 instances and 9 attributes. Each instance consists of nine cytological attributes of benign or malignant breast FNA [51, 52]. 16 instances are not used for analysis due to lack of the some features.

3.3.3 Diabetes Mellitus Data Set

Diabetes mellitus is very dangerous disease which effects all blood vessels particularly capillaries [53]. There exists two major different type of diabetes mellitus. Type I diabetes is commonly diagnosed in young adults and children [54] whose pancreas can not produce insulin due to loss of the beta cells of the pancreas. Hence these patients require the exogenous administration of insulin. On the other hand, secretion of insulin is very limited in type II diabetes, which is not enough to decrease blood glucose level compared to subjects with healthy body [55]. Beside insulin resistance occurs in most type II diabetes.

The purpose of the Pima Indians Diabetes (PID) Data Set is to forecast whether patients are diabetes or not from several demographic, detected, and measured patient features. All patients in this data set are from Phoenix, Arizona, USA. This data set is received from Data Mining Repository of University of California, Irvine (UCI) [35, 36]. The data set contains 768 instances and 8 attributes [35].

3.4 Results

In this chapter, the performance of the DNN classifier is measured by 7 medical data sets. The obtained results related with accuracy rates for each data sets are compared with the state-of-the-art classifiers. At the same time, PCA is utilized to observe the ability of the dimensional reduction of the AE.

The comparative results of the DNN classifier with state-of-the-art classifiers are reported in Sect. 3.4.1 for data sets related heart diseases, Sect. 3.4.2 for data sets related with breast cancer and in Sect. 3.4.3 for data set related with diabetes.

Table 4 Comparison of designed DNN with state-of-the-art methods for Cleveland heart disease data set

Methods	Accuracy rate (%)
Designed DNN	**85.2**[a]
Navia Bayes [56]	83.8[a]
SMO [56]	84.4[a]
IBK [56]	76.9[a]
AdaBoostM1 [56]	83.5[a]
J48 [56]	76.5[a]
PART [56]	81.5[a]
NPC [57]	84.0[a]
HNPC (minimum) [58]	84.4[a]
HNPC (product) [58]	85.0[a]
Modified HNPC (minimum) [59]	77.4[a]
Modified HNPC (product) [59]	**86.7**[a]
AdaBoostM1 [11]	82.2[a]
Bagging [11]	83.7[a]
BayesNet [11]	82.2[a]
Dagging [11]	82.2[a]
DecisionTable [11]	83.3[a]
DTNB [11]	82.5[a]
FT [11]	82.2[a]
LMT [11]	82.2[a]
Logistic [11]	83.7[a]
MultiClassClassifier [11]	83.7[a]
NaiveBayes [11]	83.3[a]
NaiveBayesSimple [11]	82.9[a]
NveBayesUpdateable [11]	83.3[a]
RandomCommittee [11]	82.2[a]
RandomForest [11]	83.7[a]
RandomSubSpace [11]	82.2[a]
RBFNetwork [11]	84.0[a]
RotationForest [11]	82.5[a]
SimpleLogistic [11]	82.2[a]
SMO [11]	83.3[a]
DAM [11]	83.7[a]
Original SAM [60]	75.7[b]
GSAM [60]	78.0[b]
IT2FLS-KMIP [61]	80.7[b]
IT2FLS-GCCD [61]	81.0[b]

[a]10-fold cross validation
[b]5-fold cross validation

Table 5 Comparison of designed DNN with state-of-the-art methods for Hungarian heart disease data set

Methods	Accuracy rate (%)
Designed DNN	**83.5**[a]
J48 [12]	78.2[a]
Random forest [12]	80.7[a]
ABM1 [12]	76.8[a]
SMO [12]	82.0[a]
Bagging [12]	81.9[a]
Naive Bayes [12]	82.8[a]
FBCDSS [62]	50.5[a]
NNBDSS [62]	46.4[a]
Genfis2 [13]	39.5[a]
Rough-Fuzzy classifier [13]	42.4[a]
Weighted fuzzy rules and decision tree rules [63]	49.9[a]
$L_{1/2}$ [14]	**86.8**[b]
Lasso [14]	**85.9**[b]
SCAD [14]	**86.1**[b]
MCP [14]	**85.6**[b]
Elastic net [14]	**86.4**[b]

[a] 10-fold cross validation
[b] Average of 50 runs (training %60 test %40)

Table 6 Comparison of designed DNN with state-of-the-art methods for long beach heart disease data set

Methods	Accuracy rate (%)
Designed DNN	**84.0**[a]
$L_{1/2}$ [14]	81.6[b]
SCAD [14]	81.2[b]
Lasso [14]	80.4[b]
MCP [14]	81.8[b]
Elastic net Hungrain_zhang2014	80.1[b]
FDSS [9]	75.0[c]
MLP-ANN [9]	76.0[c]
k-NN [9]	85.0[c]
C4.5 [9]	53.0[c]
RIPPER [9]	41.0[c]
DT [64]	68.4[a]
KNN [64]	81.1[a]
SVM [64]	83.4[a]

[a] 10-fold cross validation
[b] Average of 50 runs (training %60 test %40)
[c] Unknown

3.4.1 Results of Heart Diseases Data Sets

The results of 10-fold cross validation for DNN and state-of-the-art methods are presented in Tables 4, 5, 6 and 7 for each heart disease data set. The new features generated by each AE is visualized in Figs. 8, 9, 10 and 11, respectively, for each data set.

As can be seen in Tables 5, 6 and 7, the designed DNN is the best classifier with accuracy rate (83.5), (92.6) and (92.6) for Hungarian heart disease data set, Long Beach heart disease data set and Switzerland heart disease data set, respectively, compared with the state-of-the-art methods. On the other hand, according to Table 4, the designed DNN with accuracy rate (85.2) is the second classifier among the others for Cleavland CAD data set. Modified HNPC (product) with accuracy rate (86.7) is the best one. On the other hand, accuracy rate of designed DNN is lower than $L_{1/2}$, Lasso, SCAD and MCP methods for Hungarian heart disease data set. However, the average accuracies of 50 runs are reported without using 10 cross validation technique.

In order to understand the dimensional reduction mechanism of the AE, output of hidden layer of the AE is visualized by using PCA for each of the data sets including Cleveland heart disease, Hungarian heart disease data set, Long Beach heart disease data set and Switzerland heart disease data set. The DNN contains two hidden layers, each with four neurons.

AEs are used to reduce the dimensionality in order of $\mathbb{R}^{14} \rightarrow \mathbb{R}^4$, $\mathbb{R}^{11} \rightarrow \mathbb{R}^4$, $\mathbb{R}^{10} \rightarrow \mathbb{R}^4$ and $\mathbb{R}^{10} \rightarrow \mathbb{R}^4$ for each CAD data set. The changing of performance of the AE is recorded in Figs. 8, 9, 10 and 11 one by one when the dimensionality of the new representation $\mathbb{R}^4$ varies. As clearly seen in these figures, the effect of AE can be observed. The AE embeds raw features to a new space where features are defined in a clearer manner than before to classify the data sets.

Table 7 Comparison of designed DNN with state-of-the-art methods for Switzerland heart disease data set

Methods	Accuracy rate (%)
Designed DNN	**92.6**[a]
Genfis2 [13]	62.3[a]
Rough-Fuzzy classifier [13]	79.8[a]
weighted fuzzy rules and decision tree rules [63]	52.2[a]
FDSS [9]	70.0[b]
MLP-ANN [9]	81.0[b]
k-NN [9]	62.0[b]
C4.5 [9]	53.0[b]
RIPPER [9]	41.0[b]
DT [64]	86.5[a]
KNN [64]	88.5[a]
SVM [64]	92.2[a]

[a] 10-fold cross validation
[b] Unknown

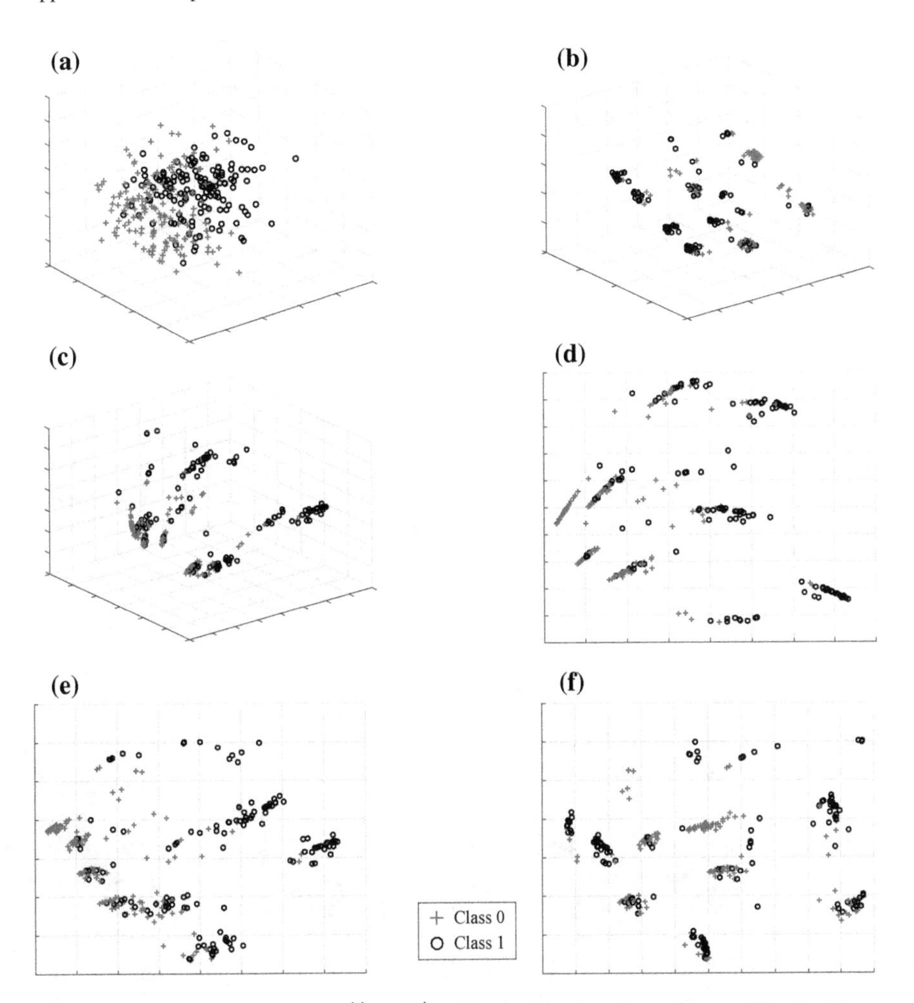

Fig. 8 Dimensionality reduction: $\mathbb{R}^{14} \rightarrow \mathbb{R}^{4}$: **a** 3D visualization of raw features Cleveland heart disease data set, **b** 3D visualization of new features generated by the first autoencoder **c** 3D visualization of new features generated by the second autoencoder **d** 2D visualization of new features generated by second autoencoder projected on X-Y plane **e** 2D visualization of new features generated by second autoencoder projected on X-Z plane **f** 2D visualization of new features generated by second autoencoder projected on Y-Z plane

3.4.2 Results of Breast Cancer Data Sets

10-fold cross validation of test accuracies of DNN and state-of-the-art methods for each of the breast cancer data sets are reported in Tables 8 and 9. Effect of AE on reducing dimension of the features is visualized in Figs. 12 and 13 for each data set, respectively.

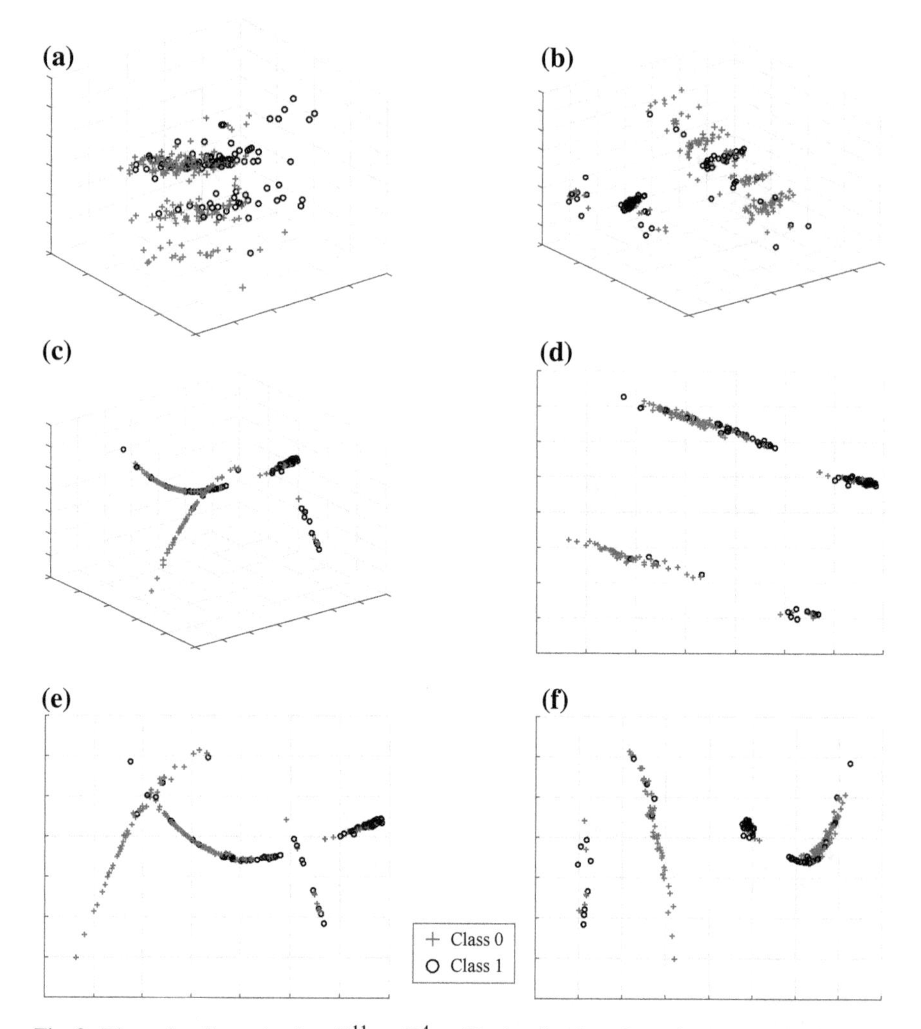

Fig. 9 Dimensionality reduction: $\mathbb{R}^{11} \rightarrow \mathbb{R}^4$: **a** 3D visualization of raw features Hungarian heart disease data set, **b** 3D visualization of new features generated by the first autoencoder **c** 3D visualization of new features generated by the second autoencoder **d** 2D visualization of new features generated by second autoencoder projected on X-Y plane **e** 2D visualization of new features generated by second autoencoder projected on X-Z plane **f** 2D visualization of new features generated by second autoencoder projected on Y-Z plane

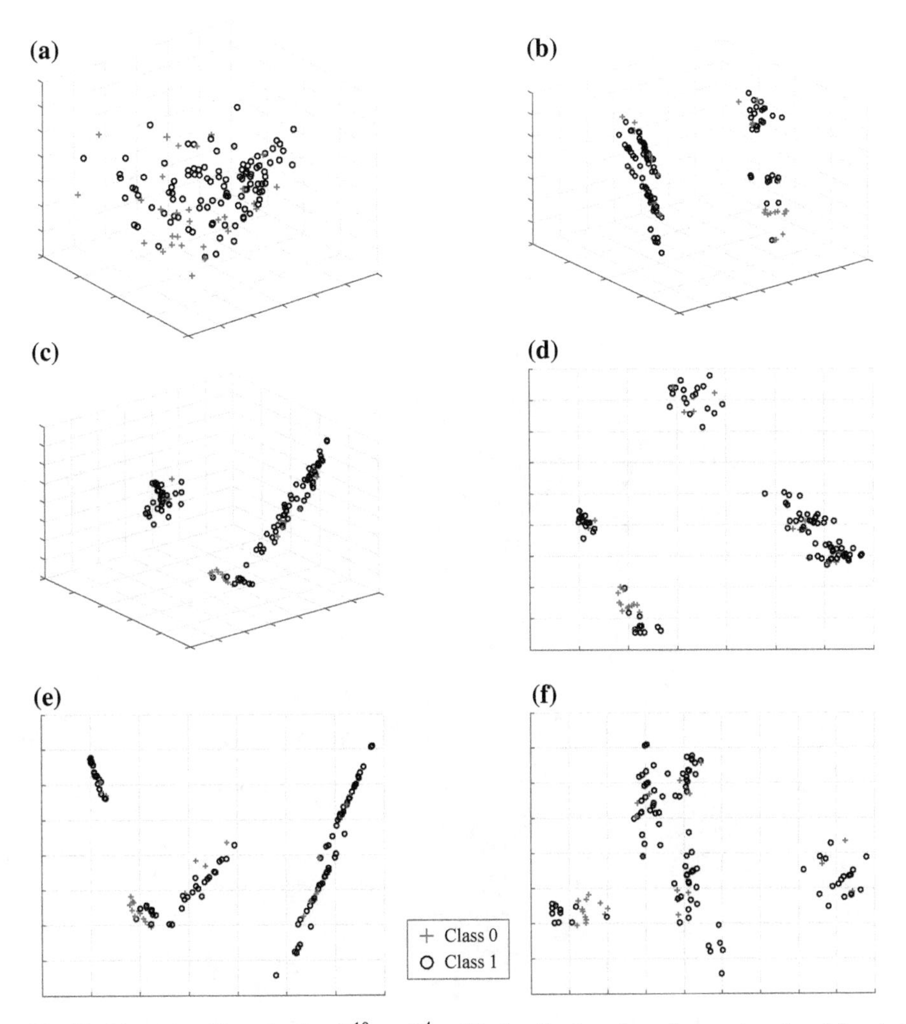

Fig. 10 Dimensionality reduction: $\mathbb{R}^{10} \rightarrow \mathbb{R}^{4}$: **a** 3D visualization of raw features long beach heart disease data set, **b** 3D visualization of new features generated by the first autoencoder **c** 3D visualization of new features generated by the second autoencoder **d** 2D visualization of new features generated by second autoencoder projected on X-Y plane **e** 2D visualization of new features generated by second autoencoder projected on X-Z plane **f** 2D visualization of new features generated by second autoencoder projected on Y-Z plane

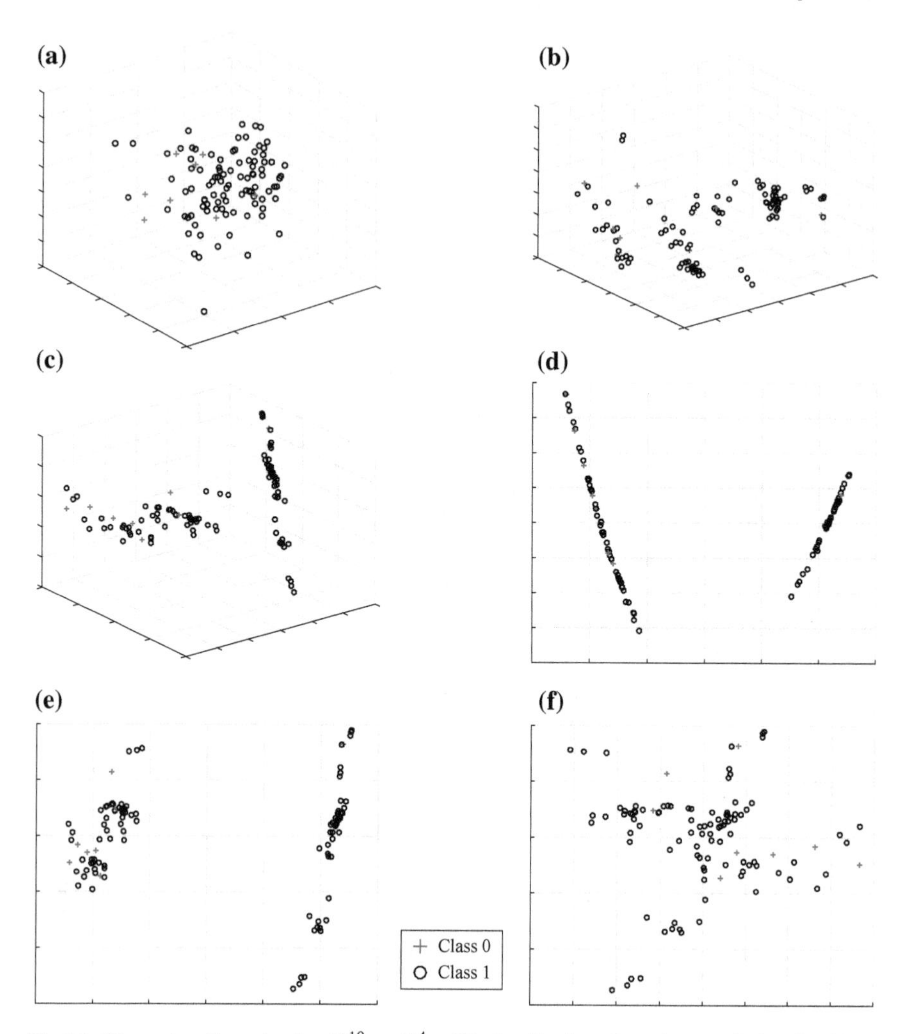

Fig. 11 Dimensionality reduction: $\mathbb{R}^{10} \rightarrow \mathbb{R}^{4}$: **a** 3D visualization of raw features Switzerland heart disease data set, **b** 3D visualization of new features generated by the first autoencoder **c** 3D visualization of new features generated by the second autoencoder **d** 2D visualization of new features generated by second autoencoder projected on X-Y plane **e** 2D visualization of new features generated by second autoencoder projected on X-Z plane **f** 2D visualization of new features generated by second autoencoder projected on Y-Z plane

Table 8 Comparison of designed DNN with state-of-the-art methods for breast cancer data set

Methods	Accuracy rate (%)
Designed DNN	**75.8**[a]
MLP [10]	**76.1**[a]
RBF [10]	72.3[a]
PNN [10]	24.6[a]
SOM [10]	29.8[a]
SVM [10]	71.3[a]
kNN [10]	72.1[a]
nB [10]	68.8[a]
MLP/GA [10]	**80.4**[a]
PCNN [10]	**77.6**[a]

[a] 10-fold cross validation

According to Table 8, designed DNN is the fourth best classifier with accuracy rate (75.8) after MLP/GA with accuracy rate (80.4), PCNN with accuracy rate (77.6) and MLP with accuracy rate (76.1) for Breast Cancer data set. On the other hand, it is the fifth best algorithm with accuracy rate (97.2) after GA-KDE, PSO-KDE, DAM, DTNB with accuracy rates (98.5), (98.5), (97.8) and (97.5), respectively, for classification of Wisconsin Breast Cancer data set. All of them are run by using 10-fold cross validation technique.

The DNN is designed as two hidden layers with four neurons. AEs are used to reduce the dimensionally $\mathbb{R}^9 \rightarrow \mathbb{R}^4$ for breast cancer data set. The effect of AE can be observed form these figures. The AE maps raw features to a new space where features are defined in a clearer manner than before to classify the data sets appropriately.

3.4.3 Results of Diabetes Data Sets

10-fold cross validation of accuracy of the DNN and state-of-the-art classifiers for PID are listed in Table 10. Effect of each AE on reducing dimension is visualized in Fig. 14 for effective classification. According to Table 10, designed DNN is the best algorithm with accuracy rate (78.1).

In order to observe the effect of hidden layers on the performance of AE over PID, the DNN is designed as two hidden layer with eight and four neurons respectively. AEs are used to degrade the dimensionally $\mathbb{R}^8 \rightarrow \mathbb{R}^4$ for PID data set. The changing of performance of the AE is reported in Fig. 14 when the dimensionality of the new representation as $\mathbb{R}^4$.

The effect of AE can be observed from Fig. 14. The AE maps raw features to a new space where features are defined in a clearer manner than before to classify the data sets appropriately.

Table 9 Comparison of designed DNN with state-of-the-art methods for Wisconsin breast cancer data set

Methods	Accuracy rate (%)
Designed DNN	**97.2**[a]
MLP [10]	95.8[a]
RBF [10]	95.2[a]
PNN [10]	82.5[a]
SOM [10]	90.1[a]
SVM [10]	96.9[a]
kNN [10]	94.1[a]
nB [10]	96.3[a]
MLP/GA [10]	91.4[a]
PCNN [10]	94.2[a]
AdaBoostM1 [11]	95.6[a]
Bagging [11]	96.1[a]
BayesNet [11]	97.2[a]
Dagging [11]	96.7[a]
DecisionTable [11]	95.7[a]
DTNB [11]	**97.5**[a]
FT [11]	96.9[a]
LMT [11]	96.4[a]
Logistic [11]	96.6[a]
MultiClassClassifier [11]	96.6[a]
NaiveBayes [11]	96.1[a]
NaiveBayesSimple [11]	96.3[a]
NveBayesUpdateable [11]	96.1[a]
RandomCommittee [11]	96.4[a]
RandomForest [11]	97.0[a]
RandomSubSpace [11]	95.5[a]
RBFNetwork [11]	95.9[a]
RotationForest [11]	**97.2**[a]
SimpleLogistic [11]	96.6[a]
SMO [11]	96.9[a]
DAM [11]	**97.8**[a]
PSO-KDE [47]	**98.5**[a]
GA-KDE [47]	**98.5**[a]
Self-training [65]	85.8[a]
Random co-training [65]	90.5[a]
Rough co-training [65]	92.3[a]
AR1+NN [66]	**97.4**[b]
AR2+NN [66]	95.6[b]
AR1+AR2+NN [66]	**98.4**[b]

(continued)

Table 9 (continued)

Methods	Accuracy rate (%)
GA-MOO-ANN [67]	**98.1**[c]
WN Bayes [68]	**98.5**[d]

[a]10-fold cross validation
[b]3-fold cross validation
[c]Average of 10 runs (training %50, validation %25 and test %25)
[d]5-fold cross validation

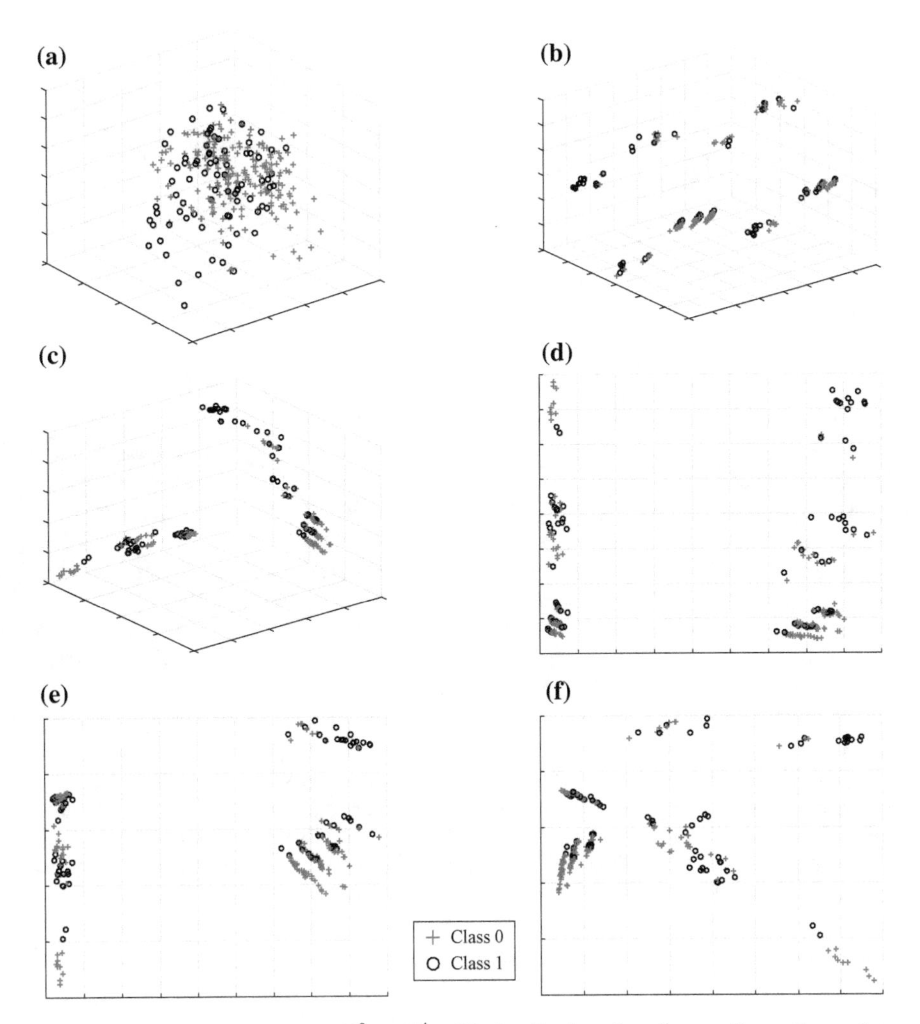

Fig. 12 Dimensionality reduction: $\mathbb{R}^9 \rightarrow \mathbb{R}^4$: **a** 3D visualization of raw features Breast Cancer data set, **b** 3D visualization of new features generated by first autoencoder **c** 3D visualization of new features generated by second autoencoder **d** 2D visualization of new features generated by second autoencoder based X-Y **e** 2D visualization of new features generated by second autoencoder based X-Z **f** 2D visualization of new features generated by second autoencoder based Y-Z

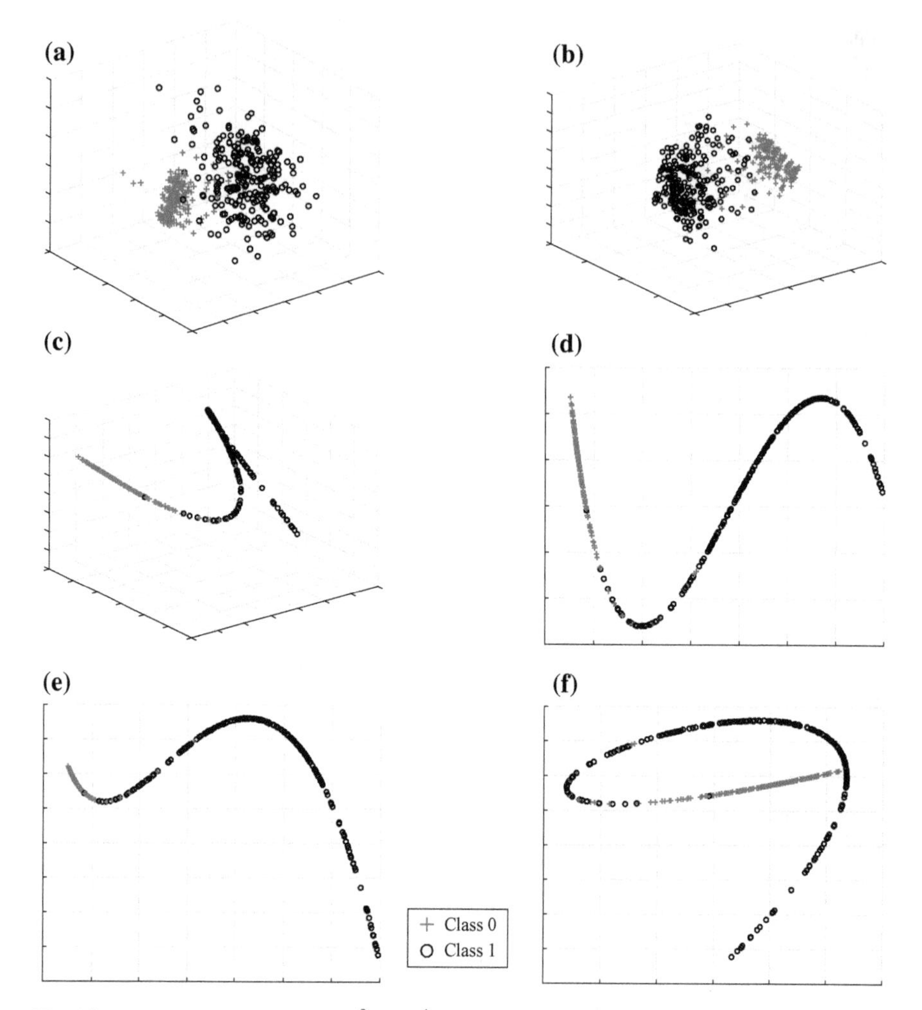

Fig. 13 Dimensionality reduction: $\mathbb{R}^9 \rightarrow \mathbb{R}^4$: **a** 3D visualization of raw features Wisconsin Breast Cancer data set, **b** 3D visualization of new features generated by the first autoencoder **c** 3D visualization of new features generated by the second autoencoder **d** 2D visualization of new features generated by second autoencoder projected on X-Y plane **e** 2D visualization of new features generated by second autoencoder projected on X-Z plane **f** 2D visualization of new features generated by second autoencoder projected on Y-Z plane

Table 10 Comparison of designed DNN with state-of-the-art methods for PID data set

Methods	Accuracy rate (%)
Designed DNN	**78.1**[a]
AdaBoostM1 [11]	74.3[a]
Bagging [11]	74.6[a]
BayesNet [11]	74.3[a]
Dagging [11]	74.0[a]
DecisionTable [11]	71.2[a]
DTNB [11]	73.8[a]
FT [11]	77.3[a]
LMT [11]	77.4[a]
Logistic [11]	77.2[a]
MultiClassClassifier [11]	77.2[a]
NaiveBayes [11]	76.3[a]
NaiveBayesSimple [11]	76.3[a]
NveBayesUpdateable [11]	76.3[a]
RandomCommittee [11]	75.2[a]
RandomForest [11]	72.3[a]
RandomSubSpace [11]	75.2[a]
RBFNetwork [11]	75.3[a]
RotationForest [11]	76.8[a]
SimpleLogistic [11]	77.4[a]
SMO [11]	77.3[a]
DAM [11]	70.3[a]
CoABCMiner [69]	75.5[a]
CORE [69]	74.3[a]
PGIRLA [69]	73.1[a]
C4.5Rules [69]	71.7[a]
SIA [69]	71.4[a]
HIDER [69]	70.9[a]
LDWPSO [69]	69.6[a]
ABCMiner [69]	65.9[a]
LUkEYK [15]	76.0[b]
LUkEYK+SVM [15]	76.0[b]
SVM [15]	73.0[b]
LDA [15]	72.0[b]
Bayes [15]	71.0[b]
kEYK [15]	64.0[b]

[a]10-fold cross validation
[b]Unknown

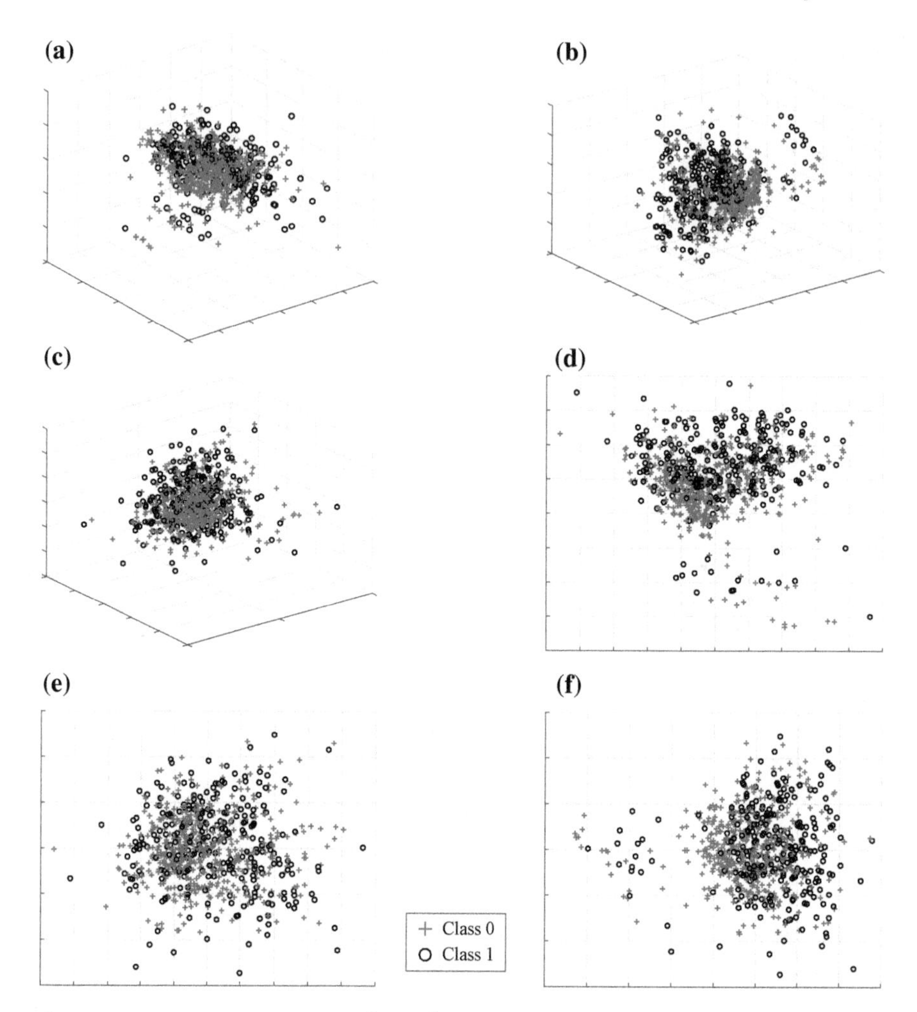

Fig. 14 Dimensionality reduction: $\mathbb{R}^8 \rightarrow \mathbb{R}^4$: **a** 3D visualization of raw features PID Data set, **b** 3D visualization of new features generated by the first autoencoder **c** 3D visualization of new features generated by the second autoencoder **d** 2D visualization of new features generated by second autoencoder projected on X-Y plane **e** 2D visualization of new features generated by second autoencoder projected on X-Z plane **f** 2D visualization of new features generated by second autoencoder projected on Y-Z plane

4 Conclusion

In this chapter, we present a DNN based classifier with very good accuracy for classification of medical data sets, including heart disease, breast cancer and diabetes mellitus. We demonstrate this fact by reporting the results of a rigorous experimental testing procedure that we employ for comparing the performance of the presented

DNN classifier with representative conventional as well as state-of-the-art classification methods from the literature. Specifically, we evaluate the performance on three different diseases, on a number of different data sets for each disease and by conducting 10-fold cross validation for each disease/data set combination.

We also demonstrate the effectiveness of the use of AEs to reduce the dimensionality of the input data. We visualize this by using the PCA technique and show that the use of AEs may be a very efficient strategy to reduce the dimensionality of the input data. This allows the design of a classifier with lower complexity, which then results in easier training.

The presented DNN classifier provides the highest accuracy rates in almost all of the classification experiments with the exception of a few cases. Therefore, the presented DNN based approach to disease diagnosis may constitute a strong alternative to currently available methods. As we mention before, this approach is gaining increasing popularity as it is easier, cheaper, quicker and non-invasive. We therefore conclude that the presented DNN based classifier may be used as a different alternative to currently existing disease diagnosis approaches.

References

1. Y. LeCun, Y. Bengio, G. Hinton, Deep learning. Nature **521**(7), 436–444 (2015)
2. W.-B. Huang, F.-C. Sun et al., Building feature space of extreme learning machine with sparse denoising stacked-autoencoder. Neurocomputing **174**, 60–71 (2016)
3. J. Xu, X. Luo, G. Wang, H. Gilmore, A. Madabhushi, A deep convolutional neural network for segmenting and classifying epithelial and stromal regions in histopathological images. Neurocomputing
4. J. Xu, L. Xiang, Q. Liu, H. Gilmore, J. Wu, J. Tang, A. Madabhushi, Stacked sparse autoencoder (SSAE) for nuclei detection on breast cancer histopathology images
5. H. Badem, A. Caliskan, A. Basturk, M.E. Yuksel, Classification and diagnosis of the parkinson disease by stacked autoencoder, in *9th National Conference on Electrical and Electronics Engineering, ELECO 2016*, 2016, pp. 499–502
6. H. Badem, A. Caliskan, A. Basturk, M.E. Yuksel, Classification of human activity by using a stacked autoencoder, in *Medical Technologies National Conference (TIPTEKNO 2016)*, 2016, pp. 370–373
7. A. Caliskan, M.E. Yuksel, H. Badem, A. Basturk, A deep neural network classifier for decoding human brain activity based on magnetoencephalography. Elektron. ir Elektrotech. **23**(2), 63–67 (2017)
8. A. Caliskan, H. Badem, A. Basturk, M.E. Yuksel, Diagnosis of the parkinson disease by using deep neural network classifier. Istanb. Univ. J. Electr. Electron. Eng. **17**, 3311–3318 (2017)
9. N.A. Setiawan, P. Venkatachalam, A.F.M. Hani, Diagnosis of coronary artery disease using artificial intelligence based decision support system, in *Proceedings of the International Conference on Man-machine Systems (ICoMMS)*, Batu Ferringhi, Penang (2009)
10. F. Gorunescu, S. Belciug, Evolutionary strategy to develop learning-based decision systems. Application to breast cancer and liver fibrosis stadialization. J. Biomed. Inform. **49**, 112–118 (2014)
11. M. Aldape-Pérez, C. Yáñez-Márquez, O. Camacho-Nieto, I. López-Yáñez, A.-J. Argüelles-Cruz, Collaborative learning based on associative models: application to pattern classification in medical datasets. Comput. Hum. Behav. Part B **51**, 771–779 (2015)

12. K. Singh, J. Rong, L. Batten, Sharing sensitive medical data sets for research purposes–a case study, in *International Conference on Data Science and Advanced Analytics (DSAA)*, vol. 2014, pp. 555–562 (2014)
13. K. Srinivas, G.R. Rao, A. Govardhan, Rough-fuzzy classifier: a system to predict the heart disease by blending two different set theories. Arab. J. Sci. Eng. **39**(4), 2857–2868 (2014)
14. B. Zhang, H. Chai, Z. Yang, Y. Liang, G. Chu, X. Liu, Application of L 1/2 regularization logistic method in heart disease diagnosis. Bio-med. Mater. Eng. **24**(6), 3447–3454 (2014)
15. H.S. Bilge, Y. Kerimbekov, Classification with Lorentzian distance metric, in *23rd Signal Processing and Communications Applications Conference (SIU)*, 2015, pp. 2106–2109
16. J.M. Keller, M.R. Gray, J.A. Givens, A fuzzy k-nearest neighbor algorithm. IEEE Trans. Syst. Man Cybern. SMC **15**(4), 580–585 (1985)
17. L. Rokach, O. Maimon, Top-down induction of decision trees classifiers-a survey. IEEE Trans. Syst. Man Cybern. Part C Appl. Rev. **35**(4), 476–487 (2005)
18. J.R. Quinlan, Induction of decision trees. Mach. Learn. **1**(1), 81–106 (1986)
19. J.R. Quinlan, *C4.5: Programs for Machine Learning* (Elsevier, 2014)
20. S. Grumbach, T. Milo, Towards tractable algebras for bags, in *Proceedings of the Twelfth ACM SIGACT-SIGMOD-SIGART Symposium on Principles of Database Systems*, ACM, 1993, pp. 49–58
21. G. Bagallo, D. Haussler, Boolean feature discovery in empirical learning. Mach. Learn. **5**(1), 71–99 (1990)
22. Q. Lu, L. Getoor, Link-based classification, in *Proceedings of ICML*, vol. 3, 2003, pp. 496–503
23. C.S. Leslie, E. Eskin, W.S. Noble, The spectrum kernel: a string kernel for SVM protein classification, in *Pacific Symposium on Biocomputing*, vol. 7, 2002, pp. 566–575
24. J. Shawe-Taylor, N. Cristianini, *Kernel Methods for Pattern Analysis* (Cambridge University, 2004)
25. C.J. Burges, A tutorial on support vector machines for pattern recognition. Data Mining Knowl. Discov. **2**(2), 121–167 (1998)
26. H. Guan, J. Zhou, M. Guo, A class-feature-centroid classifier for text categorization, in *Proceedings of the 18th International Conference on World Wide Web*, ACM, 2009, pp. 201–210
27. H. Badem, A. Basturk, A. Caliskan, M.E. Yuksel, A new efficient training strategy for deep neural networks by hybridization of artificial bee colony and limited-memory BFGS optimization algorithms. Neurocomputing **266**(2017), 506–526 (2017)
28. A. Caliskan, M.E. Yuksel, H. Badem, A. Basturk, Performance improvement of deep neural network classifiers by a simple training strategy. Eng. Appl. Artif. Intell. **67**, 14–23 (2018)
29. G.E. Hinton, R.R. Salakhutdinov, Reducing the dimensionality of data with neural networks. Science **313**(5786), 504–507 (2006)
30. J. Ngiam, A. Coates, A. Lahiri, B. Prochnow, Q.V. Le, A.Y. Ng, On optimization methods for deep learning, in *Proceedings of the 28th International Conference on Machine Learning (ICML-11)*, 2011, pp. 265–272
31. Y. Bengio, Practical recommendations for gradient-based training of deep architectures, *Neural Networks: Tricks of the Trade* (Springer, 2012), pp. 437–478
32. N. Andrew, Sparse autoencoder (2011)
33. Y. Zhang, E. Zhang, W. Chen, Deep neural network for halftone image classification based on sparse auto-encoder. Eng. Appl. Artif. Intell. **50**, 245–255 (2016)
34. R.H. Byrd, P. Lu, J. Nocedal, C. Zhu, A limited memory algorithm for bound constrained optimization. SIAM J. Sci. Comput. **16**(5), 1190–1208 (1995)
35. C. Blake, C.J. Merz, UCI repository of machine learning databases, Department of Information and Computer Science, Irvine, CA, University of California, 55. http://www.ics.uci.edu/~mlearn/mlrepository.html
36. J.W. Smith, J. Everhart, W. Dickson, W. Knowler, R. Johannes, Using the ADAP learning algorithm to forecast the onset of diabetes mellitus, in *Proceedings of the Annual Symposium on Computer Application in Medical Care*, American Medical Informatics Association, 1988, p. 261
37. S. Matthias, Approaches to analyse and interpret biological profile data, Potsdam University

38. P. Anooj, Clinical decision support system: risk level prediction of heart disease using decision tree fuzzy rules. Int. J. Res. Rev. Comput. Sci. **3**(3), 1659–1667 (2012)
39. E. Ephzibah, Cost effective approach on feature selection using genetic algorithms and LS-SVM classifier. IJCA Spec. Issue Evol. Comput. Optim. Tech. ECOT
40. R. Detrano, A. Janosi, W. Steinbrunn, M. Pfisterer, J.-J. Schmid, S. Sandhu, K.H. Guppy, S. Lee, V. Froelicher, International application of a new probability algorithm for the diagnosis of coronary artery disease. Am. J. Cardiol. **64**(5), 304–310 (1989)
41. D. Aha, D. Kibler, Instance-based prediction of heart-disease presence with the Cleveland database, University of California
42. J.H. Gennari, P. Langley, D. Fisher, Models of incremental concept formation. Artif. Intell. **40**(1–3), 11–61 (1989)
43. A.P. Bradley, The use of the area under the ROC curve in the evaluation of machine learning algorithms. Pattern Recogn. **30**(7), 1145–1159 (1997)
44. F. Paulin, A. Santhakumaran, Classification of breast cancer by comparing back propagation training algorithms. Int. J. Comput. Sci. Eng. **3**(1), 327–332 (2011)
45. E.A. Rakha, I.O. Ellis, An overview of assessment of prognostic and predictive factors in breast cancer needle core biopsy specimens. J. Clin. Pathol. **60**(12), 1300–1306 (2007)
46. G.R. Sizilio, C.R. Leite, A.M. Guerreiro, A.D. Neto, Fuzzy method for pre-diagnosis of breast cancer from the fine needle aspirate analysis. Biomed. Eng. Online **11**(1), 83 (2012)
47. R. Sheikhpour, M.A. Sarram, R. Sheikhpour, Particle swarm optimization for bandwidth determination and feature selection of kernel density estimation based classifiers in diagnosis of breast cancer. Appl. Soft Comput. **40**, 113–131 (2016)
48. C.L. Rock, W. Demark-Wahnefried, Nutrition and survival after the diagnosis of breast cancer: a review of the evidence. J. Clin. Oncol. **20**(15), 3302–3316 (2002)
49. R.S. Michalski, I. Mozetic, J. Hong, N. Lavrac, The multi-purpose incremental learning system AQ15 and its testing application to three medical domains. Proc. AAAI **1986**, 1–041 (1986)
50. M. Tan, L. Eshelman, Using weighted networks to represent classification knowledge in noisy domains, in *Proceedings of the Fifth International Conference on Machine Learning*, 1988, pp. 121–134
51. W.H. Wolberg, O.L. Mangasarian, Multisurface method of pattern separation for medical diagnosis applied to breast cytology. Proc. Natl. Acad. Sci. **87**(23), 9193–9196 (1990)
52. J. Zhang, Selecting typical instances in instance-based learning, in *Proceedings of the Ninth International Machine Learning Conference*, 1992, pp. 470–479
53. A. Gul, M.A. Rahman, A. Jaleel, Changes in glycosylated proteins in type-2 diabetic patients with and without complications. J. Ayub Med. Coll. Abbottabad **17**(3), 33–37 (2005)
54. S.F.B. Jaafar, D.M. Ali, Diabetes mellitus forecast using artificial neural network (ANN), in *2005 Asian Conference on Sensors and the International Conference on new Techniques in Pharmaceutical and Biomedical Research*, IEEE, 2005, pp. 135–139
55. H.M. Fonseca, V.H. Ortiz, A.I. Cabrera, Stochastic neural networks applied to dynamic glucose model for diabetic patients, in *1st International Conference on Electrical and Electronics Engineering, 2004, (ICEEE)*, IEEE, 2004, pp. 522–525
56. J. Nahar, T. Imam, K.S. Tickle, Y.-P.P. Chen, Computational intelligence for heart disease diagnosis: a medical knowledge driven approach. Exp. Syst. Appl. **40**(1), 96–104 (2013)
57. M. Bounhas, K. Mellouli, H. Prade, M. Serrurier, Possibilistic classifiers for numerical data. Soft Comput. **17**(5), 733–751 (2012)
58. K. Baati, T.M. Hamdani, A.M. Alimi, Hybrid naive possibilistic classifier for heart disease detection from heterogeneous medical data, in *2013 13th International Conference on Hybrid Intelligent Systems (HIS)*, 2013, pp. 234–239
59. K. Baati, T.M. Hamdani, A.M. Alimi, A modified hybrid naive possibilistic classifier for heart disease detection from heterogeneous medical data, in *2014 6th International Conference of Soft Computing and Pattern Recognition (SoCPaR)*, 2014, pp. 353–358
60. T. Nguyen, A. Khosravi, D. Creighton, S. Nahavandi, Classification of healthcare data using genetic fuzzy logic system and wavelets. Exp. Syst. Appl. **42**(4), 2184–2197 (2015)

61. T. Nguyen, A. Khosravi, D. Creighton, S. Nahavandi, Medical data classification using interval type-2 fuzzy logic system and wavelets. Appl. Soft Comput. **30**, 812–822 (2015)
62. P.K. Anooj, Implementing decision tree fuzzy rules in clinical decision support system after comparing with fuzzy based and neural network based systems, in *2013 International Conference on IT Convergence and Security (ICITCS)*, 2013, pp. 1–6
63. P. Anooj, Clinical decision support system: risk level prediction of heart disease using weighted fuzzy rules and decision tree rules. Open Comput. Sci. **1**(4), 482–498 (2011)
64. A. Caliskan, M.E. Yuksel, Classification of coronary artery disease data sets by using a deep neural network. EuroBiotech J
65. D. Miao, C. Gao, N. Zhang, Z. Zhang, Diverse reduct subspaces based co-training for partially labeled data. Int. J. Approx. Reason. **52**(8), 1103–1117 (2011). https://doi.org/10.1016/j.ijar.2011.05.006, http://www.sciencedirect.com/science/article/pii/S0888613X11000880
66. S. Palaniappan, T. Pushparaj, A novel prediction on breast cancer from the basis of association rules and neural network. Int. J. Comput. Sci. Mob. Comput. IJCSMC **4**, 269–77 (2013)
67. F. Ahmad, N.A. Mat Isa, Z. Hussain, S.N. Sulaiman, A genetic algorithm-based multi-objective optimization of an artificial neural network classifier for breast cancer diagnosis. Neural Comput. Appl. **23**(5), 1427–1435 (2012)
68. M. Karabatak, A new classifier for breast cancer detection based on Naïve Bayesian. Measurement **72**, 32–36 (2015)
69. M. Celik, F. Koylu, D. Karaboga, Coabcminer: an algorithm for cooperative rule classification system based on artificial bee colony. Int. J. Artif. Intell. Tools **25**(01), 1550028 (2016)

Why Dose Layer-by-Layer Pre-training Improve Deep Neural Networks Learning?

Seyyede Zohreh Seyyedsalehi and Seyyed Ali Seyyedsalehi

Abstract Deep perceptron neural networks are capable of implementing a hierarchy of successive nonlinear conversions. But training these neural networks by conventional learning methods such as the error back-propagation is faced with serious obstacles owing to local minima. The layer-by-layer pre-training method has been recently proposed for training these neural networks and has shown considerable performance. In the pre-training method, the complex problem of training deep neural networks is broken down into some simple sub-problems in which some corresponding single-hidden-layer neural networks are trained through the error back-propagation algorithm. In this chapter, the theoretical principles regarding how this method effectively improves the training of deep neural networks are discussed, and the maximum discrimination theory is proposed as a proper framework for analysis of training convergence in these neural networks. Subsequently, discriminations of inputs in different layers of two similar deep neural networks, one of which is directly trained through the conventional error back-propagation algorithm and the other through layer-by-layer pre-training method, are compared, and results confirm the validity of the proposed framework.

Keywords Deep learning · Neural network · Layer-by-layer pre-training · Maximum discrimination · Back-propagation algorithm · Hyperplanes

---

S. Z. Seyyedsalehi · S. A. Seyyedsalehi (✉)
Faculty of Biomedical Engineering, Amirkabir University of Technology (Tehran Polytechnic), Hafez Ave., Tehran, Iran
e-mail: ssalehi@aut.ac.ir

S. Z. Seyyedsalehi
e-mail: z.seyyedssalehi@aut.ac.ir

V. E. Balas et al. (eds.), *Handbook of Deep Learning Applications*, Smart Innovation, Systems and Technologies 136,
https://doi.org/10.1007/978-3-030-11479-4_13

1 Introduction

Extracting nonlinear intrinsic manifolds embedded in data and their corresponding nonlinear components using neural networks (NNs) is an efficient method for analysis of nonlinear data. This usually requires deep and complex structures of NNs [1].

These deep NNs (DNNs) can learn multi-stage and complex nonlinear transformations by employing many layers of neurons [2, 3]. In recent years, these structures have been demonstrated desirable performance in a large number of applications especially for vision and human language understanding [4]. DNNs have been used for speech recognition and voice conversion [5–9], bioinformatics [10, 11], face recognition [12–18], and dimension reduction [19, 20]. In these structures, the possibility of hierarchical extraction of components is provided [21, 22], in such a way that high level components are constructed as a combination of low level components in several layers [23].

A review of some results in machine learning shows that training of these deep structures faces some serious problems in comparison with shallow structures. When an attempt is made to train all the layers of a DNN through an output cost function, the results are even lower compared to those for shallow ones [24]. This is because with an increase in the number of layers, in addition to the time-consuming of training DNNs, the problem of local minima is serious obstacle, such that in most cases, training does not converge. Studies have shown that in these structures, the number of local minima, which depends on the architecture and initial values of the DNN parameters, is high [25].

One of the important reasons for this problem is the high interaction between the DNN parameters in different layers during training. Throughout training the DNN, weights are optimized so that two conditions are simultaneously fulfilled; firstly, tuning of lower layers is done in such a way that they provide proper input for final adjustment (end of training) of deeper layers. Secondly, adjustment of deeper layers is done such that in the end of training, they take good advantage of the extracted components by lower layers.

Adjustment of the weights of deeper layers is easily performed if the weights of lower layers are determined. But what makes the problem complicated is the simultaneous adjustment of the weights of deep and lower layers, which makes the slope of the objective function become limited by the local values provided by the current adjustments of network parameters. Studies have demonstrated that the traditional method of training adjusts parameters in a region which leads to low generalization of the DNN [26].

Therefore, alongside direct and common training of multilayer NNs, which takes advantage of the error back-propagation method, there are new approaches found helpful in increasing the complexity of structures and data [27, 28]. Pre-training methods are the main solution in this area [29–33]. These methods attempt to find proper initial values for DNN weights, and in this way disentangle the training from the most of the local minima. Pre-training methods providing the proper initial knowledge for getting the desired object through initializing of weights, prepare the DNNs for the training and fine-tuning of weights.

Recently, the layer-by-layer pre-training is presented as an powerful and efficient method for solving the problem of deep autoencoders (DAEs) training which takes advantage of the idea of breaking down the complex problem of training DAEs into several sub-problems of training single-hidden-layer AEs [1, 31]. The performance of this method with comparing to previous pre-training methods, the step-by-step method [32] and the decomposition to the Boltzmann machines method [33] is studied, and it has been demonstrated to be faster and more powerful. Furthermore, this method enhances the ability of DAE to extract components with better generalization for reconstructing and recognizing of test images [1].

In the chapter, analytical fundamentals and the reasons behind capabilities of the layer-by-layer pre-training method in the convergence of DAEs are discussed. First, the performance of feed-forward NN neurons in high-dimensional spaces is investigated. In this regard, the performance of NNs with linear, step nonlinear and finally soft (sigmoid) nonlinear functions is discussed.

In the following, the training presses of some distinct input samples to DAEs is studied, and the required conditions for maximum discrimination of a neuron with the sigmoid function are determined. On this basis, it will be demonstrated why training randomly initialized DAE does not converge, and the layer-by-layer pre-training method by stepwisely setting of weights leads to desirable convergence of training. Finally, in order to evaluate the influence of the layer-by-layer pre-training method some comparative experiments are performed that their results are reported.

2 Functional Analysis of Linear and Nonlinear Feed-Forward Neural Networks in High-Dimensional Spaces

DAEs, used often for nonlinear component extraction and compression of data, aim at decomposition of input data to nonlinear components and then reconstruction of the input in the output with the least error [1]. DAEs are capable of reconstructing the input distinct samples in the output with desired accuracy only if they have no waste of information during the data analysis in the different layers. Thus, it is essential to distinctly define the different inputs in the layers, and also, reconstruct them in the output. If two distinct input samples are represented by a single and non-distinct description in a layer of the DAE, the nonlinear mappings of the input up to this layer have not been successful in one-to-one discrimination of these two samples. Therefore, a part of the discriminative information of the input samples is lost. To clarify the problem, first, the process of a feed-forward NN in mapping of the input space to the output is analyzed. Through step by step analysis of DAE neuron functions, reasons of training non-convergence are discussed and a proper solution to fix it is purposed.

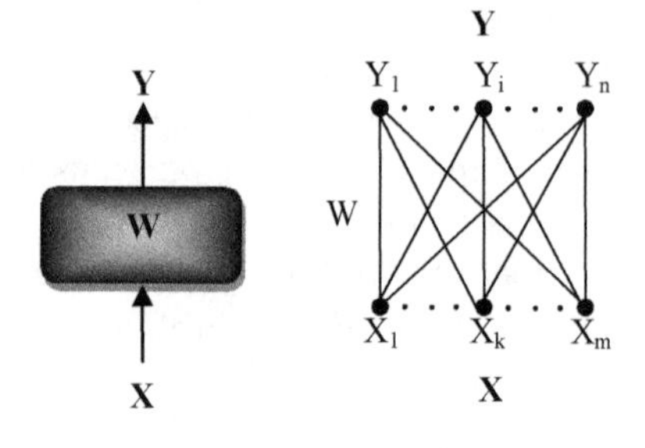

Fig. 1 A linear single-layer feed-forward NN

2.1 *Linear Neural Network (A Linear Mapping)*

First, a feed-forward single-layer NN with linear neurons without bias, i.e. $\mathbf{Y} = \mathbf{XW}$, is considered. The components of the input space ($\mathbf{X}$) are assumed orthogonal.

As illustrated in Fig. 1, $\mathbf{X}$ and $\mathbf{Y}$ are row input and output vectors and $\mathbf{W}$ is its weight matrix.

$$\begin{aligned} \mathbf{X} &= [\mathrm{X}_1, \mathrm{X}_2, \ldots, \mathrm{X}_\mathrm{m}] \\ \mathbf{Y} &= [\mathrm{Y}_1, \mathrm{Y}_2, \ldots, \mathrm{Y}_\mathrm{n}] \end{aligned} \tag{1}$$

Here, it is assumed that the weight vectors of the neurons are available according to Eq. 2, and the aim is the functional analysis of these neurons in the m-dimensional orthogonal input space $\mathbf{X}$.

$$\mathbf{W} = \begin{bmatrix} \mathrm{W}_{11} & \cdots & \mathrm{W}_{1\mathrm{n}} \\ \vdots & \ddots & \vdots \\ \mathrm{W}_{\mathrm{m}1} & \cdots & \mathrm{W}_{\mathrm{mn}} \end{bmatrix} = [\mathbf{W}_1, \mathbf{W}_2, \ldots, \mathbf{W}_\mathbf{n}] \tag{2}$$

where $\mathbf{W_i}$ is the weight vector of the ith neuron.

Figure 2 shows the vector $\mathbf{W_i}$ in the input space. Magnitude of each output component $\mathbf{Y_i}$ equals the scalar product of the input vector $\mathbf{X}$ and the weight vector of the ith linear neuron, and its direction is the same as $\mathbf{W_i}$ in the input space $\mathbf{X}$.

$$\mathbf{Y_i} = (\mathbf{X} \cdot \mathbf{W_i})\vec{\mathbf{a}}_{\mathbf{W_i}} \tag{3}$$

where $\vec{\mathbf{a}}_{\mathbf{W_i}}$ is the unit vector along $\mathbf{W_i}$.

In other words, each component $\mathbf{Y_i}$ is equal to the product of $\|\mathbf{W_i}\|$ in the $\mathbf{X}$ projected along $\mathbf{W_i}$.

$$\begin{aligned} \mathrm{Y_i} &= \left(\mathbf{X} \cdot \vec{\mathbf{a}}_{\mathbf{W_i}}\right)\vec{\mathbf{a}}_{\mathbf{W_i}} \times \|\mathbf{W_i}\| \\ &= \left(\mathbf{X} \cdot \vec{\mathbf{a}}_{\mathbf{W_i}}\right) \times \|\mathbf{W_i}\|\vec{\mathbf{a}}_{\mathbf{W_i}} \end{aligned} \tag{4}$$

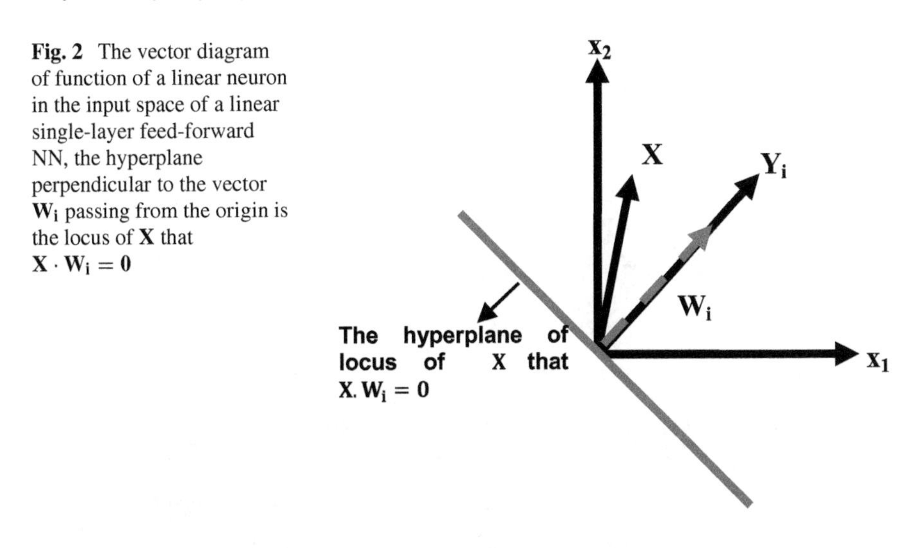

Fig. 2 The vector diagram of function of a linear neuron in the input space of a linear single-layer feed-forward NN, the hyperplane perpendicular to the vector $\mathbf{W_i}$ passing from the origin is the locus of $\mathbf{X}$ that $\mathbf{X} \cdot \mathbf{W_i} = \mathbf{0}$

where $\|\mathbf{W_i}\|$ is the magnitude or Euclidean norm of $\mathbf{W_i}$. Thus, the input vector $\mathbf{X}$ under the linear transformation $\mathbf{W}$ is converted to the vector $\mathbf{Y}$ with the following components:

$$\mathbf{Y} = [\mathrm{Y_1}, \mathrm{Y_2}, \ldots, \mathrm{Y_n}] \tag{5}$$

where $\mathrm{Y}_{\mathrm{i}|\mathrm{i}=1,2,\ldots,\mathrm{n}}$ are components of the output space $\mathbf{Y}$. Obviously, having orthogonal vectors $\mathbf{W_i}$ (the weight vectors of NN linear neurons):

$$\mathbf{W_i} \cdot \mathbf{W_r} = 0 \quad \forall \quad \mathrm{i} \neq \mathrm{r}, \quad \mathrm{i}, \mathrm{r} = 1, 2, \ldots, \mathrm{n} \tag{6}$$

leads to the orthogonal space $\mathbf{Y}$. If the orthogonality condition of components $\mathbf{Y}$ is not established, components $\mathrm{Y_i}$ will be correlated with each other and their inner product will not be zero. For each vector $\mathbf{W_i}$ of a linear neuron, the locus of the input vectors $\mathbf{X}$ whose inner product to $\mathbf{W_i}$ are zero ($\mathrm{Y_i} = 0$), is a m-1-dimensional hyperplane perpendicular to the vector $\mathbf{W_i}$ passing from the origin in the m-dimensional input space [34].

2.2 *Functional Analysis of Single-Layer Feed-Forward Neural Networks with a Step Nonlinear Activation Function and Bias*

A single-layer NN as shown in Fig. 3 is considered in which neurons have a step nonlinear function (f) and a bias B as in the following equation:

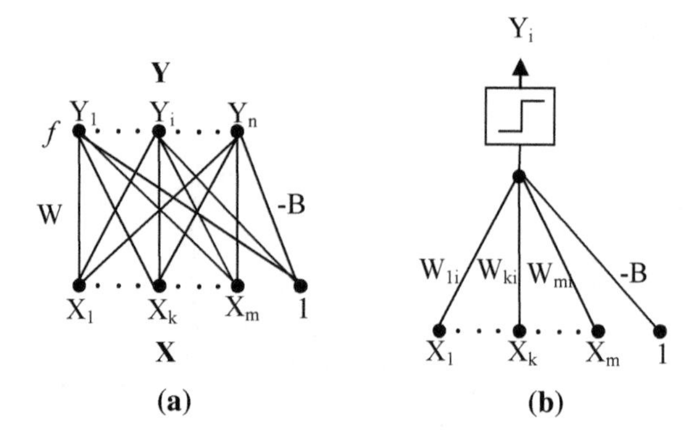

Fig. 3 **a** A single-layer feed-forward NN with a step nonlinear activation function and a bias, **b** the model of its neuron with step nonlinear function and bias

$$Y_i = f(\mathbf{X} \cdot \mathbf{W_i} - B_i), \quad f(z) = \begin{cases} 1 \; z > 0 \\ 0 \; z < 0 \end{cases} \tag{7}$$

For each neuron of the NN with the weight vector $\mathbf{W_i}$ and bias B_i, the locus $\mathbf{X}$ of the input space for which Eq. 8 is zero, i.e.:

$$(\mathbf{X} \cdot \mathbf{W_i} - B_i) = 0 \tag{8}$$

is a hyperplane that is perpendicular to the neuron's weight vector ($\mathbf{W_i}$) and passes from point $\mathbf{X_a}$ (Eq. 9) of the input space as shown in Fig. 4.

$$\mathbf{X_a} = \left(\mathbf{X} \cdot \vec{\mathbf{a}}_{\mathrm{W_i}}\right)\vec{\mathbf{a}}_{\mathrm{W_i}} \tag{9}$$

According to Eq. 8, $\|\mathbf{X_a}\|$ is calculated from Eq. 10.

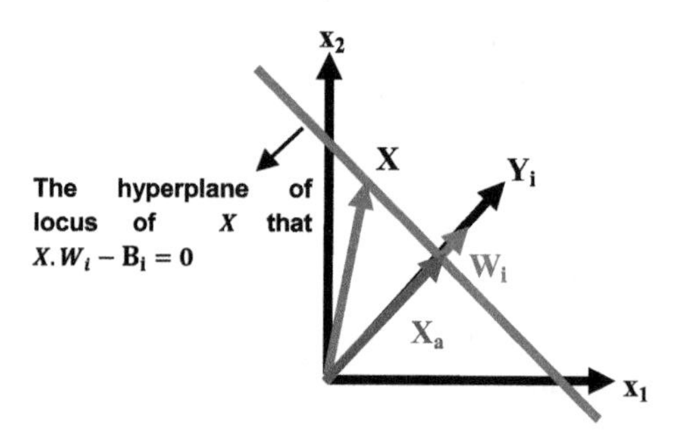

Fig. 4 The vector diagram of function of a nonlinear neuron with step activation function and bias in the input space, the hyperplane perpendicular to the vector $\mathbf{W_i}$ passing from $\mathbf{X_a}$ is the locus of $\mathbf{X}$ that $\mathbf{X} \cdot \mathbf{W_i} - \mathbf{B_i} = \mathbf{0}$ (The diagram is drawn for a two-dimensional input space ($\mathbf{m} = \mathbf{2}$), therefore, here the m-1-dimensional hyperplane is a line.)

$$(\mathbf{X} \cdot \mathbf{W_i} - B_i) = 0 \rightarrow \mathbf{X} \cdot \mathbf{W_i} = B_i \rightarrow \|\mathbf{X_a}\| = \mathbf{X} \cdot \vec{\mathbf{a}}_{W_i} = \frac{B_i}{\|\mathbf{W_i}\|} \tag{10}$$

The hyperplane is the decision boundary of the neuron in the input space **X**. Therefore, the output of the neuron for **X** of the input space at one side of the hyperplane is one and for another side is zero.

$$Y_i = 1 \quad \forall(\mathbf{X} \cdot \mathbf{W_i} - B_i) > 0 \tag{11}$$

$$Y_i = 0 \quad \forall(\mathbf{X} \cdot \mathbf{W_i} - B_i) < 0 \tag{12}$$

Thus, for each neuron with the weight vector $\mathbf{W_i}$ and the bias B_i, the location of the hyperplane in the input space (i.e. the neuron decision boundary) can be exactly defined. Also, the intersection of the hyperplane with the axis of the input space **X** ($[X_1, X_2, \ldots, X_m]$) can be calculated. As Eq. 8:

$$X_1 W_{1i} + X_2 W_{2i} + \cdots + X_m W_{mi} - B_i = 0 \tag{13}$$

Assuming orthogonality of input space **X**, the intersection of the hyperplane with the kth component of the input space as shown in Fig. 5 is:

$$\begin{aligned} X_r &= 0 \quad \forall r \neq k \\ X_k W_{ki} &= B_i \\ X_k &= \frac{B_i}{W_{ki}} \end{aligned} \tag{14}$$

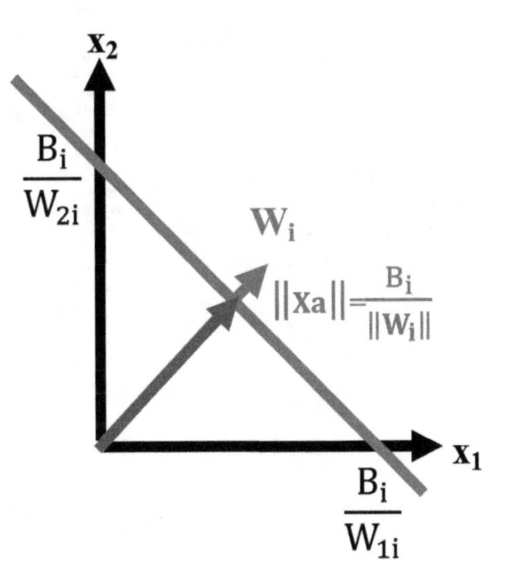

Fig. 5 Diagram of the intersections of the hyperplane with the axes of the orthogonal input space

So, for a single-layer NN, for the given input vector **X**, the output of each neuron will be zero or one depending on the location of the input **X** to the neuron hyperplane. For example, for input **X** in Fig. 6, the output of the NN is given by:

$$\mathbf{Y} = \begin{bmatrix} 0\,0 & 1\,0 \end{bmatrix} \tag{15}$$

This means that any given input sample **X** is mapped to a vector of zeros and ones in the layer **Y** that indicates where **X** is located between neuron hyperplanes in the input space. For each neuron of the NN, the location of the decision boundary hyperplane of the neuron in the input space is given by $\mathbf{W_i}$ and B_i. It should be noted that for the higher resolution of subspaces separated by the hyperplanes, corresponding to different codes in the layer **Y**, more neurons (hyperplanes) are needed.

Also, all of the input samples $\mathbf{X_p}$ located in a separated subspace by hyperplanes as shown in Fig. 7, have same representations in the layer **Y** and are inseparable (dimension reduction). For example in Fig. 7, for inputs $\mathbf{X_1}$ and $\mathbf{X_2}$, the outputs are given by:

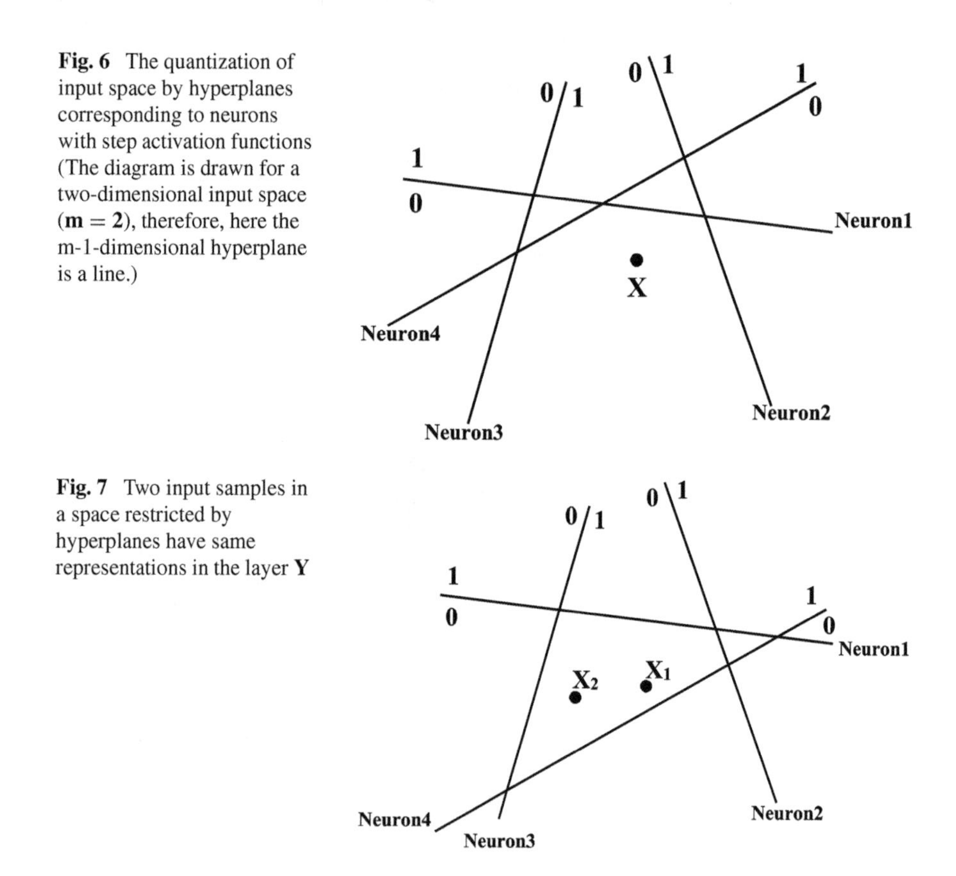

Fig. 6 The quantization of input space by hyperplanes corresponding to neurons with step activation functions (The diagram is drawn for a two-dimensional input space (**m** = **2**), therefore, here the m-1-dimensional hyperplane is a line.)

Fig. 7 Two input samples in a space restricted by hyperplanes have same representations in the layer **Y**

$$\mathbf{Y}_1 = \mathbf{Y}_2 = \begin{bmatrix} 0\,0\;\;1\;1 \end{bmatrix} \tag{16}$$

2.3 Functional Analysis of Single-Layer Feed-Forward Neural Networks with a Soft Nonlinear Activation Function (Sigmoid) and Bias

In this section, a single-layer NN in which the step functions of the neurons are replaced with soft non-linear functions (sigmoid) (as in Eq. 17) is analyzed. Sigmoid functions and also their derivative functions are given in Fig. 8.

$$\begin{aligned} \mathrm{Y_i} &= \mathrm{f}(\mathbf{X} \cdot \mathbf{W_i} - \mathrm{B_i}), \\ \mathrm{f(z)} &= \frac{1}{1 + \mathrm{e}^{-\mathrm{z}}} \end{aligned} \tag{17}$$

Here also, it is assumed that components of the input space are orthogonal. Due to the difference of step and soft (sigmoid) functions, the replacement of step functions with soft (sigmoid) functions change the hard decision boundaries to softened (fuzzy) decision boundaries. Nevertheless, the locations of the hyperplanes corresponding to them depended on $\mathbf{W_i}$ and $\mathrm{B_i}$, don't change. In other words, each neuron with sigmoid function and the fuzzy boundaries has a hyperplane corresponding to the step function whose the location depends on $\mathbf{W_i}$ and $\mathrm{B_i}$ on which $\mathbf{X} \cdot \mathbf{W_i} - \mathrm{B_i} = 0$. The sigmoid nonlinear function has the highest power of discrimination in a neighborhood of the decision boundary, nearly as the step function. This is because, the slope of the sigmoid function is maximum at this point referred to the inflection point, as in Fig. 8. However, by changing the bias ($\mathrm{B_i}$) and weights ($\mathbf{W_i}$), this inflection point is adjustable, as in Fig. 8.

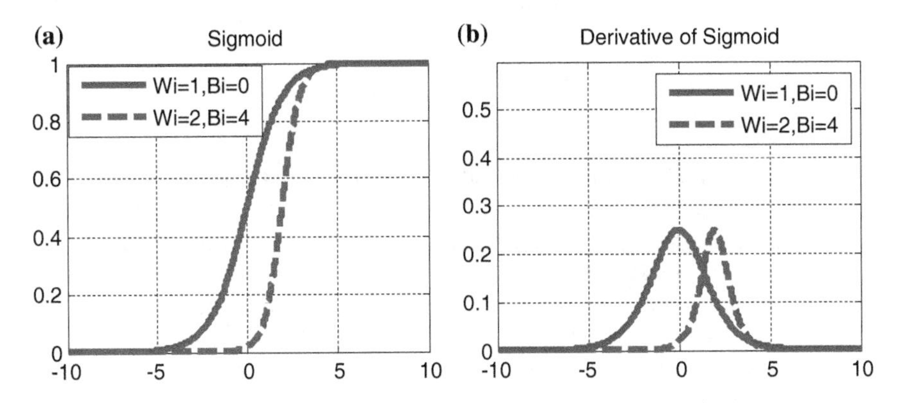

Fig. 8 Diagrams of sigmoid functions (**a**) and their derivatives (**b**)

Therefore, the output of a single-layer NN with sigmoid functions for the input $\mathbf{X}$ will be in the interval (0, 1) depending on position of $\mathbf{X}$ with respect to hyperplanes corresponding to the neurons. The output value depends on the distance $\mathbf{X}$ from the hyperplane of the neuron (the membership degree of $\mathbf{X}$ to the two decision boundaries of the neuron) where by distancing $\mathbf{X}$ from the hyperplane of each neuron, the neuron output becomes closer to one or zero (depending on $\mathbf{W_i}$ and B_i). For example, for the input $\mathbf{X}$ as Fig. 6, if functions of neurons are sigmoid, $\mathbf{Y}$ will be:

$$\mathbf{Y} = \begin{bmatrix} 0.1 & 0.3 & 0.9 & 0.4 \end{bmatrix} \tag{18}$$

In this case, the representations of inputs on the layer $\mathbf{Y}$ contain more information than the case where the functions of the neurons were step. Here, by varying $\mathbf{X}$ in an area enclosed by hyperplanes, a single value cannot be achieved in the layer $\mathbf{Y}$, but these values will be approximately close to each other. By passing a fuzzy boundary, at least one of the components of $\mathbf{Y}$ will quickly change (crossing the steepest slope of the sigmoid function). Thus, areas enclosed by the hyperplanes are discriminated.

Training of samples (inputs and desired outputs) to the NN means modifying NN weights ($\mathbf{W_i}$) and biases (B_i) for adjusting the hyperplanes corresponding to the neurons as the output global error function is optimized. In other words, the outputs of neurons become closer to the desired outputs as much as possible.

3 Analysis of Training Some Distinct Samples to a DAE with Nonlinear Activation Functions

For training of a DAE (Fig. 9) with P distinguished samples (Fig. 10 and Eq. 19), it is necessary to minimize the output global error function.

$$\{\mathbf{X_p} | \mathrm{p} = 1, 2, \ldots, \mathrm{P}\} \tag{19}$$

$$\mathrm{E} = \frac{1}{2} \sum_{\mathrm{p}=1}^{\mathrm{P}} \left\| \mathbf{Z_p} - \mathbf{X_p} \right\|^2 \tag{20}$$

where $\mathbf{X_p}$ is the input vector to DAE and also the its desired output, and $\mathbf{Z_p}$ is DAE output for the input vector $\mathbf{X_p}$. The necessary condition to be minimized the output global error function (Eq. 20) nearly zeros is maintaining the distinctions of samples in all of the hidden and also output layers (no dimension reduction). Because, if two distinct input patterns $\mathbf{X_i}$ and $\mathbf{X_j}$ are represented in DAE hidden layers by same descriptions, DAE outputs will be same for them. However, their desired outputs which are the same as inputs are different. Therefore, minimum of the error function E_{min} (Eq. 21) will not be zero, because $\mathbf{X_i} \neq \mathbf{X_j}$ while $\mathbf{Z_i} = \mathbf{Z_j}$.

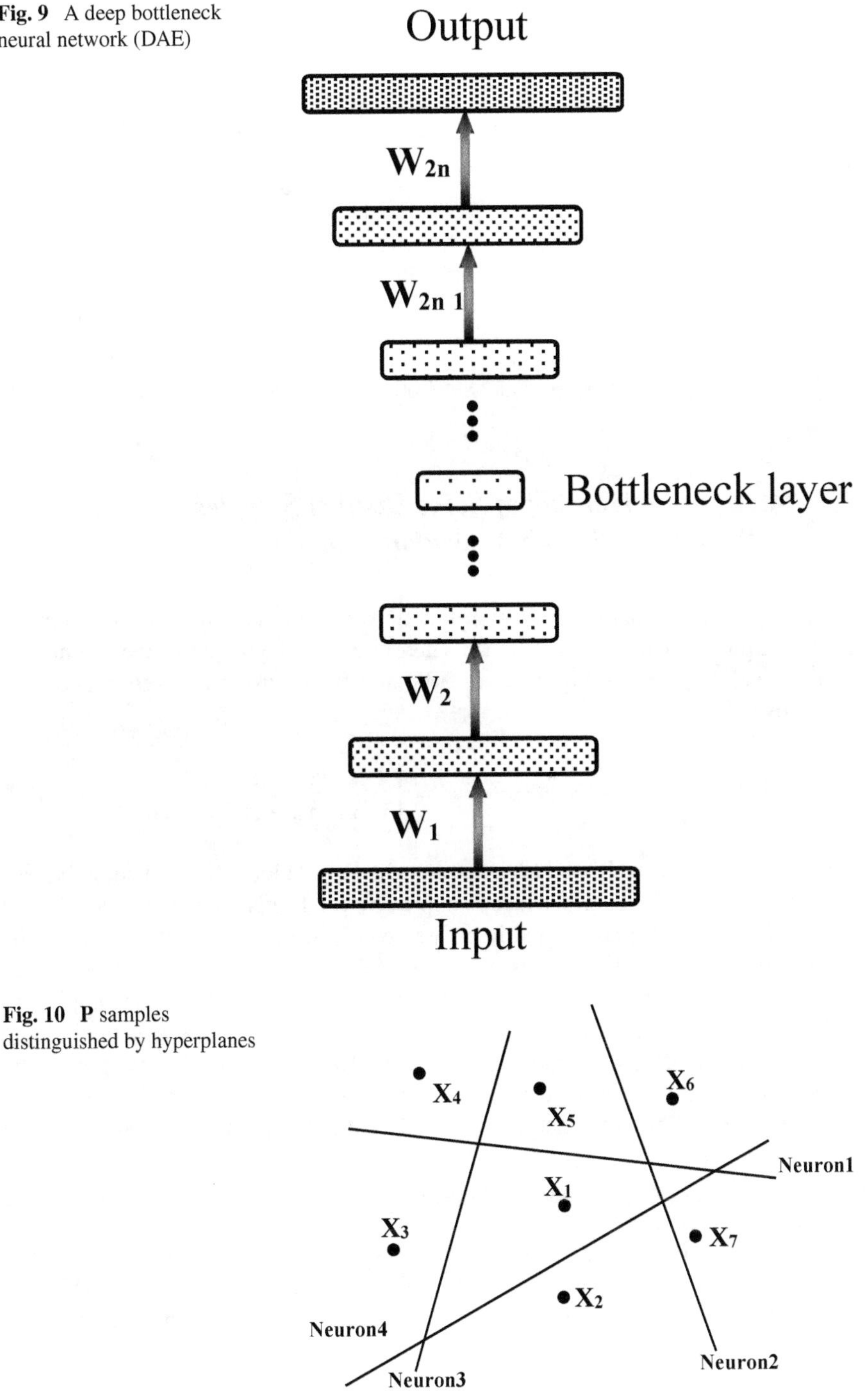

Fig. 9 A deep bottleneck neural network (DAE)

Fig. 10 **P** samples distinguished by hyperplanes

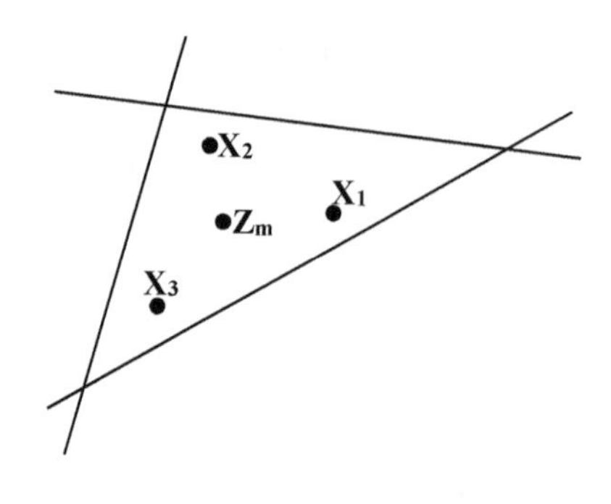

Fig. 11 Based on minimizing sum of squared error function in the output layer for samples located in one area enclosed between hyperplanes in the input space, the output error function is minimum in $\mathbf{Z_m}$

$$E_{min} = \frac{1}{2}\left[\|\mathbf{X_i} - \mathbf{Z_i}\|^2 + \|\mathbf{X_j} - \mathbf{Z_j}\|^2\right] \tag{21}$$

3.1 Analysis of Training of Some Distinct Samples to the DAE with Step Activation Functions

First, it is assumed that the activation functions of all hidden neurons are step and of the output layer neurons are linear. Therefore, Y_{ji}, the output of the i th neuron of j th layer, is given by Eq. 22 where $\mathbf{W_{ji}}$ and B_{ji} are its weight vector and bias, respectively.

$$Y_{ji} = f(\mathbf{X} \cdot \mathbf{W_{ji}} - B_{ji}) = \begin{cases} 1 & \mathbf{X} \cdot \mathbf{W_{ji}} - B_{ji} > 0 \\ 0 & \mathbf{X} \cdot \mathbf{W_{ji}} - B_{ji} < 0 \end{cases} \tag{22}$$

By modifying the weights during training, the output global error function (Eq. 20) can be minimized if all various input samples $\mathbf{X_p}$ can be distinctly represented in all the DAE layers. Samples located in an area enclosed between hyperplanes will be considered identical in later layers, and their discriminant information will be filtered (dimension reduction). Therefore, the DAE output descriptions for thesis samples are not ultimately distinguishable. It can be shown that in training based on minimizing the sum of squared error function, all of the samples in one area will be expressed by single representations in the next layers and in the output layer. For instance, their single representation in the output layer will be $\mathbf{Z_m}$, which is their mean, as shown in Fig. 11.

Therefore, the necessary condition for convergence of DAE training and getting of output error near zero is maintaining the discriminations of sample descriptions of in all layers to the output. To discriminate samples, being at least one hyperplane between each of two sample descriptions in all layers is essential, as shown in Fig. 12. Here, each hyperplane distinguishing two separate sample descriptions (not duplicate hyperplanes) in each layer is as a nonlinear binary component.

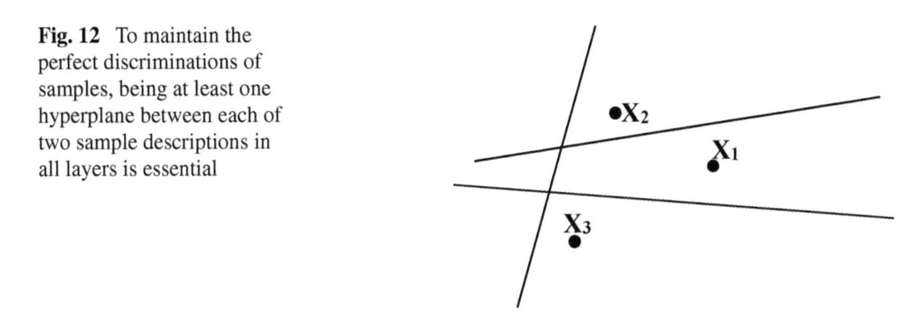

Fig. 12 To maintain the perfect discriminations of samples, being at least one hyperplane between each of two sample descriptions in all layers is essential

3.2 Analysis of Training of Some Distinct Samples to the DAE with Continuous (Sigmoid) Activation Functions

By replacing of the step functions with the sigmoid functions, Eq. 22 is modified as Eq. 23.

$$\mathrm{Y_{ji}} = f(\mathbf{X}\cdot\mathbf{W_{ji}} - \mathrm{B_{ji}}) = \frac{1}{1 + \mathrm{e}^{-(\mathbf{X}\cdot\mathbf{W}_{ji} - \mathrm{B}_{ji})}} \tag{23}$$

Here also, minimizing of the global error function by the back-propagation algorithm depends on maintaining discriminations of sample representations in all of the layers. However, due to using sigmoid nonlinear functions instead of step, maintaining the discriminations between two samples in the NN layers with step functions is converted to maximizing discriminations of input sample representations by neurons of every layer. Necessary conditions for discrimination maximization between the two samples in a neuron output are determined as following.

3.2.1 Necessary Conditions for Maximizing Discrimination of Two Distinct Input Samples by a Neuron with Sigmoid Continuous Function

In this section, we maximize the distance between representations of two samples $\mathbf{X}_1$ and $\mathbf{X}_2$ in the output of the ith neuron with distance $\Delta\mathbf{X}$ in the input space (as shown in Fig. 13).

$$\Delta\mathbf{X} = \mathbf{X}_2 - \mathbf{X}_1 \tag{24}$$

Theorem *Necessary conditions for maximizing discrimination of two distinct input samples* $\mathbf{X}_1$ *and* $\mathbf{X}_2$ *in the input space, by a neuron with the weight vector* $\mathbf{W}_i$, *the bias* B_i, *and sigmoid continuous nonlinear function are*

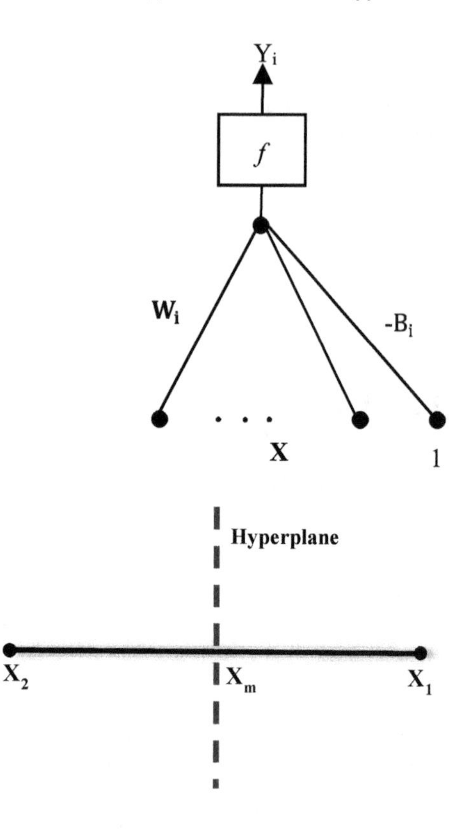

Fig. 13 The ith neuron of a NN

Fig. 14 The discrimination of the representations of two samples will be maximized in the neuron output, if the hyperplane corresponding to it is perpendicular to $\Delta\mathbf{X}$ in the point $\mathbf{X_m}$

$$\mathbf{X_m} \cdot \mathbf{W_i} - \mathrm{B_i} = 0 \tag{25}$$

where $\mathbf{X_m}$*, the mean of* $\mathbf{X_1}$ *and* $\mathbf{X_2}$*, is*

$$\mathbf{X_m} = \frac{\mathbf{X_1} + \mathbf{X_2}}{2} \tag{26}$$

and $(\Delta\mathbf{X} \cdot \mathbf{W_i})$ *is maximized (the weight vector* $\mathbf{W_i}$ *must be at the same direction of* $\Delta\mathbf{X}$*).*

Proof $\mathbf{X_1}$ and $\mathbf{X_2}$, two distinct input samples, are assumed as Fig. 14. For the ith neuron with the weight vector $\mathbf{W_i}$ and the bias $\mathrm{B_i}$ can be written:

$$\mathrm{Y_{i_1}} = f(\mathbf{X_1} \cdot \mathbf{W_i} - \mathrm{B_i}) \tag{27}$$

$$\mathrm{Y_{i_2}} = f(\mathbf{X_2} \cdot \mathbf{W_i} - \mathrm{B_i}) \tag{28}$$

$$\Delta\mathrm{Y_i} = \mathrm{Y_{i_2}} - \mathrm{Y_{i_1}} \tag{29}$$

where Y_{i_1} and Y_{i_2} are outputs of the ith neuron for inputs $\mathbf{X_1}$ and $\mathbf{X_2}$. For the neuron with the weight vector $\mathbf{W_i}$ and bias B_i, in the n-dimensional input space, in general, the Eq. 8 represents a hyperplane that is perpendicular to $\mathbf{W_i}$ and pass from point $\mathbf{X_a} = \frac{B_i}{\|\mathbf{W_i}\|}$ (see Sect. 2.2). If the sigmoid nonlinear function of the neuron is replaced with a step function, the hyperplane will be the boundary decision locus of the neuron. Therefore, for the neuron with the sigmoid function, this hyperplane is called the hyperplane corresponding to the neuron.

For highest distinction between $\mathbf{X}_1$ and $\mathbf{X}_2$ at the ith neuron output, ΔY_i should be maximized. For this purpose, using the Taylor series expansion, Eqs. (27) and (28) can be written as:

$$\begin{aligned} Y_{i_1} &= f(\mathbf{X_1} \cdot \mathbf{W_i} - B_i) \\ &= f(\mathbf{X_m} \cdot \mathbf{W_i} - B_i) - \frac{1}{2} f'(\mathbf{X_m} \cdot \mathbf{W_i} - B_i)(\Delta \mathbf{X} \cdot \mathbf{W_i}) + \cdots \end{aligned} \tag{30}$$

$$\begin{aligned} Y_{i_2} &= f(\mathbf{X_2} \cdot \mathbf{W_i} - B_i) \\ &= f(\mathbf{X_m} \cdot \mathbf{W_i} - B_i) + \frac{1}{2} f'(\mathbf{X_m} \cdot \mathbf{W_i} - B_i)(\Delta \mathbf{X} \cdot \mathbf{W_i}) + \cdots \end{aligned} \tag{31}$$

Assuming that weights $\mathbf{W_i}$ and bias B_i are too small at the beginning of training, the second and higher order terms in Eqs. 30 and 31 can be ignored (The slope variations of f is negligible in this range).

$$\Delta Y_i = Y_{i_2} - Y_{i_1} \cong (\Delta \mathbf{X} \cdot \mathbf{W_i}) f'(\mathbf{X_m} \cdot \mathbf{W_i} - B_i) \tag{32}$$

Therefore, ΔY_i (the distance of the representations of two samples in the neuron output) will be maximized if $f'(\mathbf{X_m} \cdot \mathbf{W_i} - B_i)$ and $(\Delta \mathbf{X} \cdot \mathbf{W_i})$ are maximized based on $\mathbf{W_i}$ and B_i:

$$maximized(\Delta \mathbf{X} \cdot \mathbf{W_i}) \tag{33}$$

$$maximized\left(f'(\mathbf{X_m} \cdot \mathbf{W_i} - B_i)\right) \tag{34}$$

For the purposes these are necessary:

1. The weight vector $\mathbf{W_i}$ must be at the same direction of $\Delta \mathbf{X}$.
2. $(\mathbf{X_m} \cdot \mathbf{W_i} - B_i) = 0$, whereupon the derivative of the sigmoid function is maximum.

As a result, the hyperplane corresponding to the neuron must pass from $\mathbf{X_m}$ and be perpendicular to $\Delta \mathbf{X}$ (as seen Fig. 14). Therefore, $\mathbf{X_m}$ must be the inflection point of the sigmoid function f (the maximum of the sigmoid derivative function).

After determining the conditions for maximizing the neuron's discrimination between two input samples, in the following, training some distinct samples to a DAE with soft (sigmoid) functions will be explained.

3.2.2 Stagewise Analysis of Training Some Distinct Samples to a DAE

At the beginning of DAE training, starting from small random weights, due to the small and random values of $\mathbf{W_i}$ and B_i and also, uncorrelated input and weight vectors of each neuron, the outputs of neurons of all the layers will have a value approximately equal to 0.5 (Eq. 35).

$$Y_{ji} = f\left(\mathbf{Y_{j-1}} \cdot \mathbf{W_{ji}} - \mathbf{B_{ji}}\right) \cong 0.5 \quad j = 1, 2, \ldots, 2n \tag{35}$$

(The value of phrase in parentheses is nearly equal to zero.) where $\mathbf{Y_{j-1}}$ is the output vector of the previous layer neurons, and $\mathbf{Y_0}$ is the input vector $\mathbf{X}$. Therefore, the distinct inputs $\mathbf{X}$ will be represented by nearly the same values and not distinguished in the next layers. With the start of training, weights are modified to minimize the error function E in Eq. 22. However, the weights of the last layer $\mathbf{W_{2n}}$ are modified much faster than the weights of the previous layers. Comparing the delta error signals computed in the back-propagation algorithm (Eqs. 36 and 37) for the output and previous layers show that by back-propagating and multiplying the delta error signal vectors by the weight vectors with small and random values (uncorrelated with $\boldsymbol{\delta}$), their discrimination information is extremely weakened and filtered. Specifically, it is more severe for earlier layers.

$$\boldsymbol{\delta}_{\mathbf{Z}} = \boldsymbol{\delta}_{\mathbf{Y_{2n}}} = f'(\mathbf{Y_{2n-1}} \cdot \mathbf{W_{2n}} - B_{2n})(\mathbf{X} - \mathbf{Z}) \tag{36}$$

$$\boldsymbol{\delta}_{\mathbf{Y_{2n-1}}} = f'(\mathbf{Y_{2n-2}} \cdot \mathbf{W_{2n-1}} - B_{2n-1})\left(\boldsymbol{\delta}_{\mathbf{Z}} \cdot \mathbf{W_{2n}^T}\right) \tag{37}$$

where $\mathbf{X}$ is the desired output of the DAE (It is the same as input) and $\mathbf{Z}$ is the DAE output. $\boldsymbol{\delta}_{\mathbf{Z}}$ is the error signal vector at the output layer. Therefore, in order to be rapidly minimized E, $\mathbf{W_{2n}}$, the weight vectors of the neurons of the last layer, are modified so that all distinct inputs represented with nearly the same representations in the $2n - 1$th layer are mapped to a single output $\mathbf{Z}$ and also E is minimized at the beginning of training. Due to the minimum of the error function (Eq. 21) for the same outputs of different samples is at the mean of them $\mathbf{Z_m}$, the single output $\mathbf{Z}$ will converge to $\mathbf{Z_m}$, firstly. For example, as shown in Fig. 15, the output of the DAE for the input samples $\mathbf{X_1}, \ldots, \mathbf{X_4}$ will converge to $\mathbf{Z_m}$, firstly.

Then, by back-propagating error to deeper layers, DAE attempts to create the first discrimination between samples by modifying weights of deeper layers to the

Fig. 15 The minimum of the error function for the same outputs of different samples is at the mean of them $\mathbf{Z_m}$

input. This is performed by back-propagating $\boldsymbol{\delta}_{\mathbf{Z}}$ to prior layers. The formation of the output distinctions depends on forming discriminant information of the input representations (as discriminant nonlinear components for data), primarily in the first hidden layer $\mathbf{Y}_1$ and then in higher layers to the output. The first discrimination is largely formed aligned with the first linear principal component of data (the largest variance) during the training, as shown in Fig. 16. Because of the first linear principal component of some distinct samples functions as $\Delta\mathbf{X}$ for two distinct samples during training to a DAE.

To create the discrimination in the first hidden layer, the delta error signal vector in this layer $\boldsymbol{\delta}_{\mathbf{Y}_1}$ should have significant amounts i.e. $\|\boldsymbol{\delta}_{\mathbf{Y}_1}\| \neq 0$ back-propagated layer-by-layer from the DAE output $\boldsymbol{\delta}_{\mathbf{Z}}$ to $\mathbf{Y}_1$. However, during back-propagating of the error signals $\boldsymbol{\delta}$ in layers, due to the small and random weights, the discriminant information of $\boldsymbol{\delta}_{\mathbf{Z}}$ is filtered and in the lower layers is extremely weakened. While, in the single-hidden-layer NNs, discriminant information of $\boldsymbol{\delta}_{\mathbf{Z}}$ immediately influences the weights of the first hidden layer $\mathbf{W}_1$ and by modifying of them different representations in layer $\mathbf{Y}_1$ for distinct input samples are provided. Therefore, the single-hidden-layer NNs can be used to rapidly initialize DAE weights.

Based on this analysis, to train DAEs, by the layer-by-layer pre-training, the complex problem of training the deep structure of DAE is broken to some simpler sub-problems; training of single-hidden-layer NNs. Then, their weights are used in the DAE weights as initial values [1]. In the next section, a summary of the method is given.

4 Layer-by-Layer Pre-training

In the layer-by-layer pre-training method, the DAE, including 2n layers, is decomposed into n single-hidden-layer AEs. Then, each of the AEs is trained to minimize its loss of input discrimination information, and therefore the outputs are distinctly reconstructed.

Figure 17 shows how the DAE is decomposed into several single-hidden-layer AEs. The weights of mapping and demapping sections of the first AE are $\mathbf{W}_1$ and $\mathbf{W}_{2n}$ in the first and last layers of the DAE, respectively. The first AE is trained by the training data in an auto-associative manner using the back-propagation algorithm;

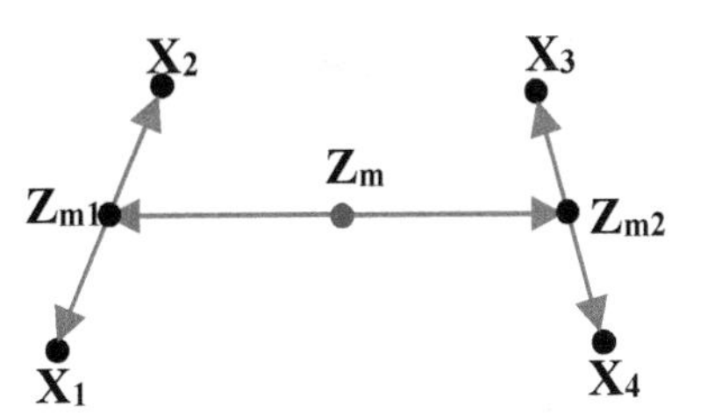

Fig. 16 For the discrimination between samples, the first discrimination is largely formed aligned with the first linear principal component of data

which yields the initial weights of the first ($\mathbf{W_1}$) and last ($\mathbf{W_{2n}}$) layers. Afterwards, the non-linear projections of the inputs in the bottleneck (hidden) layer space of the first AE are calculated, and applied as the inputs for the second AE to determine the initial values of $\mathbf{W_2}$ and $\mathbf{W_{2n-1}}$. This process continues in this way [1].

After the pre-training phases, the obtained weights are considered as the initial weights in the integrated DAE, and then fine-tuned using the error back-propagation algorithm over it.

5 Experimental Results and Discussion

In this section, we study the performances of the pre-training method to maintain distinguishing information of data. In this way, the ability of the method to train DAEs to extract nonlinear principal components of faces in Bosphorus and ORL databases is evaluated. In the following, these databases are introduced.

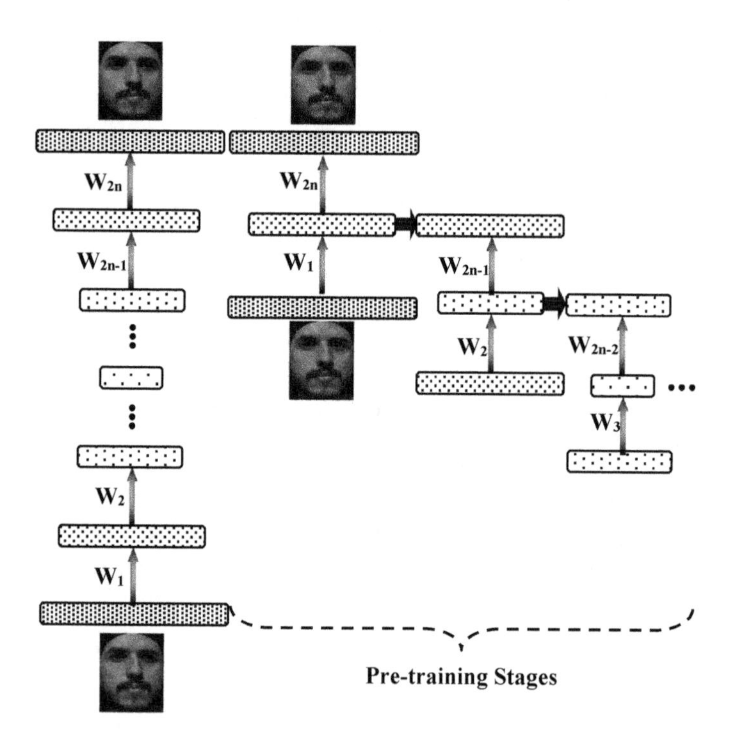

Fig. 17 The manner of breaking down DAE to some AEs in the layer-by-layer pre-training method [1]

Fig. 18 Random samples from face databases, top to bottom: Bosphorus.E, Bosphorus.P and ORL

5.1 Databases

5.1.1 Bosphorus Database

The Bosphorus data consists of multi-expression (Bosphorus.E) and multi-pose (Bosphorus.P) 3D and 2D face images in which the 2D images of 95 subjects from this collection were selected randomly for this study [35]. These images are used in the gray scale format with 256 levels of gray. Moreover, to reduce the computation load, the resolution of these images was decreased to 92 * 114.

For each part of the database, two images for test, two images for validation and the remaining images for training were used for everyone. At the end, the training, validation and test sets were consisted of 2358, 190 and 190 images for the Bosphorus.E dataset and 949, 190 and 190 images for the Bosphorus.P dataset, respectively.

5.1.2 ORL Database

The ORL database contains images of 40 individuals, each person having 10 different images. The size of each image is 92 * 112 pixels, with 256 grey levels per pixel [36]. For every subject, 2 images for test, 2 images for validation and the remaining images for training were used. Finally, the training, validation and test sets were consisted of 240, 80 and 80 images, respectively. Figure 18 illustrates examples of images of the used databases.

Table 1 DAE parameters

The number of hidden layers	7
The number of hidden layer neurons	1000-400-200-100-200-400-1000
The activation function of hidden neurons	Sigmoid
Learning rate	0.001–0.0001
Momentum rate	0.7

5.2 *Training DAE*

The DAE structure studied in this research is introduced in Table 1. For any dataset, the DAE initial weights were assigned by the following two methods before training occurred: (1) random values, (2) values obtained from the layer-by-layer pre-training method. In the layer-by-layer pre-training method, the seven-hidden-layer DAEs were broken down into four single-hidden-layer BNNs. The first BNN was pre-trained with the input samples, and the 1000-dimensional representations extracted at its hidden layer were used to pre-train the second BNN with the 1000-400-1000 structure; and the same procedure was continued. In this way, the initial weights of the DAEs were obtained. Then, the resulting weights were substituted in the DAEs and fine-tuned.

5.3 *The Effect of the Layer-by-Layer Pre-training Method on Convergence Improvement of Training the DAEs*

During the training of DAEs, reconstruction errors of validation datasets were traced and not improving of them were placed as criteria (the model generalization) for stopping the train.

Table 2 compares global reconstruction errors (E_{MSRE} [1]) of train and validation datasets and epochs needed for acceding them by the two methods; randomly initialized and pre-trained DAEs. For each dataset, as Table 2 indicates, the trainings of pre-trained DAEs were rapidly converged and the training and validation errors were lower. The weights of the training phase were used in two subsequent experiments.

5.4 *Evaluation of Locations of Hyperplanes Corresponding to Neurons of DAE Hidden Layers*

As mentioned earlier, to maintain the full discrimination of samples, being at least one hyperplane between the representations of the both samples is essential in every layer. In this section, to evaluate this case, the number of distinct subspaces formed

Table 2 E_{MSRE} for DAEs on three databases

Dataset	Methods	Epochs	Train (E_{MSRE})	Validation (E_{MSRE})
Bosphorus.E	Randomly initialized DBNN	1043	0.048	0.048
	Pre-trained DBNN	91	0.021	0.026
Bosphorus.P	Randomly initialized DBNN	2175	0.075	0.085
	Pre-trained DBNN	37	0.027	0.053
ORL	Randomly initialized DBNN	1106	0.131	0.135
	Pre-trained DBNN	3	0.037	0.09

by hyperplanes corresponding to neurons of each layer containing only one sample representation, have been compared for the randomly initialized and pre-trained DAEs. A discrimination parameter is defined for this evaluation by Eq. 38.

$$D(j) = \frac{h(j)}{N} * 100 \tag{38}$$

where D(j) is the discrimination percentage in the jth layer, N is the total number of samples and $h(j)$ is the number of subspaces enclosed by hyperplanes of the jth layer in which only one example exists. Therefore, larger D(j) for a layer is indicative of the better protection of distinctions between samples in the layer.

In Fig. 19, D(j) for the different layers of the randomly initialized and pre-trained DAEs for the training (the first column), validation (the second column) and test (the third column) data of the Bosphorus.E (the first row), the Bosphorus.P (the second row), the ORL (the third row) databases are shown. As can be seen for each of the datasets, discrimination parameters D(j) is 100% in all layers of pre-trained DAE. This means that at least one hyperplane is taken between every two samples at every layer. In the other words all samples are distinctively represented in all layers. However, the discrimination percentages are decreased for the randomly initialized DAE in different layers, and the loss for the third to last hidden layers is more prominent. With a step forward in the layers, due to the improper training of randomly initialized DAE, some of the distinguishing information between the samples were filtered by multiplying in the weights.

Comparing Fig. 19a–c shows more fall in D(j) of deeper layers in (a). This is due to the high density of samples in training dataset which makes the number of subspaces with only one sample to be low. Therefore, the discrimination percentages of the different layers for the training dataset are lower than validation and test datasets.

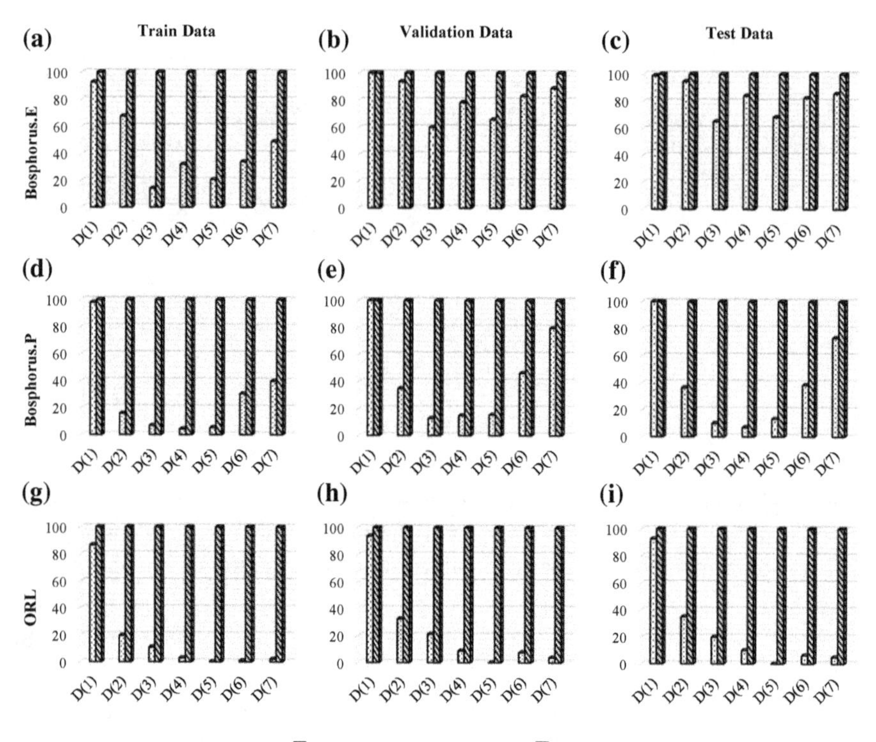

Fig. 19 The discrimination rates (**D(j)**) of the different layers of the randomly initialized and pre-trained DAEs for the training (the first column), validation (the second column) and test (the third column) datasets of the Bosphorus.E (the first rows), the Bosphorus.P (the second rows), the ORL (the third rows) databases

A confusion may be occurred here, that why the discrimination percentage of some layers of randomly initialized DAEs are more than the previous layers. If the neurons with step functions had been used, all of the samples located in a subspace enclosed by hyperplanes would have a single representation. Therefore, they are identically represented in this layer and the later layers. However, in our experiments, neuron functions of the DAEs were the sigmoid continues function. So, the samples located in a subspace enclosed by hyperplanes corresponding to the neurons do not necessarily have the exact same representations, but those are close to each other. As a result, the distinctions of their expressions may be strengthened or weakened in the next layer. Therefore, due to the strengthening of their slight distinctions, increasing the discrimination percentage are shown in some later layers.

Diagrams of rows 2 and 3 of Fig. 19 show the results of these evaluations for Bosphorus.P and ORL databases, respectively. The results for these databases are similar to Bosphorus.E and the discrimination rates of the different layers of the pre-

Table 3 Identity recognition rate for images of the Bosphorus.E

Hidden layers	Methods			
	Randomly initialized DAE		Pre-trained DAE	
	Validation	Test	Validation	Test
Layer 1	93.7	91.6	98.9	97.9
Layer 2	85.3	79.5	99.5	97.9
Layer 3	78.4	73.7	98.9	97.9
Layer 4	76.8	71.6	99.5	98.4
Layer 5	77.4	68.9	99.5	97.9
Layer 6	75.3	67.4	99.5	97.9
Layer 7	75.8	66.8	98.4	97.9

trained DAEs are 100% for three sections of each database. However, the discrimination rates D(j) for the randomly initialized DAEs are very lower on all datasets.

Overall, these experiments showed that in pre-trained DAEs, samples were located in more discriminated subspaces in all the layers. In other words, through layer-by-layer setting hyperplanes and accurate tuning of them, more discriminated representations of all the samples in all the layers are provided.

E. Evaluation of Identity Discriminability of Representations in the Different Layers of the DAEs

In this section we evaluate discriminability of sample representations in different layers of the DAEs. For this purpose, we recognize the identities of the face images using their representations extracted by DAE layers. The K-Nearest Neighbor (KNN) classifier is used for recognizing in which its training datasets were representations of training images and its test datasets were representations of validation or test images. Therefore, in each hidden layer, for the input images, the output vectors of the neurons are computed and used for recognition.

Tables 3, 4 and 5 show recognition rates for images mapped in different layers of the randomly initialized and pre-trained DAEs for Bosphorus.E, Bosphorus.P, and ORL databases, respectively. As illustrated, by going to outputs of randomly initialized DAEs, distinguishing information of identity for validation and test datasets are weakened. Whilst, the pre-trained DAEs maintain this information, so that identity recognition rates at the different layers of DAEs are almost constant and significantly higher. Low results in Table 4 for Bosphorus.P are due to the dataset. This database contains images of faces under rotation at 0–90°, which makes the recognition difficult. However, it is shown that the recognition rates are significantly higher for pre-trained DAEs.

Table 4 Identity recognition rate for images of the Bosphorus.P

Hidden layers	Methods			
	Randomly initialized DAE		Pre-trained DAE	
	Validation	Test	Validation	Test
Layer 1	20	16.3	34.2	34.7
Layer 2	11.0	7.9	35.3	36.3
Layer 3	7.4	5.8	37.9	38.4
Layer 4	7.4	5.8	35.8	36.8
Layer 5	7.4	3.7	36.8	35.8
Layer 6	7.4	4.7	37.4	36.3
Layer 7	5.8	4.2	32.6	35.8

Table 5 Identity recognition rate for images of the ORL

Hidden layers	Methods			
	Randomly initialized DAE		Pre-trained DAE	
	Validation	Test	Validation	Test
Layer 1	67.5	61.2	96.2	93.7
Layer 2	40	26.2	95	95
Layer 3	25	15	95	95
Layer 4	22.5	13.7	97.5	95
Layer 5	23.7	12.5	97.5	95
Layer 6	23.7	12.5	95	95
Layer 7	23.7	10	95	93.7

6 Conclusion

In this chapter, the theoretical framework of the layer-by-layer pre-training method for DAE training convergence was presented. In this regard, NNs with linear, step nonlinear, and soft nonlinear (sigmoid) activation functions in high-dimensional spaces were studied and analyzed.

Subsequently, autoassociative training of some samples into a DAE was analyzed and requirements for maximum discrimination by each neuron of the DAE with sigmoid function were determined. Accordingly, it was shown why the training of randomly initialized DAEs do not converge and the layer-by-layer pre-training leads to the training convergence by adjusting the weights layer-wisely.

Our experiments indicated that representations of distinct samples were located in discriminated subspaces enclosed by hyperplanes corresponding to the neurons in all layers pre-trained DAEs. Also, it was shown that in different layers of pre-trained DAEs, sample discriminations were extremely preserved. While in randomly initialized DAEs, by going to outputs distinguishing information for validation and test datasets were weakened.

References

1. S.Z. Seyyedsalehi, S.A. Seyyedsalehi, A fast and efficient pre-training method based on layer-by-layer maximum discrimination for deep neural networks. Neurocomputing **168**, 669–680 (2015)
2. S.Z. Seyyedsalehi, S.A. Seyyedsalehi, New fast pre-training method for training of deep neural network. Int. J. Signal Data Process. **10**(1), 13–26 (2013). (In persian)
3. L. Szymanski, B. McCane, Deep networks are effective encoders of periodicity. IEEE Trans. Neural Netw. **25**(10), 1816–1827 (2014)
4. M. Bianchini, F. Scarselli, On the complexity of neural network classifiers: a comparison between shallow and deep architectures. IEEE Trans. Neural Netw. **25**(8), 1553–1565 (2014)
5. P.Z. Eskikand, S.A. Seyyedsalehi, Robust speech recognition by extracting invariant features. Procedia Soc. Behav. Sci. **32**, 230–237 (2012)
6. Z. Ansari, S.A. Seyyedsalehi, Toward growing modular deep neural networks for continuous speech recognition. Neural Comput. Appl. 1–20 (2017)
7. H. Behbood, S.A. Seyyedsalehi, H.R. Tohidypour, M. Najafi, S. Gharibzadeh, A novel neural-based model for acoustic-articulatory inversion mapping. Neural Comput. Appl. **21**(5), 935–943 (2012)
8. Z. Ansari, S.A. Seyyedsalehi, Proposing two speaker adaptation methods for deep neural network based speech recognition systems, in *7th International Symposium on Telecommunications (IST)*, pp. 452–457. IEEE (2014)
9. T. Nakashika, T. Takiguchi, Y. Ariki, Voice conversion using RNN pre-trained by recurrent temporal restricted Boltzmann machines. IEEE Trans. Audio Speech Lang. Process **23**(3), 580–587 (2015)
10. S. Babaei, A. Geranmayeh, S.A. Seyyedsalehi, Towards designing modular recurrent neural networks in learning protein secondary structures. Exp. Syst. Appl. **39**(6), 6263–6274 (2012)
11. M. Spencer, J. Eickholt, J. Cheng, A deep learning network approach to ab initio protein secondary structure prediction. IEEE Trans. TCBB **12**(1), 103–112 (2015)
12. S.Z. Seyyedsalehi, S.A. Seyyedsalehi, Simultaneous learning of nonlinear manifolds based on the bottleneck neural network. Neural Process. Lett. **40**(2), 191–209 (2014)
13. S.Z. Seyyedsalehi, S.A. Seyyedsalehi, Bidirectional pre-training method for deep neural network learning. Comput. Intell. Electr. Eng. **6**(2), 1–10 (2015). (In persian)
14. S. Gao, Y. Zhang, K. Jia, J. Lu, Y. Zhang, Single sample face recognition via learning deep supervised autoencoders. IEEE Trans. Inf. Forensics Secur **10**(10), 2108–2118 (2015)
15. S.Z. Seyyedsalehi, S.A. Seyyedsalehi, Improving the nonlinear manifold separator model to the face recognition by a single image of per person. Int. J. Signal Data Process. **12**(1), 3–16 (2015). (In persian)
16. S.Z. Seyyedsalehi, S.A. Seyyedsalehi, Attractor analysis in associative neural networks and its application to facial image analysis. Comput. Intell. Electr. Eng. (2018). (In persian)
17. M. Hayat, M. Bennamoun, S. An, Deep reconstruction models for image set classification. IEEE Trans. Pattern Anal. Mach. Intell. **37**(4), 713–727 (2015)
18. S.M. Moghadam, S.A. Seyyedsalehi, Nonlinear analysis and synthesis of video images using deep dynamic bottleneck neural networks for face recognition. Neural Netw. **105**, 304–315 (2018)
19. S.H. Lee, C.S. Chan, S.J. Mayo, P. Remagnino, How deep learning extracts and learns leaf features for plant classification. Pattern Recogn. **71**, 1–13 (2017)
20. P. Eulenberg, N. Köhler, T. Blasi, A. Filby, A.E. Carpenter, P. Rees, F.J. Theis, F.A. Wolf, Reconstructing cell cycle and disease progression using deep learning. Nat. Commun. **8**(1) (2017)
21. H. Goh, N. Thome, M. Cord, J.-H. Lim, Learning deep hierarchical visual feature coding. IEEE Trans. Neural Netw. **25**(12), 2212–2225 (2014)
22. R. Salakhutdinov, J. Tenenbaum, A. Torralba, Learning with hierarchical-deep models. IEEE Trans. Pattern Anal. Mach. Intell. **35**(8), 1985–1971 (2013)

23. Y. Bengio, Learning deep architectures for AI. Found. Trends Mach. Learn. **2**, 1–127 (2009)
24. G. Ian, Y. Bengio, A. Courville, *Deep Learning* (MIT press, 2016)
25. N. Plath, Extracting low-dimensional features by means of deep network architectures, Ph.D. dissertation, Technische Universität Berlin, Apr 2008
26. Y. Bengio, Evolving culture versus local minima, in *Growing Adaptive Machines* (Springer, Berlin, Heidelberg, 2014), pp. 109–138
27. A.S. Shamsabadi, M. Babaie-Zadeh, S.Z. Seyyedsalehi, H.R. Rabiee, A new algorithm for training sparse autoencoders, in *2017 25th European Signal Processing Conference (EUSIPCO)*. IEEE, pp. 2141–2145 (2017)
28. G.E. Hinton, S. Osindero, Y.-W. Teh, A fast learning algorithm for deep belief nets. Neural Comput. **18**(7), 1527–1554 (2006)
29. N. Jaitly, P. Nguyen, A. W. Senior, V. Vanhoucke, Application of pretrained deep neural networks to large vocabulary speech recognition, in *Proceedings of Interspeech* (2012)
30. G.E. Dahl, D. Yu, L. Deng, A. Acero, Context-dependent pre-trained deep neural networks for large-vocabulary speech recognition. IEEE Trans. Audio, Speech Lang. Process. **20**(1), 30–42 (2012)
31. S.Z. Seyyedsalehi, S.A. Seyyedsalehi, New fast pre training method for deep neural network learning, in *Proceedings of 19th ICBME* (2012). (In Persian
32. I. Nejadgholi, S.A. Seyyedsalehi, S. Chartier, A brain-inspired method of facial expression generation using chaotic feature extracting bidirectional associative memory. Neural Process Lett. **46**(3), 943–960 (2017)
33. G.E. Hinton, R.R. Salakhutdinov, Reducing the dimensionality of data with neural networks. Science **313**(5786), 504–507 (2006)
34. S.A. Seyyedsalehi, S.A. Motamedi, M.R. Hashemigolpayegany, M.H. Ghasemian, Towards describing function of the human brain and neural networks using nonlinear mappings in highdimensional spaces. J. Daneshvar **11&12**, 1–10 (1996). (In Persian)
35. A. Savran, N. Alyüz, H. Dibeklioğlu, O. Çeliktutan, B. Gökberk, B. Sankur, L. Akarun, Bosphorus database for 3D face analysis. Biom. Identity Manag. **5372**, 47–56 (2008)
36. F.S. Samaria, A.C. Harter, Parameterisation of a stochastic model for human face identification. in *Proceedings of the Second IEEE Workshop on Applications of Computer Vision*, pp. 138–142 (1994)

Springer: Deep Learning in eHealth

Peter Wlodarczak

Abstract In recent years, a lot of advances have been made in data analysis using Machine Learning techniques, and specifically using Deep Learners. Deep Learners have been performing particularly well for multimedia mining tasks such as object or face recognition and Natural Language Processing tasks such as speech recognition and voice commands. This opens up a lot of new possibilities for medical applications. Deep Learners can be used for medical imaging, behavioral health analytics, pervasive sensing, translational bioinformatics or predictive diagnosis. This chapter provides an introduction in Deep Learning for health informatics and presents some of its applications for eHealth.

Keywords Deep learning · Predictive analytics · Convolutional Neural Networks · eHealth · Biomedicine · Correlation analysis

1 Introduction

In all western societies, the health care systems face many challenges. Increasing health care costs, an aging population, the growing burden of chronic diseases (Alzheimer's disease, cancer, Parkinson's disease, hypertension etc.) and a shortage of specialists are only some of the problems that seek for a solution. An ever increasing demand for services, underfunding and understaffing, and pressures inflicted by state and federal government health regulations, hospital based services are forced to become more efficient in how they provide their services [1]. New approaches in medical diagnosis, patient monitoring and personalized treatment promise to mitigate some of the issues that are currently unsolved.

Electronic health (eHealth) has become an active study field since it promises to alleviate some of the problems health care systems face today. There is no consistent definition of eHealth in literature. eHEalth denotes the use of electronic communication and information technology in the health care sector [2]. Sometimes it also

P. Wlodarczak (✉)
University of Southern Queensland, West Street, Toowoomba, QLD 4350, Australia
e-mail: wlodarczak@gmail.com

V. E. Balas et al. (eds.), *Handbook of Deep Learning Applications*,
Smart Innovation, Systems and Technologies 136,
https://doi.org/10.1007/978-3-030-11479-4_14

includes the transmission and storage of clinical data. eHealth makes it possible, at least partly, to monitor patients in real-time, offer personalized treatments, optimize resource management, detect emergencies early or before they occur, predict treatment outcomes or improve treatments. Key enablers for eHealth solutions are Big Data, wireless communication, the Internet of Things (IoT) and Artificial Intelligence (AI).

1.1 Internet of Things

The basic idea of IoT is to connect all physical things in the world including machines and humans to the Internet [3]. Things can be physical such as smart devices, or virtual such as services. In health, IoT can interconnect medical devices for the purpose of providing real time information about the health state of patients. IoT produces large amounts of data called "Big Data". Big Data needs effective tools for processing and analyzing and requires a lot of storage and processing power.

1.2 Big Data

Big Data refers to large volumes of data, produced at high velocity in a variety of formats such as text, image, video or audio. Big Data is the ability of society to discover actionable knowledge in novel ways to produce useful insights, goods and services and new business models of significant value [4]. In the health domain the analysis of Big Data allows to contextualize clinical data to obtain a better picture of the medical problem and find optimal solutions.

1.3 Artificial Intelligence

Most early research in Artificial Intelligence (AI) was aiming at imitating human intelligence by giving machines the capability to reason and deduce facts. This is sometimes called symbolic processing because the computer manipulates symbols that reflect the environment [5]. Other capabilities of AI include the ability to learn and self-correct. Due to the ever increasing volumes of data and growing computer power, AI algorithms can automatically learn rules that are too complex or too numerous for a developer to program. Automatic learning, or Machine Learning (ML) has been applied to many data analysis problems that were too complex to be solved in the past. Using ML computers modify or adapt their actions (whether these actions are making predictions, or controlling a robot) so that these actions get more accurate. Accuracy is measured by how close the chosen actions reflect the correct ones [5]. Within AI, ML has emerged as the method of choice for developing practical solutions for computer vision, robot control, speech recognition, natural language processing, and other applications [6]. ML has also been applied to generate more accurate solutions for medical problems.

In recent years, so-called deep learning approaches to machine learning have had a major impact on speech recognition and computer vision [7]. While ML using shallow learners has been explored for a while for medical applications, using Deep Learning is a recent phenomenon in clinical research. Deep learners (DL) are a type of artificial neural networks (ANN) with many data processing layers that learn representations by increasing the abstraction level from every layer to the next [8]. Each layer consists of many neurons or perceptrons. While using many neurons in every layer allows for an extensive coverage of the raw data at hand, the layer-by-layer pipeline combines their outputs in a nonlinear way and generates a lower dimensional projection of the input space [9]. DLs take advantage of the hierarchical structures found in nature. For instance, edges form shapes, shapes form parts, parts form objects. If a DL is presented with an image for object recognition, it "decomposes" the image by increasing the abstraction level at each layer. DLs render a high level abstraction of the raw input data, representing an automatic feature set. The ability to automatically create a feature set is one of the big advantages of DL over shallow learners, where a handcrafted feature set is required. Feature extraction is a laborious task and DL can automatically create much more sophisticated feature sets than manual feature engineering can provide.

DL have been used in many medical applications. They have been used to:

- Automatically detect structures such as organs or deformations in medical imaging
- Classify radiographs or CAT scans of tumors
- Reduce radiation dose from X-rays
- Predict the results of a cancer therapy
- Improve the design of biomarkers
- Personalize the treatment for patients with chronic disease such as cancer or HIV
- Predict the effect of new drugs

to name a few.

The biggest disadvantage of DLs is the carnivorous demand for training data that is often difficult to get in a medical context. DLs also require a lot of processing power and the recent advances in the efficient use of Graphics Processing Units (GPU) and parallelization algorithms has fostered the dissemination of DLs in medical applications. DL architectures can be highly parallelized by transferring algebraic calculations to the GPU.

This chapter describes some popular DLs and their applications in eHealth.

2 Deep Learners

Deep learning (DL) is an unsupervised feature learning approach that can be used to derive higher-level features from low-level data and thus helps to avoid expensive and time-consuming feature engineering tasks to identify relevant features [10].

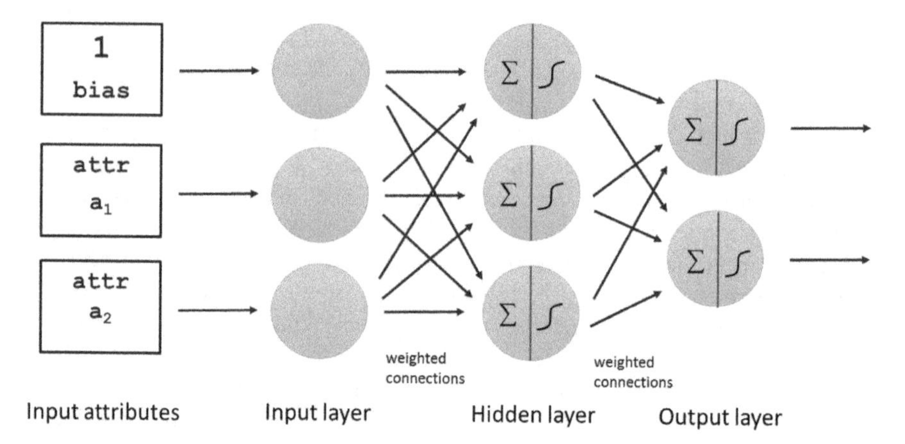

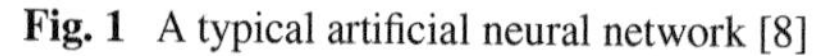
Fig. 1 A typical artificial neural network [8]

Unsupervised learning does not require labeled data for learning, for instance X-ray images with or without a medical condition such as fibrosis or cancer. However, DL can also be trained using labeled data, which is called supervised learning.

There are many different types of DLs. They are well documented in literature: [7–9, 11] and to describe them is beyond the scope of this chapter. This section gives an overview of the basic functioning of DLs.

DLs are in essence Artificial Neural Networks (ANN) with multiple hidden layers. There is no agreed upon definition of when a learner is a shallow learner or a DL. Some authors consider ANNs with more than two layers DLs [9].

A typical ANN consists of perceptrons, the neurons, organized into layers. The layers consist of one input and one output layer with one or more hidden layers in between. It should be noted that having more layers does not necessarily improve the DLs results. Adding too many layers and perceptrons can lead to an overfitted learner. An overfitted learner captures noise instead of the features of interest and can result in poor accuracy. Figure 1 shows a typical ANN.

Some DLs such as Convolutional Neural Networks (CNN) take a fixed size input vector, for instance a pixel array of a histopathological image, and map it to an output, e. g. a type of tumor. During training, the DL is presented with different images with tumors. Learning happens by assigning and adjusting the weights of the connections between the perceptrons. After each learning cycle, the error is calculated using a loss function. The error is then backpropagated and the weights are adjusted until the loss function converges. This type of ANN is called a feed forward network since the signal passes from the input layer to the output layer in one direction. Other types of ANN have foreward and backward connections. They are called Recurrent Neural Networks (RNN). To decrease the loss function, typically a mechanism called gradient descent is used.

2.1 *Types of Deep Learners*

There are many different DL architectures and it is often not obvious from the beginning, which architecture works best for a specific problem. Figure 2 shows some Learning architectures that have been applied in the area of health informatics.

2.1.1 Convolutional Neural Networks

Convolutional Neural Networks (CNN) are inspired by the human visual cortex and are probably the most popular DLs. They are a type for feed forward network and consist of many layers. Its architecture can be defined as an interleaved set of feed-forward layers containing convolutional filters followed by rectification, reduction or pooling layers [9]. As the signal passes through the network, each layer extracts a higher level abstraction of the input from the previous layer. The layer-by-layer pipeline generates a representation of lower dimensionality (dimensionality reduction) of the input space. The output provides a high level abstraction or automatic feature set that would otherwise require handcrafted features that are less sophisticated. This is advantageous for instance for medical imaging, where irregularities in tissue morphology can be used to classify tumors, or in translational bioinformatics, where a specific sequence of nucleotides can replicate a specific protein. They are capable of detecting patterns that are difficult do find by human experts, for instance early stages of diseases in tissue samples.

2.1.2 Deep Recurrent Networks

Recurrent Neural Networks (RNN) can analyze streaming data. A CNN uses fixed size input vectors. Contrarily, RNN can be used on data such as text, speech or DNA sequences where the output depends on previous inputs. RNNs maintain an inner state using a recurrent layer where perceptrons have an additional connection with themselves, acting as memory that can be used for consecutive inputs. For instance, a word in a text might only makes sense in the context of a phrase or the whole sentence. Figure 3 shows a RNN with one recurrent hidden layer.

In eHealth RNNs can be used to analyze medical texts such as anamnesis. For instance, different patients with the same disease might have different symptoms. RNNs can scan through thousands of text documents and find similarities to support a physician in diagnosing an illness.

Architecture	Description	Key Points
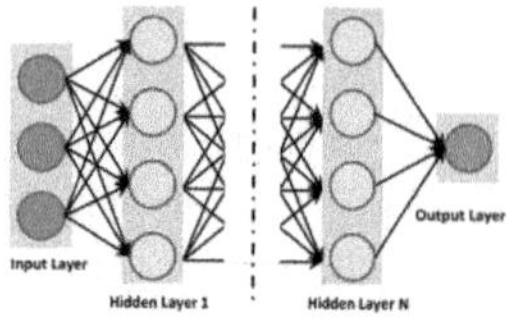 Input Layer, Hidden Layer 1, Hidden Layer N, Output Layer	**Deep Neural Network** • General deep framework usually used for classification or regression • Made of many hidden layers (more than 2) • Allows complex (non-linear) hypotheses to be expressed	**Pros:** • Widely used with successes in many areas **Cons:** • Training is not trivial because once the errors are back-propagated to the first few layers they become minuscule • The learning process can be very slow
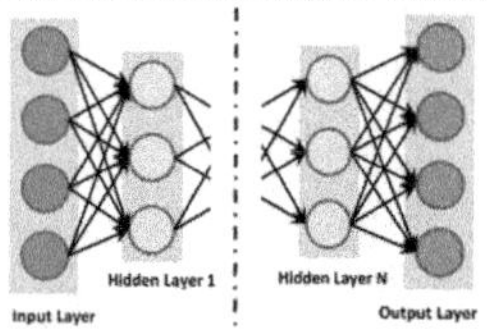 Input Layer, Hidden Layer 1, Hidden Layer N, Output Layer	**Deep Autoencoder** • Proposed in [5] and is mainly designed for feature extraction or dimensionality reduction • Has the same number of input and output nodes • Aims to recreate the input vector • Unsupervised learning method	**Pros:** • Does not require labelled data for training • Many variations have been proposed to make the representation more robust: Sparse AutEnc. [6], Denoising AutEnc. [7], Contractive AutEnc. [8], Convolutional AutEnc. [9] **Cons:** • Requires a pre-training stage • Training can also suffer from vanishing of the errors
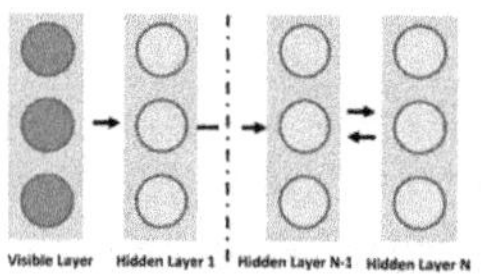 Visible Layer, Hidden Layer 1, Hidden Layer N-1, Hidden Layer N	**Deep Belief Network** • Proposed in [10] is a composition of RBM where each sub-network's hidden layer serves as the visible layer for the next • Has undirected connections just at the top two layers • Allows unsupervised and supervised training of the network	**Pros:** • Proposes a layer-by-layer greedy learning strategy to initialize the network • Inferences tractable maximizing the likelihood directly **Cons:** • Training procedure is computationally expensive due to the initialization process and sampling
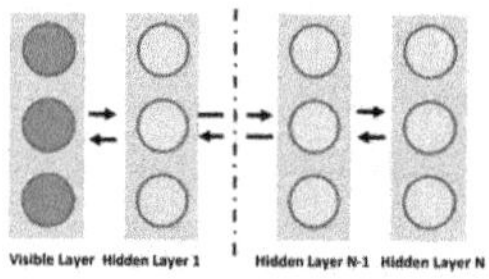 Visible Layer, Hidden Layer 1, Hidden Layer N-1, Hidden Layer N	**Deep Boltzmann Machine** • Proposed in [11] is another approach based on the Boltzmann family • Possesses undirected connections (conditionally independent) between all layers of the network • Uses a stochastic maximum likelihood [12] algorithm to maximize the lower bound of the likelihood	**Pros:** • Incorporates top-down feedback for a more robust inferences with ambiguous inputs **Cons:** • Time complexity for the inference is higher than DBN • Optimization of the parameters is not practical for large datasets
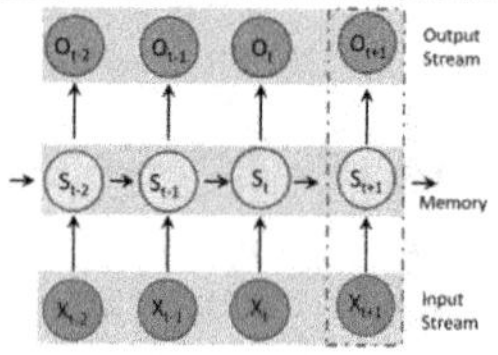 O_{t-2}, O_{t-1}, O_t, O_{t+1} Output Stream; S_{t-2}, S_{t-1}, S_t, S_{t+1} Memory; X_{t-2}, X_{t-1}, X_t, X_{t+1} Input Stream	**Recurrent Neural Network** • Proposed in [13] is a NN capable of analyzing stream of data • Useful in applications where the output depends on the previous computations • Shares the same weights across all steps	**Pros:** • Can memorize sequential events • Can model time dependencies • Has shown great success in many Natural Language Processing applications **Cons:** • Learning issues are frequent due to the vanishing gradient and exploding gradient problems
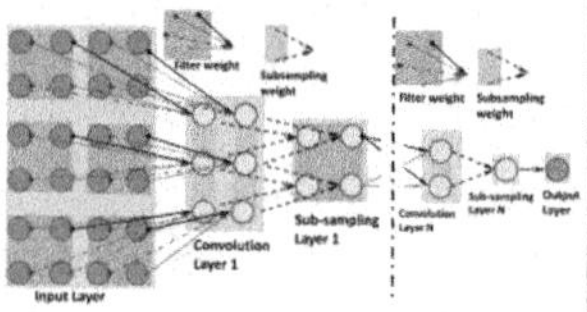 Input Layer, Convolution Layer 1, Sub-sampling Layer 1, Output Layer	**Convolutional Neural Network** • Proposed in [14], it is well suited for 2D data such as images • Every hidden convolutional filter transforms its input to a 3D output volume of neuron activations • Inspired by the neurobiological model of the visual cortex [15]	**Pros:** • Few neuron connections required with respect to a typical NN • Many variants have been proposed: AlexNet [16], Clarifai [17], and GoogLeNet [18] **Cons:** • It may require many layers to find an entire hierarchy of visual features • It usually requires a large dataset of labelled images

Fig. 2 Popular Deep Learning architectures [9]

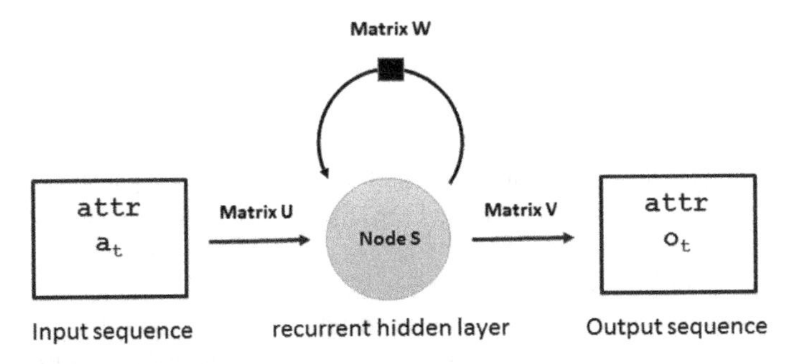

Fig. 3 Recurrent neural network [8]

3 Medical Applications of Deep Learners

Whereas ML techniques have been successfully applied to many biomedical applications, DLs are a recent field of research and the adoption of DL in medicine is slow and still faces many challenges. However, there has been rapid progress in some areas and DLs have proven to yield promising results. Medical applications of DL can be roughly divided in three areas:

- Prediction of biological processes, e. g. for predicting the efficiency of a treatment or designing new drugs [12, 13]
- Medical decision support, e. g. through early disease detection or DNA analysis [14–16]
- Personalized treatments, e. g. chemotherapy that has been tailored for a specific patient [9, 17].

There are other applications of DL in public health or assistive device support [18, 19]. Some of the DL research in biomedicine is still in its infancy and can at best support medical decision making, but an expert opinion is still needed. Automatic diagnosis and treatment is still a matter of the future, but in the area of personalized treatment, treatment effectiveness prediction as well as drug design we are likely to see a lot of progress.

3.1 Predictive Health Care

Predictive health care has many applications. Predictive health care aims to detect diseases before they break out or to detect them early and start treatment before more serious conditions occur. Other areas include predicting the effectiveness of new drugs or treatments and determine the disease process and patient prognosis. Some diseases such as Alzheimer's Disease (AD), are very difficult to detect in the early pathogeny and an accurate and objective diagnosis is problematic due to their mild symptoms [15]. DL can be used in such cases since they can detect anomalies even if they are difficult to see for the human eye, for instance in radiographs or

CAT scans. Anomaly detection is a task that DL are very suitable for since they can detect small deviations that remain often undetected by humans in early stages. Also DL can be trained using, for instance, radiographs of healthy tissue to detect what is pathological. Since medical images of healthy tissue is generally easier to obtain, training data is more available and mitigates the problem of sparse data sets. Behavioral changes in patients can also be used as early indicators for illness, for instance to determine if a patient becomes symptomatic. DL can tap into these different data sources and combine them in their prediction model to obtain a more complete picture of the medical problem at hand.

Predictive health care is also used to predict the efficiency of drug treatments. Discovery and development of new drugs is a costly and time consuming task. Often the results are disappointing and during development new approaches have to be found. In silico drug development has evolved over the past decades and provided targeted and more cost effective approaches for drug development. Modern approaches use now Data Mining techniques such as DL for predicting the effectiveness of a drug or treatment. More recently, DL has also been used for drug repurposing. Drug repurposing or target extension allows prediction of new potential applications of medications or even new therapeutic classes of drugs using gene expression data before and after treatment (e.g., before and after incubation of a cell line with multiple drugs) [12].

3.2 Medical Decision Support

Medical decision support is likely to be one of the main application fields of DL in medicine in the near future. DLs support the medical expert in various stages of a medical case. This can span the whole illness life cycle from diagnosing a medical condition, proposing a treatment, determining the healing progress up to proposing post disease therapy or life style change suggestions.

DL are often superior in image analysis and disease prediction than humans [15]. DL has also been applied to biomedical text analysis [20]. Since DL schemes are domain independent, they can analyze and correlate any kind of data.

Correlation analysis can use different data sources, for instance past anamnesis, the genome and symptoms of other patients, to provide a more accurate diagnosis. DL are also effective in finding correlations in single data sources, for instance Magnetic Resonance Imaging (MRI). They have been used to correlate different brain regions [15], or for histology image segmentation. Isolating glands from the background tissue is important for analyzing and diagnosing histological images [21]. Usually CNN architectures are used for correlation analysis. They create abstractions of the input signals even if the data sources are heterogeneous, that can be more easily correlated to detect anomalies such as cancer diagnosis. A medical expert cannot efficiently scan through the amounts of data that a DL can, which makes medical decision support a likely candidate for early adoption. However, since it is often difficult to reproduce how a DL came to a conclusion, a physician will likely trust his intuition and experience and not leave the ultimate decision to a machine.

3.3 Personalized Treatments

Closely related to medical decision support are personalized treatments. DL can provide decision-making support based on the prediction and precision of care adapted to the needs of individual patients and thus provide personalized treatments or stratified medicine. DL has paved the way for personalized treatments by offering an unprecedented efficiency and power in sifting through large, multimodal unstructured data stored in hospitals, at cloud providers and research organization [9]. Personalized treatments can be based on various information. It can use behavioral health analytics [19] or DNA analysis and genome mining [22] to determine biomarkers. A biomarker is a measurable indicator of a biological state such as a disease. Biomarkers allow clinicians and industry professionals to advance more accurate diagnoses and prognostications by determining the probability that a patient will develop a disease. Deep Neural Networks demonstrate state-of-the-art performance in classifying cancer using gene expression data, in extracting features from sparse transcriptomics data (both mRNA and miRNA data) and in predicting splicing code patterns [12].

Due to the versatility of CNNs, they have been used in many personalized treatment applications. CNNs have outperformed baseline approaches not only in classical deep learning applications, such as image recognition (gene expression annotation and microarray segmentation), but also in annotation of sequence polymorphism data in tools such as DeepBind, DeepSEA and Basset [17].

Often many factors influence the outbreak and development of a disease. Genetic factors, exposure to germs and parasites but also environmental factors contribute to diseases such as cancer. Genomics aims to identify gene alleles that influence the development of an illness. Pharmacogenomics evaluates variations in an individuals drug response to treatment brought about by differences in genes [9]. It is used to reduce side effects or reduce dosage levels. Combining all the information available is a laborious task and DLs can effectively analyze the different data sources and provide personalized medical therapy.

4 Challenges

The adoption of DL and AI in general face many challenges in medicine. On the one hand, this is due to the nature of medical applications, where high requirements for security, availability, reliability, privacy and efficiency are demanded. For instance, a home care application needs to work without interruption so it can alert a health care center in case of an emergency. On the other hand, research in DL for eHealth is still facing many technical problems that have to be overcome. DLs can analyze massive amounts of data and detect useful patterns that cannot effectively be interpreted by humans. The massive scale of modern biological data is simply too large and complex for human-led analysis [17]. Due to advances in storage and processor technologies, DLs can scan massive amounts of data and automatically develop rules that are too complex for a developer to implement. But since the rules are generated automatically, DL also means a loss of control. Also, ML and DL schemes are prone to overfitting. Overfitting happens when a learner becomes too complex and starts capturing noise. This can result in false positives, for instance a falsely diagnosed disease.

DLs need carnivorous amounts of data for training and testing. Since there are high privacy requirements when it comes to medical data, the required volumes are one of difficult to obtain. Also, some diseases are rare and only scarce medical datasets are available. There are learning techniques that can be adopted for scarce datasets, also in some cases the data sets can be enhanced with synthetic data. Transfer learning is a technique that applies knowledge from a source domain to a target domain with considerably less data points. A pre-trained DL can be used and fine tuned using for instance MRI scans, if the image data set is sparse. But the availability of suitable data sets remains one of the biggest challenges for the adoption of DL in medicine.

One of the greatest challenges for AI in the health research community still remains translating research from labs to everyday use in hospitals and medical practice [1]. In medicine emergencies happen that require immediate action and there is little scope for experimentation. Also wrong treatments can end in permanent damage or death. In some cases simulators can be a way around this problem, but to date there are few simulators suitable for DL in the medical context.

5 Future Research

Modern biology has now entered the era of Big Data. Data sets are too large, high-dimensional, and complex for classical computational methods in biology [12]. Due to the big progress in ML and more recently the advances using DL schemes for analyzing these large data sets, particularly CNN for multimedia and RNN for speech and text analysis, a lot of research in the area of biomedicine and eHealth using DL has been conducted. There are lots of discussions about whether AI will replace many traditional brain jobs and cut jobs. We will probably still see much progress in the areas of AI and DL. But it is to be expected that soon the limitations of DL will become obvious. ANN are trying to imitate human learning. But ANNs are at best inspired by nature. There is no evidence of such mechanisms such as back-propagation or gradient descent in the brain. The inner working of the human brain is still largely unknown. We do not know how we recall episodes from the past. Also humans do a lot subconsciously. We do not need to focus to do things such as walking or moving our arms. We do all these things subconsciously. Yet the big achievement of the human brain is not our ability to reason. According to Moravecs paradox contrary to traditional assumptions, high-level reasoning (playing chess and checkers, abstract thinking, performing Intelligence tests, using mathematics, engineering etc.) requires low computation power compared to low-level sensorimotor skills that require enormous computational resources [23]. How this is achieved is largely unknown today. To be able to simulate human brain functionality on a computer a lot of research is still needed in the area of neuroscience. Trying to imitate human thinking on a computer can help understand the problems nature has solved throughout evolution. More research that aims to reproduce brain functionality can also improve our understanding of how the brain works, but up to date it remains largely a "black box".

Whereas the internals of the brain are mostly unknown, the inner workings of DLs are also difficult to interpret and DLs themselves are a "black box" for researchers [24]. That is a problem when DLs need to be optimized since it is not clear, which parts of the DL contribute to what extend to the result.

Automatic feature extraction is one of the predominant feature that makes them the technique of choice over shallow learners in areas such as natural language processing and object recognition. However, there are many different DL architectures and it is often difficult to decide which architecture works best. Often several different DL schemes are trained and compared and the best performing architecture is then selected for production. This is time and resource consuming. Automatic selection of shallow learning schemes is already in place, but DL are considerably more complex and more research is needed to find optimal architectures for DLs.

In the medical context, obtaining enough training data is often a problem. Sparse and noisy data sets result in considerable drop in performance indicating that there are several challenges to be addressed [9]. Also, testing a trained DL is often a problem since humans are not feasible objects for experimenting. Developing effective simulators and generating synthetic data can mitigate the problem of training and testing using sparse data sets. But to date there are no tools in place that would provide satisfying results.

Currently, DL are trained for a specific purpose, for instance detecting anomalies in MRI images or blood samples. Future research should focus on creating learners that can detect many different types of symptoms. Ultimately we want to have DLs that can detect any kind of disease and propose treatments. This leads to general purpose learners that do not serve a specific purpose. This area is sometimes called Artificial General Intelligence (AGI) [25] and some research in this area has already been conducted, however we need to develop a better understanding of how intelligence works until we will see anything close to human intelligence on a computer.

References

1. S. Khanna, A. Sattar, D. Hansen, Artificial intelligence in health the three big challenges. Aust. Med. J. **6**(5), 315–7 (2013)
2. T. Pteri, N. Varga, L. Bokor, A survey on multimedia quality of experience assessment approaches in mobile healthcare scenarios, in *eHealth 360: International Summit on eHealth, Budapest, Hungary, 14–16 June 2016, Revised Selected Papers*, ed. by K. Giokas et al. (Springer International Publishing, Cham, 2017), pp. 484–491
3. T. Chun-Wei, L. Chin-Feng, C. Ming-Chao, L.T. Yang, Data mining for internet of things: a survey. IEEE Commun. Surv. Tutor. **16**(1), 77–97 (2014)

4. V. Mayer-Schonberger, K. Cukier, *Big Data: A Revolution That Will Transform How We Live, Work, and Think* (Houghton Mifflin Harcourt Publishing Company, New York, USA, 2013)
5. S. Marsland, *Machine Learning: An Algorithmic Perspective*, 2nd edn. (Chapman & Hall/CRC, 2014)
6. M.I. Jordan, T.M. Mitchell, Machine learning: Trends, perspectives, and prospects. Science **349**(6245), 255–60 (2015)
7. I.H. Witten, E. Frank, M.A. Hall, C.J. Pal, *Data Mining: Practical Machine Learning Tools and Techniques*, 4th edn. (Morgan Kaufmann Publishers Inc., 2016)
8. P. Wlodarczak, J. Soar, M. Ally, Multimedia data mining using deep learning, in *Fifth International Conference on: Proceedings of Digital Information Processing and Communications (ICDIPC), IEEE Xplore, Sierre* (2015), pp. 190–196
9. D. Rav, C. Wong, F. Deligianni, M. Berthelot, J. Andreu-Perez, B. Lo, G.Z. Yang, Deep learning for health informatics. IEEE J. Biomed. Health Inf. **21**(1), 4–21 (2017)
10. E. Portmann, U. Reimer, G. Wilke, From a data-driven towards a knowledge-driven society: making sense of data, in *The Application of Fuzzy Logic for Managerial Decision Making Processes: Latest Research and Case Studies* (2017), p. 93
11. I. Goodfellow, Y. Bengio, A. Courville, *Deep Learning* (The MIT Press, 2016)
12. A. Aliper, S. Plis, A. Artemov, A. Ulloa, P. Mamoshina, A. Zhavoronkov, Deep learning applications for predicting pharmacological properties of drugs and drug repurposing using transcriptomic data. Mol. Pharm. **13**(7), 2524–30 (2016)
13. H. Zhong, J. Xiao, Enhancing health risk prediction with deep learning on big data and revised fusion node paradigm, in *Scientific Programming*, vol. 2017 (2017), p. 18
14. J. Budaher, M. Almasri, L. Goeuriot, Comparison of several word embedding sources for medical information retrieval, in *CLEF (Working Notes): Proceedings of theCLEF (Working Notes)* (2016), pp. 43–46
15. C. Hu, R. Ju, Y. Shen, P. Zhou, Q. Li, Clinical decision support for Alzheimer's disease based on deep learning and brain network, in *2016 IEEE International Conference on Communications (ICC): Proceedings of the 2016 IEEE International Conference on Communications (ICC)* (2016), pp. 1–6
16. Y. Xu, Y. Li, Y. Wang, M. Liu, Y. Fan, M. Lai, E. Chang, Gland instance segmentation using deep multichannel neural networks. IEEE Trans. Biomed. Eng. (2017)
17. P. Mamoshina, A. Vieira, E. Putin, A. Zhavoronkov, Applications of deep learning in biomedicine. Mol. Pharm. **13**(5), 1445–54 (2016)
18. M. Poggi, S. Mattoccia, A wearable mobility aid for the visually impaired based on embedded 3D vision and deep learning, in *2016 IEEE Symposium on Computers and Communication (ISCC): Proceedings of the 2016 IEEE Symposium on Computers and Communication (ISCC)* (2016), pp. 208–213
19. T. Zebin, P.J. Scully, K.B. Ozanyan, Inertial sensor based modelling of human activity classes: feature extraction and multi-sensor data fusion using machine learning algorithms, in *eHealth 360: International Summit on eHealth, Budapest, Hungary, 14–16 June 2016, Revised Selected Papers*, ed. by K. Giokas et al. (Springer International Publishing, Cham, 2017), pp. 306–314
20. F. Li, M. Zhang, B. Tian, B. Chen, G. Fu, D. Ji, Recognizing irregular entities in biomedical text via deep neural networks. Pattern Recogn. Lett. (2017)
21. Y. Xu, Z. Jia, Y. Ai, F. Zhang, M. Lai, E.I.C. Chang, Deep convolutional activation features for large scale Brain Tumor histopathology image classification and segmentation, in *2015 IEEE International Conference on Acoustics, Speech and Signal Processing (ICASSP): Proceedings of the 2015 IEEE International Conference on Acoustics, Speech and Signal Processing (ICASSP)*, pp. 947–951
22. P. Wlodarczak, J. Soar, M. Ally, Genome mining using machine learning techniques, in *Inclusive Smart Cities and e-Health*, vol. 9102, ed. by A. Geissbhler et al. (Springer International Publishing, 2015), ch. 39, pp. 379–384

23. V.S. Rotenberg, Moravec's paradox: consideration in the context of two brain hemisphere functions. Act. Nerv. Super. **55**(3), 108 (2013)
24. Y. Nohara, Y. Wakata, N. Nakashima, Interpreting medical information using machine learning and individual conditional expectation, in *MedInfo: Proceedings of the MedInfo* (2015), p. 1073
25. Y. Li, Deep reinforcement learning: an overview (2017). arXiv:1701.07274

Deep Learning for Brain Computer Interfaces

Ankita Bose, Sanjiban Sekhar Roy, Valentina Emilia Balas and Pijush Samui

Abstract From playing games with just the mind to capturing and re-constructing dreams, Brain computer Interfaces (BCIs) have turned fiction into reality. It has set new standards in the world of prosthetics, be it hearing aids or prosthetic arms, legs or vision, helping paralyzed or completely locked-in users. Not only can one get a visual imprint of their own brain activity but the future of BCI will make sharing someone else's experience possible. The entire functioning of the BCI can be segmented into acquiring the signals, processing it, translation of signals, device that gives the output and the protocol in operation. The translation algorithms can be classical statistical analysis or non-linear methods such as neural networks. Deep learning might serve as one of the translation algorithms that converts the raw signals from the brain into commands that the output devices follow. This chapter aims to give an insight into the various deep learning algorithms that have served in BCI's today and helped enhance their performances.

Keywords The brain · Brain computer interface · Electroencephalography signals · Deep learning · Convolutional neural networks

A. Bose · S. S. Roy (✉)
School of Computer Science and Engineering, Vellore Institute Technology, Vellore, India
e-mail: sanjibanroy09@gmail.com

A. Bose
e-mail: ankita.bose2015@vit.ac.in

V. E. Balas
Automation and Applied Informatics, Aurel Vlaicu University of Arad, Arad, Romania
e-mail: balas@drbalas.ro

P. Samui
Department of Civil Engineering, NIT Patna, Patna, India
e-mail: pijushsamui@gmail.com

V. E. Balas et al. (eds.), *Handbook of Deep Learning Applications*,
Smart Innovation, Systems and Technologies 136,
https://doi.org/10.1007/978-3-030-11479-4_15

1 Introduction

The human brain has stood the test of time with its extreme adaptability, surviving and excelling in both extreme and moderate conditions since the beginning of existence. When talking about adaptability nothing can serve better than neural networks (elaborated later). Understanding the brain is quite a challenge but with recent technology, nothing seems impossible. It is a known fact that the brain is entirely made up of neurons that carry charges throughout the brain and also to and from the receptors to the brain. Modern devices, like the Scalp electrodes for electroencephalography (EEG), electrode implants that penetrate the cortex, electrodes placed exactly on the layer of the cortex—Electrocorticography (ECoG) [1], help to capture the brain activity in the form of brainwaves that are converted into signals. These signals are passed through the relevant translation algorithms that convert them into directives for the output devices. Different parts of the brain process signals from different receptors in the body, as shown in Fig. 1.

The current applications of BCI range from moving cursors on the screen without touching [2], playing games with just the mind [3], accessing ones concentration levels [4], measures of excitement, focus or calm, moving robotic arms or legs by just thinking about it [5], typing on screen by thinking about pressing a key [6], controlling appliances with the mind. Hearing aids and visual aids using BCI are being worked upon and the possibilities of sharing experiences between humans over a network, communication without talking, capturing dreams seem a thing of just the near future. Any BCI can be framed by putting together these parts—acquiring of signals, processing it, translation of signals, device that delivers output and the protocol in operation [1]. BCIs have to cater to demands like speed, amount of training required, accuracy, attention requirements, therefore, based on varying

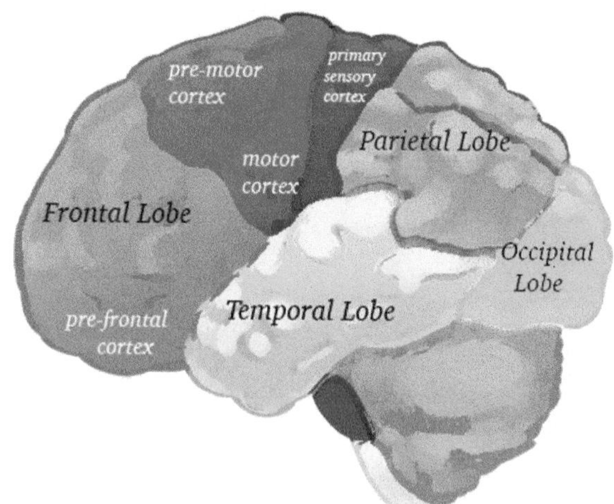

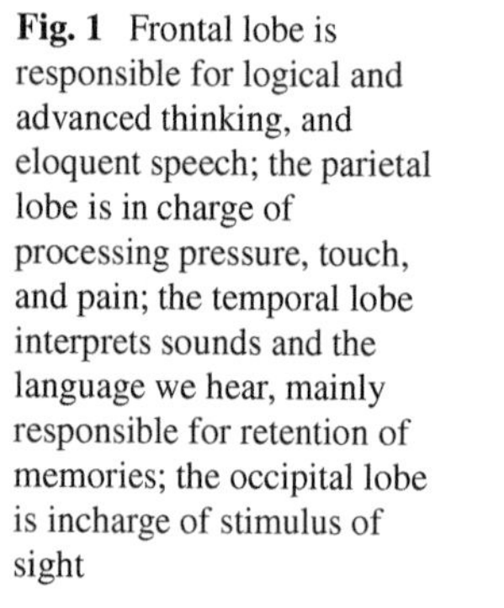
Fig. 1 Frontal lobe is responsible for logical and advanced thinking, and eloquent speech; the parietal lobe is in charge of processing pressure, touch, and pain; the temporal lobe interprets sounds and the language we hear, mainly responsible for retention of memories; the occipital lobe is incharge of stimulus of sight

levels of these with respect to different applications, the translation algorithm and other parameters are decided [7]. Since every action we carry out is a result of signals due to activities in the brain, the task of deciding the output signals can as well be considered to be a pattern recognition problem [8]. In BCI, many times a linear classifier is used which works good if some assumptions hold true otherwise results might be highly unsatisfactory. Therefore, non-linear translation algorithms like neural networks—Recurrent Neural Networks (RNNs), Convolutional Neural Networks (CNNs) and other forms of Deep Neural Networks (DNNs) are used. CNNs have mostly been used to classify the generated motor images that captured the user brain activity to perform a certain task. The rest of the chapter focusses on throwing light into these arenas.

2 Different Parts of a BCI

What a BCI essentially does is, captures the brain activity, converts it into signals, processes these signals, identifies the action and directs the output device accordingly. Functioning of the brain is traceable to a large extent, due to the waves it generates which is caused by the firing up of neurons inside. These wave patterns can be categorized under five heads—delta waves, theta waves, alpha waves, beta waves and gamma waves. Human behavior is explained by the concoction of these waves at any given point, wherein only one is dominant while others, though present, are not as marked.

- *Delta Waves*: Being the slowest brain waves, they lie in the frequency range of 0–4 Hz. They are related with the most profound levels of unwinding and therapeutic, recuperating rest and are mainly discovered regularly in newborns and youngsters. Aging reduces the production of delta waves even during periods of deep sleep. They are also associated with involuntary bodily actions such as digestion of food, and heart beat. Whereas excessive amount of delta waves might cause inability to learn and think properly and might be a result of brain injuries, fewer amounts of it, resulting due to lack of proper sleep, might cause problems in rejuvenation and healing. Sufficient creation of delta waves gives a feeling of complete revival, after waking up from proper sleep.
- *Theta Waves*: Lying in the frequency range between 4–8 Hz, these waves are mainly generated during periods of sub-consciousness, sleep and when profound and crude feelings are encountered. They are essential for stirring up innovation and intuition. Though theta wave range is a helpful state, its excessive production might cause a person to be depressed, extremely hyper or impulsive whereas lack of theta waves might cause increasing amount of stress and anxiety.
- *Alpha Waves*: These waves have a moderate frequency range of 8–12 Hz and crosses over any barrier between our cognizant reasoning and intuitive personality. It helps to attain a state of calm when essential. Too much of stress might cause the beta waves to suppress the dominance of alpha waves.

- *Beta Waves*: These waves have high frequency, 12–40 Hz, and are prevalent in states of high focus and consciousness. They are associated with cognizant idea, legitimate reasoning, and have a tendency to have an invigorating effect. Having excessive amounts of beta waves causes inability to stay calm and causes huge amounts of stress and anxiety whereas less amounts of it causes poor thinking capabilities, depression and lack of consciousness or focus. Therefore, beta waves are mostly predominant during the day where cognitive tasks like studying, presenting etc. are involved.
- *Gamma Waves*: Belonging to the highest frequency range of 40–100 Hz, these waves are associated with advanced cognitive functions, having clear perception, and intellectual abilities. People who are mentally disabled usually lack gamma dominance. While an excess of gamma waves might lead to increased excitement and stress, too little of it may cause difficulty in learning. High occurence of such waves are usually noticed in people who meditate.

The process of gaining the accurate data from the brain for a particular action is called Forward Modelling, and determining the action from a given set of data is called Inverse Modelling [9]. The overall BCI system is demonstrated in Fig. 2.

Data Acquisition is acquiring signals from the brain and digitizing them. The digitized data is straight away passed on to the Signal Processing unit.

Data from the brain can be acquired in the following ways:

- Electrode implants in the cortex: they give the clearest and best signals but these implants can be very risky.
- Electroencephalography (EEG): placing electrodes on the scalp that give indistinct data that may not be very good to give accurate results but involves no risk at all.
- Electrocorticography (ECoG): electrodes placed exactly on the layer of the cortex which is not that harmful and serves as a balance between risk and clarity.

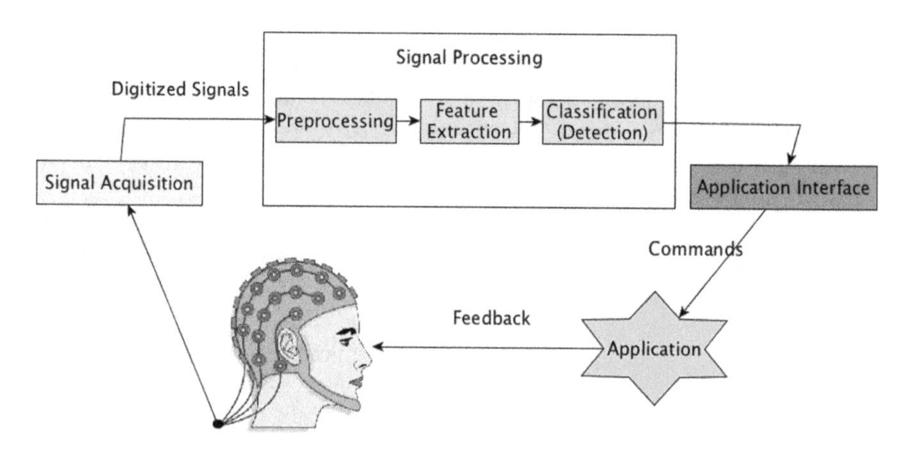

Fig. 2 Generic brain computer interface [1]

The *Signal Processing* module converts the brain signals into commands for the output device [1, 10]. This task is carried forward by feature extraction, the target being to reduce the noise to signal ratio, and translation of the extracted features. Firing rates of neurons or amplitudes of waves might serve as important features. Each translation algorithm is characterized by transfer function, adaptive capacity, and output. These algorithms can be linear or non-linear based on the features extracted and the kind of output required. The linear transfer function could be statistical methods or simple linear equations whereas the non-linear transfer functions could be deep neural networks. In many system, a combination of linear and non-linear transfer functions are used.

The *User Application* module receives the ultimate commands from the signal processing module and performs actions such as moving the cursor, playing games, moving a prosthetic arm, or leg, typing on a screen without physically touching a keyboard. The applications of BCIs can be classified into *Exogenous* and *Endogenous*. In endogenous systems the trained user has direct control over the environment and responses are generated with his/her own intent and hence serves better than exogenous systems that although, do not require extensive training but are constrained with a structured environment and the user is subjected to external stimuli.

The *Operator module* makes controls easy for the experimenter by defining the system parameters, providing a graphical interface and marking the beginning and end of any operation. It also displays information received from any other module of the BCI, including real time analysis results which makes it better to visualise the activities.

3 Deep Neural Networks for Brain Computer Interfaces

Any network with an input layer, more than one hidden layers and an output layer is said to be a DNN [11]. Deep Learning algorithms are advanced machine learning algorithms that help achieve better classification or regression accuracies in case of problems where a linear classifier does not suffice [12]. Deep Belief Networks have been used in areas of classifying images for calamities [13], DNNs have been used in the field of security [14], and it has wider applications in the field of text classifications in the form of spams [15] and the like. DNNs act as non-linear classification or regression transfer functions in the signal processing module. BCIs work with *motor imagery* [16] in which the user rehearses a particular task mentally without actually implementing it. In motor imagery the actions of the subject and the computer's interpretation of the input signals matter. Classification algorithms using CNN architecture as demonstrated in Fig. 3, have been applied to new representations of the motor imagery data. Devising optimum Deep Learning algorithms is challenging due the high dimensionality of EEG data, channel correlation and presence of noise. Over the years CNNs have emerged as the best classification algorithm when dealing with images [17].

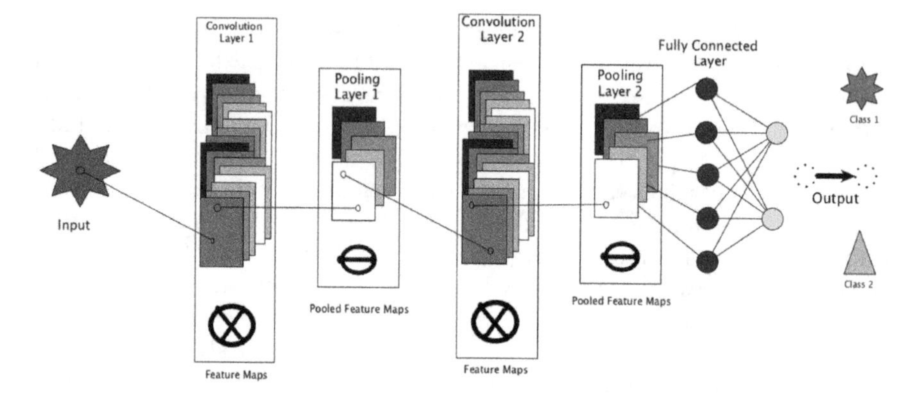

Fig. 3 A convolutional network with 3 intermediate layers

To apply CNNs to a data, the correct representation of EEG has to be designed first. In recent studies about learning *Temporal Information for BCI using CNNs* [17], the Filter-Bank Common Spatial Patterns (FBCSP) along with extraction of the temporal features, have been found to yield considerably good results. These features are then fed into the CNN layer [18], that is customized to optimize the performance, as described. In a particular scenario, the features selected are assumed to share a common morphology even if they vary intrinsically, therefore a common convolution kernel is used to learn by discriminating the classes but preserving the channels and reducing the temporal dimension and hence called Channel-Wise CNN (CW-CNN).

In another situation when the FBCSP inputs are apriori concoction of the original EEG signals then a convolution kernel size equal to the number of channels yields a redundant channel mixing so it is better to use with CW-CNN.

In a third scenario the feature channels are first convoluted and then mixed which yields a single channel feature. This ultimate signal consists of the summation of all individual convoluted channels and resembles the 2D CNN architecture.

Like in all other algorithms this technique also has to be optimised by tuning in the hyperparameters like number of different pooling layers, convolutional layers, kernel size, dropout and number of hidden nodes.

In another study, Fourier Transform has been embedded with CNNs for the classification of EEG signals for *Steady State Visual Evoked Potential (SSVEP)* [19]. This work has been accomplished in three stages—creation of different channels, signal transformations to relevant frequency domain and concluded by the classification step. CNNs allow classification of data without requiring prior knowledge about it but the addition of a signal processing module in between the hidden layers introduces some prior domain knowledge. In this approach 2 intermediate layers have been used. In between the two intermediate layers lies the layer which is the result of Fourier Transformation on the first hidden layer. This step helps in transitioning from the temporal domain to the frequency domain. Backpropagation along with gradient descent has been used as the mode of learning. Classification accuracies on test data range from 95 to 97%.

In many BCI signal processing units the spatial information is ignored, contrary to that, in [20], a novel approach has been proposed where the EEG signals have been converted into multi-spectral images that preserve topology. A Deep Recurrent Convolutional Network has been chosen to learn EEG representations, that help in finding features which are more resistant to spatial, spectral and temporal variations. This serves as a solution to the challenges faced by modern BCI that struggles to find representations that do not vary in case of different subjects or in different situations for the same subject. Two major approaches have been evaluated—Single frame approach where a single image was constructed from measurements over the complete trial and fed into the CNN, and a Multi frame approach where the entire trial was divided into frames and images were generated over each frame and fed into the CNN. In both the cases multiple configurations of CNNs constituting varying numbers of convolution and pooling layers were used. In case of multi frame approach Long Short Term Memory (LSTM) layers were added after the Convolution Network architecture, since images are passed in sequence over the entire trial it can carry important information in the patterns.

A study, by Guan and Sakhavi [21], makes efforts to reduce or nullify the time taken by each subject to train the BCI device. It would be easier if the knowledge gained by a BCI from one user could be transferred to another user to reduce the training time and promote transfer learning. Here also the FBCSP method has been used for feature extraction from the motor images. To train the CNN model, data from many subjects has been used to generalise the algorithm. The CNN has been used alongside a Multilayer Perceptron Layer (MLP) and features from both are then concatenated. The K L Divergence loss function is used along with adam optimizer. The error is backpropagated and weights belonging to both CNN and MLP are updated simultaneously.

The use of Multilayer Perceptrons has been demonstrated in a study by [22] where data was fetched from positions C3, C4, P3, P4, O1, and O2 of four subjects using the EEG cap as shown in Fig. 4, while each one of them performed five different tasks mentally.

In this study AR coefficients [23] have been used as features to the MLP. Since there were five tasks, there are 5 classes. Two hidden layers were used with the minimum squared error loss function. The accuracies obtained with different subjects varied with the maximum accuracy being 70% with one of the subjects.

The quest for better algorithms lead to the formation of learning algorithms which when combined with the standard CNNs gave better results as shown in the study by [24] which focusses on mapping information related to movements to the brain signals with the help of EEG. The deep CNNs decoding performance has been elevated by a cropped training strategy combined with batch normalization along with exponential linear units, to the extent that it surpassed the FBCSP techniques. The inputs to the CNNs were 2D arrays with width as count of steps in time and height as count of electrodes. A number of CNN architectures have been explored in this study—Deep CNNs, Shallow CNNs, Hybrid CNNs and Residual CNNs for raw

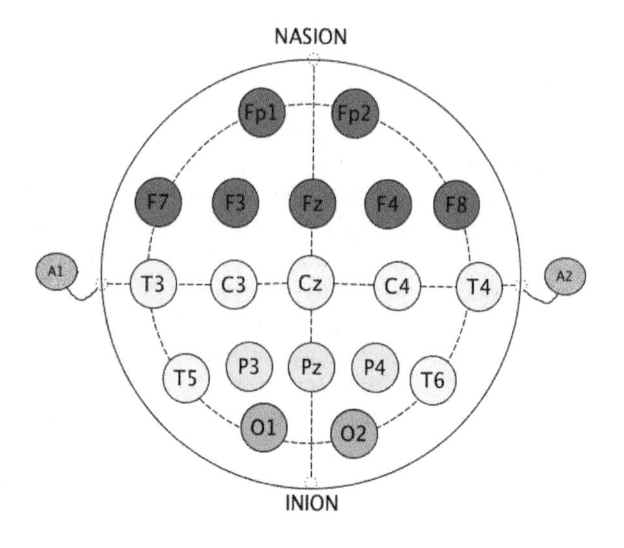

Fig. 4 Electrode positions in an EEG cap

EEG signals. The Deep CNN had 4 max-pooling blocks and the dense layer used was softmax layer for classification. In the shallow CNN there were only two convolution layers with larger kernel sizes, squaring nonlinearity, a mean pooling layer and a logarithmic activation function. The Hybrid CNN was designed by merging the deep CNN and the shallow CNN. Residual CNNs are different in that they have a good number of intermediate layers and the input of a CNN layer is connected to its own output, so the convolutional layer outputs a residual that changes the previous layers output. In addition a number of training algorithms have been proposed, for more details refer [24].

In most of the proposed algorithms cross validation is used to improve the accuracy results as it includes hyperparameter tuning and building different hypothesises. The use of CNNs is well pronounced as EEG still remains one of the main methods of data acquisition [25–30].

4 Applications of BCI

One of the main applications of BCI's so far has been taking control over external devices like appliances, onscreen keyboards, cursors and the like. Major contributions have also been made in the field of prosthetics, making crucial changes in the main principals of signal acquisition. From allowing completely locked in patients to communicate, to making hearing through the skin possible, BCI's can reach crazy limits, to the extent that someday "hacking into the mind" will be made possible. This could resolve all problems where language acts as a barrier to effective communication- deaf and dumb people could communicate using their minds, parents would be able to understand exactly why their babies cry, lack of vocabulary to express oneself would no longer degrade effective understanding. It would be lot more easier for crime detectors to detect lies when hacking into the mind will be made possible in the near future. This section aims to elaborate more on few of the successful BCI implementations so far.

- *Onscreen cursor movement using EEG signals*: In this case signals are mainly drawn from the sensorimotor cortex. Users can move the cursor, after some amount of training, in either 1 dimensional space or 2 dimensional space and the function underlying it may differ in being either linear or non-linear, thus allowing 3 different combinations i.e.: 1 dimensional linear, 2 dimensional linear and 1 dimensional non-linear. All these methods give best results with 10–20 features and the best results are given by the later two. Control of the amplitude of the mu and beta frequency bands are learned by the users while they try to move the cursor on the screen around 3 boxes which are aligned vertically and performance is judged by the number of correct and incorrect boxes selected by the user.
- *Game playing*: Games have different genres—action, sports, strategic, role-playing, puzzle, adventure or simulations. The different control strategies involved are as follows:

  – Motor Imagery: this involves detection and monitoring of the wave patterns in the brain that is generated due to imagination of movement of different parts of the body thus resulting in differing sesorimotor oscillations.
  – Neurofeedback: in some cases it monitors the mental recreational state of the subject whereas sometimes it also monitors parameters like the fluctuations in heart rate etc. This is mainly used for passive BCI architectures and it works by mainly inspecting the theta, alpha, SMR, and the beta waves.
  – Visual Evoked Potential (VEP): a measure of the change in electric potential caused due some alterations in the visual environment and is suitable due to multiple factors one of which is that the frequencies are unaltered by blinking of the eyes.

  Out of these, motor imagery has been the most widely used control strategy followed by Neurofeedback and VEP. For action games, where motor imagery is used, the subjects task is to move an on-screen object without any constraints in dimensions. VEP is also used in action games where the subject tries to balance an avatar on-screen. It has been observed that VEP gives about 90% accuracy. BCI strategic games include chess or cards where P300 is used as a control method. While it is not possible to create a fully functioning BCI for role-playing games because it is a concoction of many aspects like exploration, planning etc., the aspect of agitation and relaxation of a subject is monitored by passive BCI's. Simulation and Puzzle games both are very ideal for BCI's because they are more focussed on making the users more suited to the BCI environment. In these genres, Motor imagery and P300 are used respectively.
- *Prosthetics*: Most of the existing prosthetic devices today work by deriving sensory inputs from the nerve endings at the remnants of the injured part, say the thigh stump or the shoulder. This principle will fail in cases of total paralysis or cases

where the nerve endings are damaged or don't even exist. In such cases taking direct inputs from the EEG signals generated by the brain works best. These signals, drawn by placing electrodes over the sensorimotor cortex which is mainly involved in movement of different body parts, are passed through a classification algorithm that helps translate the signals into valid output commands to be followed by the external prosthetic device. There have been successful implementations of prosthetics aided by BCI like prosthetic arms and legs. Extensions to hearing and visual aids are also being worked upon.

Mentioned above are few among the innumerable BCI applications [11]. Channelizing the brain maps to different applications is just a matter of the computer learning to recognise the brain waves as they are being emitted. The computer is trained to classify the signals by providing it task labels corresponding to each waves and training it over a period. So next time when similar waves are generated, it knows exactly what kind of task is required to be done.

5 Conclusion and Future Work

BCIs are gaining more attention at a fast pace due to its enormous uses in the world of prosthetics. Invasive techniques such as, implanting a programmable silicon mesh into the brain which will expand inside and cross the blood brain barrier and give the purest information about neuron activity, or *Neural Dust* which are extremely tiny silicon sensors which could be spread across the cortex and communicate with an external device through ultrasound, are still in the development stage because the working of the brain is an extremely complex mechanism and there are problems related to lasting of external sensors in the body, neurosecurity, among others, to cater to. Non-invasive techniques such as MRI and EEG are in use today but the extent to which they are able to measure the crude and accurate brain activity is still not satisfactory since many a times they generate false alarms. Researchers are more focussed on extending the number of applications of BCI rather than optimising each algorithm to make it work in real time. The response time, accuracy, amount of training are the main parameters among the others that has to be optimised. Improvements require work in all the three domains—more accurate signal acquisition, better methods of representations and signal processing, optimised and efficient translation algorithms. Machine Learning techniques such as Support Vector Machines with different kernels, Radial Basis Functions, Linear Discriminant Analysis, Independent Component Analysis are already used but since a very long time not much attention had been paid to Deep Learning for classification purpose and it is still not that widely used though it has commendable potentials that is yet to be explored. With more applications, it can be hoped that the powers of Deep Learning are exploited. BCIs will not only serve in prosthetics but will also help in every day education, monitoring and controlling conscious and sub-conscious states of the mind, communication over a network through just the brain, sharing of knowledge and experiences and "disability" might only be a term in books.

References

1. G. Schalk, G. Schalk, D.J. McFarland et al., BCI2000: a general-purpose brain-computer interface (BCI) system. IEEE Trans. Biomed. Eng. **51**, 1034–1043 (2004). https://doi.org/10.1109/tbme.2004.827072
2. J.R. Wolpaw, D.J. McFarland, G.E. Fabiani et al., Conversion of EEG activity into cursor movement by a brain-computer interface (BCI). IEEE Trans. Neural Syst. Rehabil. Eng. **12**, 331–338 (2004). https://doi.org/10.1109/tnsre.2004.834627
3. D. Marshall, D. Marshall, D. Coyle et al., Games, gameplay, and BCI: the state of the art. IEEE Trans. Comput. Intell. AI Games **5**, 82–99 (2013). https://doi.org/10.1109/tciaig.2013.2263555
4. K.R. Müller, M. Tangermann, G. Dornhege et al., Machine learning for real-time single-trial EEG-analysis: from brain–computer interfacing to mental state monitoring. J. Neurosci. Methods **167**, 82–90 (2008). https://doi.org/10.1016/j.jneumeth.2007.09.022
5. C. Guger, G. Pfurtscheller, W. Harkam, C. Hertnaes, Prosthetic control by an EEG-based brain-computer interface (BCI), in *Proceedings of AAATE 5th European Conference for the Advancement of Assistive Technology*, pp. 3–6 (1999)
6. F. Akram, S.M. Han, T.S. Kim, An efficient word typing P300-BCI system using a modified T9 interface and random forest classifier. Comput. Biol. Med. **56**, 30–36 (2015). https://doi.org/10.1016/j.compbiomed.2014.10.021
7. J.R. Wolpaw, N. Birbaumer, W.J. Heetderks et al., Brain-computer interface technology: a review of the first international meeting. IEEE Trans. Rehabil. Eng. **8**, 164–173 (2000). https://doi.org/10.1109/tre.2000.847807
8. F. Lotte, M. Congedo, A. Lécuyer et al., A review of classification algorithms for EEG-based brain–computer interfaces. J. Neural Eng. **4**, R1–R13 (2007). https://doi.org/10.1088/1741-2560/4/2/r01
9. V. Shenoy Handiru, V.S. Handiru, A.P. Vinod, C. Guan, EEG source space analysis of the supervised factor analytic approach for the classification of multi-directional arm movement. J. Neural Eng. **14**, 046008 (2017). https://doi.org/10.1088/1741-2552/aa6baf
10. J.R. Wolpaw, G. Pfurtscheller, N. Birbaumer et al., Brain–computer interfaces for communication and control. Clin. Neurophysiol. **113**, 767–791 (2002). https://doi.org/10.1016/s1388-2457(02)00057-3
11. I. Sutskever, A. Krizhevsky, G.E. Hinton, ImageNet classification with deep convolutional neural networks. Commun. ACM **60**, 84–90 (2017). https://doi.org/10.1145/3065386
12. S.S. Roy, R. Roy, V.E. Balas, Estimating heating load in buildings using multivariate adaptive regression splines, extreme learning machine, a hybrid model of MARS and ELM. Renew. Sustain. Energ. Rev. **82**, 4256–4268 (2018)
13. S.S. Roy, P. Kulshrestha, P. Samui, Classifying images of drought-affected area using deep belief network, kNN, and random forest learning techniques, *Deep Learning Innovations and Their Convergence With Big Data*, pp. 102–119. IGI Global
14. S.S. Roy, A. Mallik, R. Gulati, M.S. Obaidat, P.V. Krishna, A deep learning based artificial neural network approach for intrusion detection, in *International Conference on Mathematics and Computing* (Springer, Singapore, 2017)
15. S.S. Roy, V.M. Viswanatham, Classifying spam emails using artificial intelligent techniques. Int. J. Eng. Res. Africa **22** (2016)
16. G. Pfurtscheller, C. Neuper, Motor imagery and direct brain-computer communication. Proc. IEEE **89**, 1123–1134 (2001). https://doi.org/10.1109/5.939829

17. A. Bandhu, S.S. Roy, Classifying multi-category images using deep learning: a convolutional neural network model, in *2017 2nd IEEE International Conference on Recent Trends in Electronics, Information & Communication Technology (RTEICT)*, pp. 915–919, May, 2017. IEEE
18. S. Sakhavi, S. Sakhavi, C. Guan et al., Learning temporal information for brain-computer interface using convolutional neural networks. IEEE Trans. Neural Netw. Learn. Syst. 1–11. https://doi.org/10.1109/tnnls.2018.2789927
19. H. Cecotti, A. Graeser, Convolutional neural network with embedded Fourier transform for EEG classification. IEEE, pp. 1–4 (2008)
20. P. Bashivan, I. Rish, M. Yeasin, N. Codella, Learning representations from EEG with deep recurrent-convolutional neural networks (2015). arXiv:1511.06448
21. C. Guan, S. Sakhavi, Convolutional neural network-based transfer learning and knowledge distillation using multi-subject data in motor imagery BCI. IEEE, pp. 588–591 (2017)
22. C.W. Anderson, Z. Sijercic, Classification of EEG signals from four subjects during five mental tasks, pp. 407–414 (1996)
23. Z.A. Keirn, Z.A. Keirn, J.I. Aunon, J.I. Aunon, A new mode of communication between man and his surroundings. IEEE Trans. Biomed. Eng. **37**, 1209–1214 (1990). https://doi.org/10.1109/10.64464
24. R.T. Schirrmeister, J.T. Springenberg, L.D.J. Fiederer et al., Deep learning with convolutional neural networks for EEG decoding and visualization. Hum. Brain Mapp. **38**, 5391–5420 (2017). https://doi.org/10.1002/hbm.23730
25. H. Cecotti, H. Cecotti, A. Graser, A. Gräser, Convolutional neural networks for P300 detection with application to brain-computer interfaces. IEEE Trans. Pattern Anal. Mach. Intell. **33**, 433–445 (2011). https://doi.org/10.1109/tpami.2010.125
26. H. Cecotti, A time–frequency convolutional neural network for the offline classification of steady-state visual evoked potential responses. Pattern Recogn. Lett. **32**, 1145–1153 (2011). https://doi.org/10.1016/j.patrec.2011.02.022
27. S. Stober, D.J. Cameron, J.A. Grahn, Using convolutional neural networks to recognize rhythm stimuli from electroencephalography recordings, in *Neural Information Processing Systems (NIPS)*, pp. 1–9 (2014)
28. M. Hajinoroozi, Z. Mao, Y. Huan, Prediction of driver's drowsy and alert states from EEG signals with deep learning. IEEE, pp. 493–496 (2015)
29. H. Yang, Y. Huijuan, S. Sakhavi et al., On the use of convolutional neural networks and augmented CSP features for multi-class motor imagery of EEG signals classification. IEEE, pp. 2620–2623 (2015)
30. A. Phadtare, A. Bahmani, A. Shah, R. Pietrobon, Scientific writing: a randomized controlled trial comparing standard and on-line instruction. BMC Med. Educ. **9**, 27 (2009). https://doi.org/10.1186/1472-6920-9-27

Reducing Hierarchical Deep Learning Networks as Game Playing Artefact Using Regret Matching

Arindam Chaudhuri and Soumya K. Ghosh

Abstract Human behavior prediction in strategic scenarios has been addressed through hierarchical deep learning networks in the recent past. Here we present a mathematical framework towards reduction of hierarchical deep learning network as game centric object. Considering simple game, we show the equivalence between training problem's global minimizers and Nash equilibria. Then we extend the game to hierarchical deep learning networks where the correspondence revolving Nash equilibria and network's critical points are addressed. With respect to these connections other learning methods are investigated. The experiments are done considering the artificial datasets which are developed from RPS game, CT experiments, Poker variant games as well as real MNSIT dataset. The experimental evaluation shows proposed framework's efficiency. It is concluded that regret matching achieves good training performance than other deep learning networks.

Keywords Game theory · Hierarchical deep learning networks · Reduction · Global minimizers · Nash equilibria

1 Introduction

In this chapter we revisit the human behavior prediction problem in strategic setups using hierarchical fuzzy deep learning networks (HFDLN) by [1]. A new approach is envisaged whereby HDFLN is reduced as game playing artefact. Giving due considerations to the well-known reductions [2–4] in literature, duality is not considered here as a possible candidate even though it allows a flexible perspective. The no-regret

A. Chaudhuri (✉)
Samsung R & D Institute Delhi, Noida 201304, India
e-mail: arindam_chau@yahoo.co.in; arindamphdthesis@gmail.com

S. K. Ghosh
Department of Computer Science Engineering, Indian Institute of Technology Kharagpur, Kharagpur 701302, India
e-mail: skg@iitkgp.ac.in

V. E. Balas et al. (eds.), *Handbook of Deep Learning Applications*, Smart Innovation, Systems and Technologies 136,
https://doi.org/10.1007/978-3-030-11479-4_16

strategies have been used in the past successfully [5]. They have effective training methods towards the learning algorithms [6]. In this direction regret matching [7] has emerged as an efficient stochastic optimization in terms of performance.

Some of the important research work for using regret minimization to solve offline optimization problems are available in [8–11]. Two methods worth mentioning are adaptive sub-gradient methods and traditional stochastic gradient descent. The loss simplification appears in either batch gradient or incremental gradient approaches. In regret minimization the simplification of the losses class by choosing minimizer from any particular functions family can be found in [12]. Any game can be used to solve optimal coloring [13]. The regret minimization has also been used in game theory [14]. Some of the other research works revolving around deep neural networks are available in [15, 16].

The prima face of this research work is to highlight how HDFLN can be presented as a game playing proposition with associated player actions and utilities. Looking at the different variations of the learning problem with or without considering the regularization we coin handshakes through critical points and Nash equilibria in strategic setups. It has been argued in the past that deep learning-based games are not straightforward because training deep models approximately is difficult in worst case [17]. It may be stated that regret matching has offered good training performance in comparison with deep learning heuristics providing sparse results. HDFLN reduction here breaks this juxtaposition. It addresses new situations for training any hierarchical based deep learning networks that have not been explored earlier.

This chapter has the following structure. In Sect. 2 HFDLN's reduction is presented. This follows the experimental evaluation of the proposed mathematical framework in Sect. 3. Finally, conclusions are illustrated in Sect. 4.

2 HFDLN Reduction Computational Framework

Here the mathematical proposition for HFDLN as game centric object is presented.

2.1 Problem Definition

This research work revisits the prediction of human behavior in strategic scenarios. The proposed HFDLN predictor in [1] is sketched here as a game playing entity which highlights several interesting insights. Considering the convex problem, the equivalence between training problem's global minimizers and game's Nash equilibria is established. Then the game is extended to hierarchical deep learning networks where the correspondence between network's critical points and Nash equilibria is established. The other learning methods are explored and regret matching achieves competitive training performance for HFDLN. The simulation experiments are performed over the RPS game [18], CT experiments [19], Poker variant game [20] and real MNIST dataset [21]. The entire framework ensures decision-based competitiveness.

2.2 Datasets Used

The experiments are performed on both artificial and real datasets. The artificial datasets are prepared from three different sources viz RPS game [18], CT experiments [19] and Poker variant game [20]. Besides this real MNSIT dataset [21] is also used for performing the experimental evaluations.

RPS is modeled as two-player zero-sum symmetric game. It is prepared from 60 one-shot game threads. The short delay of 9 s was taken into account. For non-reaction default gesture is adopted. Thus, we have $60 \times 9\,\text{s} = 9$ min duration threads. The game had one mixed strategy-based equilibrium with identical probability distribution. Here 700 computer science postgraduates are considered. Their average age was 21 years with 70% male population. The thread was played twice against test person. The other games were played between two threads. Each victorious player received $5 and $2 for neutral results. The players are stationed in different locations. For gestures the players used normal procedures. The information gathered had player's last and actual choice, opponent player's terminal choice, time factor and money received.

CT experiments dataset contains 7096 single human arguments of 200 participating topics. The responder's positive decision edits both players payoff and everything remains same for negative responds. The responder payoff ranged between $20.70 and $–20.86. For 700 cases there are zero responder edits. Only those proposals are accepted which increases player's payoff regardless of his opponent's payoff.

Poker variant game dataset is prepared by considering the counterfactual regret minimization to calculate an equilibrium solution in poker's domain. The poker variant focusses on tops-up limit Texas Hold'em used at AAAI Computer Poker Competition [20]. It is case of zero-sum and two-player game with four rounds of dealt cards, four betting rounds with under 2018 game states. The game is first abstracted and an equilibrium is populated for the abstracted game. We merge the information sets and bucket the card sequences. The results quality near equilibrium solution depends on the abstraction coarseness. As less abstraction is used there is higher quality of the resulting strategy. Hence solving larger game leads to less abstraction requirement.

2.3 Simple Games with Embedded Learning

We start our exploration by considering a simple one-layer learning game. This allows us to highlight the basic aspects which are then extended to hierarchical deep learning networks. Let us consider the standard supervised learning problem with a paired dataset $\{(a_i, b_i)\}_{i=1}^{N}$ such that $(a_i, b_i) \in A \times B$. Here one desires to learn a predictor $p: A \rightarrow B$. For the sake of simplicity, it is assumed that $A = \mathbb{R}^m$ and $B = \mathbb{R}^n$. Any standard generalized linear model takes the form:

$$p(a) = \varphi(\theta a) \tag{1}$$

Equation (1) is always valid for some output transfer function $\varphi: \mathbb{R}^n \rightarrow \mathbb{R}^n$ where matrix $\theta \in \mathbb{R}^{n \times m}$ denote the model's training parameters. Inspite the transfer function's presence φ the objective which has convexity in $c = \theta a$ is minimized and then the models are trained.

2.3.1 Learning Problem with One-Layer

Consider the loss function:

$$\upsilon: \mathbb{R}^n \times \mathbb{R}^n \rightarrow \mathbb{R} \tag{2}$$

Equation (2) is convex in the first argument. Let us assume $\upsilon_i(c) = \upsilon(c, b_i)$ and $V_i(\theta) = \upsilon_i(\theta a_i)$. The training problem minimizes $V(\theta) = N^{-1} \sum_{i=1}^{N} V_i(\theta)$ considering the parameters θ. Now we need to figure out a situation where the Nash equilibria corresponds to global minima. This fundamental correspondence creates bridge revolving game playing through deep learning. The one-layer case paves the path to highlight basic concepts which are used with deep learning networks. Another aspect viz one-shot simultaneous move is specified through players' set, actions' set and utility functions' set which specifies each player value with respect to the joint action selection. Next, we proceed to define the one-layer learning game.

2.3.2 Learning Game with One-Layer

Consider players as protagonist *prt* and antagonist *ant*. The protagonist takes the parameter matrix $\theta \in \mathbb{R}^{m \times n}$. The antagonist takes N vectors set and scalars $\{x_i, y_i\}_{i=1}^{N}, x_i \in \mathbb{R}^n, y_i \in \mathbb{R}^n$ such that $x_i^{\mathrm{T}} c + y_i \leq \upsilon_i(c) \forall c \in \mathbb{R}^n$. For each training example the antagonist considers an affine minorant of the local loss. So, both players consider their action choice without having any information regarding the other player's choice.

Considering the joint action selection $(\theta, \{x_i, y_i\})$, the antagonist's utility is defined as:

$$W^{ant} = N^{-1} \sum_{i=1}^{N} x_i^{\mathrm{T}} \theta a_i + y_i \tag{3}$$

The protagonist's utility is defined as:

$$W^{prt} = -W^{ant} \tag{4}$$

This leads us to the situation with continuous actions.

2.3.3 Nash Equilibrium

Nash equilibrium is specified through the joint actions situation. Thus if $\sigma^{prt} = \theta$ represents the protagonist's action choice and $\sigma^{ant} = \{x_i, y_i\}$ represents the antagonist's action choice then the joint action $\sigma = \left(\sigma^{prt}, \sigma^{ant}\right)$ represents the Nash equilibrium if:

$$W^{prt}\left(\tilde{\sigma}^{prt}, \sigma^{ant}\right) \le W^{prt}\left(\sigma^{prt}, \sigma^{ant}\right) \forall \tilde{\sigma}^{prt} \wedge$$

$$W^{ant}\left(\sigma^{prt}, \tilde{\sigma}^{ant}\right) \le W^{ant}\left(\sigma^{prt}, \sigma^{ant}\right) \forall \tilde{\sigma}^{ant} \tag{5}$$

With this background we determine the bijection between the one-layer learning game's Nash equilibria and the one-layer learning problem's global minimizers. If $(\theta^*, \{x_i, y_i\})$ represents the one-layer learning game's Nash equilibrium then θ^* should be the one-layer learning program's global minimum. Putting things, the other way around if θ^* represents the one-layer learning program's global minimizer then there is an antagonist strategy $\{x_i, y_i\}$ such that $(\theta^*, \{x_i, y_i\})$ represents the one-layer learning game's Nash equilibrium. The model's complexity is generally controlled through regularizing θ, though it is convenient to use constraint $\theta \in \Theta \exists \Theta$. If $(\theta^*, \{x_i, y_i\})$ represents the one-layer learning game's Nash equilibrium then θ^* should be the one-layer learning program's constrained global minimum. Putting things, the other way around if θ^* represents the one-layer learning program's constrained global minimizer then there is an antagonist strategy $\{x_i, y_i\}$ such that $(\theta^*, \{x_i, y_i\})$ represents the one-layer learning game's Nash equilibrium.

There are alternative training approaches for finding Nash equilibria which arises from the earlier stated connection. An appreciable progress has been made towards on-line algorithms for Nash equilibria [5, 22–24]. Considering the two-person zero-sum scenario, games are resolved through pitching regret-minimizing learning algorithms against one another. It is to be noted that both players should have $\varepsilon/2$ regret rate such that their strategies lead to ε-Nash equilibrium [14]. Here action of protagonist is $\theta \in \Theta$ and action of antagonist is σ_{ant}. The game playing is performed in rounds with joint action as:

$$\sigma^{(j)} = \left(\theta^{(j)}, \sigma_{ant}^{(j)}\right) \text{ on round } j \tag{6}$$

The utility function for each player W^i with $i \in \{prt, ant\}$ takes the affine form considering which the opponent player chooses. In this way each player finds himself in an online convex optimization problem [12]. The total regret of player is defined considering their utility function after J rounds as:

$$Reg^{p}\left(\sigma^{(1)}\ldots\sigma^{(J)}\right)=\max_{\theta\in\Theta}\sum_{j=1}^{J}W^{p}\left(\theta,\sigma_{ant}^{(j)}\right)-W^{p}\left(\theta^{(j)},\sigma_{ant}^{(j)}\right) \tag{7}$$

It may be argued that nature can be inducted here considering random training example. This calls for the regret to be defined in expectations' terms. In order to bring regularization here certain constraints Θ are imposed. A situation happens when:

$$\Theta=\{\theta\colon\|\theta\|_{1}\leq\gamma\} \tag{8}$$

Here L_1 ball constraint imposes L_1 regularization. The L_1 ball constraint enforces solution sparsity. Any polytope constraint leads to constrained online convex optimization reduction towards expert advice learning considering a finite number of experts [12]. This reduction permits expert advice learning towards L_1 constrained online convex optimization. In this direction, two algorithms are worth mentioning viz the normalized exponentiated weight algorithm [25, 26] and regret matching algorithm [7]. These algorithms generally operate through gradient's stochastic sample in order to achieve their updates. The exponentiated weight algorithm has better regret bounds. The regret matching algorithm has good regret bounds [27, 28]. Another notable algorithm is the projected stochastic gradient descent and it has similar regret bound [29, 30]. The utility of the discussion in this section is further highlighted through experimental evaluation on synthetic and real datasets in Sect. 3.

2.4 HFDLN as Deep Learning Game

With respect to our discussion in previous sections we are now in a position to approach the problem of training the HFDLN for human behavior prediction to be reduced to game playing. This reduction leads further insights towards training complex deep models used for solving massive-scale games. It may be mentioned at the outset that the feedforward neural network training problem has been studied by [31]. The feedforward neural network is one of the most commonly used neural networks during past few decades. It is specified through directed acyclic graph where vertices and edges are bridged by objects. The training of the network is done by assuming following facts:

(a) the input features are mapped to the vertices
(b) the output dimensions are mapped to output vertices

It is generally preferable to go for additional inputs' bias with (0, 1) input features. The identity activation functions are considered here. Consider a training input datapoint $a_t \in \mathbb{R}^m$ and the network $N = (V, E, I, O, F)$ (with V as vertices set, $E \subseteq V \times V$ is edges set, $I \subset V$ is input vertices set, $O \subset V$ is output vertices set, F is activation function set with $f_w: \mathbb{R} \to \mathbb{R}$ and training parameters $\theta: E \to \mathbb{R}$) is computed through circuit value function cv_i considering the vertices' partial order such that:

$$cv_i(p_k, \theta) = f_{p_k}(a_{ik}) \text{ for } p_k \in I \quad (9)$$

$$cv_i(w, \theta) = f_w\left(\sum_{v:(v,w)\in E} cv_i(v, \theta)\theta(v, w)\right) \text{ for } w \in V - I \quad (10)$$

Let the vector of values at the output be denoted by $cv_i(\text{o}, \theta)$. As such each f_w is differentiable, the output $cv_i(\text{o}, \theta)$ is also differentiable with respect to θ. When the constraints are imposed on θ it is assumed that vertices have the constraints factored. For each $w \in V - I$, θ is restricted to P_w and are placed in $\Theta_w \subseteq \mathbf{R}^{P_w}$ and $\Theta = \prod_{w \in V-I} \Theta_w$. The network is unconstrained for $\Theta = \mathbf{R}^P$ and the network is bounded if Θ is bounded.

2.4.1 HFDLN Learning Problem

Consider a loss function $\upsilon(c, b)$ which is convex in the first argument such that $0 \leq \upsilon(c, b) < \infty \forall c \in \mathbb{R}^n$. Also let us define define $\upsilon_i(c) = \upsilon(c, b_i)$ and $V_i(\theta) = \upsilon_i(cv_i(\text{o}, \theta))$. Here the problem is to figure out $\theta \in \Theta$ minimizing $V(\theta) = N^{-1} \sum_{i=1}^{N} V_i(\theta)$.

2.4.2 HFDLN Learning Game

Now we present one-shot simultaneous game situation having infinite action sets.

Players: This comprises of protagonist prt $\forall w \in V - I$, antagonist ant and zannis z_w set $\forall w \in V$.

Actions: Here protagonist vertex w selects a parameter function $\theta_w \in \Theta_w$. The antagonist chooses N vectors' set and scalars $\{x_i, y_i\}_{i=1}^{N}, x_i \in \mathbb{R}^n, y_i \in \mathbb{R}^n$ such that $x_i^{\mathrm{T}} c + y_i \leq \upsilon_i(c) \forall c \in \mathbb{R}^n$. In this way the antagonist selects for each training exemplar local loss's affine minorant. Each zannis z_w selects $2N$ scalars set $(r_{wi}, h_{wi}), r_{wi} \in \mathbb{R}^n, h_{wi} \in \mathbb{R}^n$ such that $r_{wi}c + h_{wi} \leq f_w(c) \forall c \in \mathbb{R}$. Thus, for each training example zanni selects affine minorant corresponding to f_w. It is to be noted that players select their action without another player's knowledge.

Utilities: Considering the joint action $\sigma = (\theta, \{x_i, y_i\}, \{r_{wi}, h_{wi}\})$, the zannis' utilities can be specified recursively considering the vertices partial order. For each $w \in I$ the zanni z_w utility on training exemplar i is:

$$W_{wi}^{z}(\sigma) = h_{wi} + r_{wi}a_{wi} \tag{11}$$

For each $w \in V - I$ the zanni z_w utility on exemplar i is:

$$W_{wi}^{z}(\sigma) = h_{wi} + r_{wi} \sum_{v:(v,w)\in E} W_{iv}^{z}(\sigma)\theta(v, w) \tag{12}$$

Each zanni z_w total utility is given by:

$$W_{w}^{z}(\sigma) = \sum_{i=1}^{N} W_{wi}^{z}(\sigma) \text{ for } w \in V \tag{13}$$

The antagonist *ant* utility is then given by:

$$W^{ant} = N^{-1} \sum_{i=1}^{N} W_{i}^{ant} \text{ with } W_{i}^{ant}(\sigma) = y_i + \sum_{k=1}^{n} x_{ki} W_{O_k i}^{z}(\sigma) \tag{14}$$

Likewise, all protagonists maintain the same utility such that:

$$W^{prt}(\sigma) = -W^{ant}(\sigma) \tag{15}$$

Considering protagonist action θ there exists all agents' unique joint action $\sigma = (\theta, \{x_i, y_i\}, \{r_{wi}, h_{wi}\})$ such that zannis and antagonist play positively. Since $W^{prt}(\sigma) = -L(\theta)$ and $\nabla_{\theta} W^{prt}(\sigma) = -\nabla L(\theta)$. Now with some protagonist $w \in V - I$ all other agents' strategies are held fixed $W^{prt}(\sigma)$. Here σ is the joint action expansion for θ. For each parameter, current cost and partial derivatives are computed such that affine function for each agent comes into place. This is discussed as we proceed further. The KKT point satisfies the KKT conditions [32, 33]. This is either a critical point with zero gradient or boundary point Θ with gradient points out of Θ perpendicularly.

2.4.3 HFDLN Learning Game Nash Equilibrium

The joint action $\sigma = (\theta, \{x_i, y_i\}, \{r_{wi}, h_{wi}\})$ is a Nash equilibrium of HFDLN learning game. With this we can argue that if deep network is not bounded, joint action $\sigma = (\theta, \{x_i, y_i\}, \{r_{wi}, h_{wi}\})$ is a Nash equilibrium of HFDLN learning game when it is θ's joint action expansion with θ as HFDLN learning problem's critical point.

Sometimes it is required to add constraints connecting edges on different nodes. Considering any deep neural network with edges ed and $ed^{'}$ there may be a constraint satisfying $\theta_{ed} = \theta_{ed'}$. If any two players act simultaneously in a game it becomes impossible situation where a player's actions depend on their opponent. Hence, as the parameters become constrained, any of the players' can take the control. This is true for both smooth and non-smooth activation functions through certain differentiable approximations.

Articulating HFDLN as a deep learning game paves the path towards equilibrium finding methods. Considering earlier reduction towards intelligent algorithms L_1 ball constraint $\Theta_w = \{\theta_w\colon \|\theta_w\|_1 \leq \gamma\}$ is used at each vertex w. For HFDLN the approach adopted through training protagonist versus antagonist and zannis [34]. It is to be noted that best response cases should be selected here. Here it is possible build new learning strategies based on learning algorithms from certain advice from experts. A non-convex optimization problem is placed considering protagonist's action θ_w. As a result of this the convergence is not expected to be a joint, globally optimal strategy for protagonists. There are generic approaches towards utilising the game in learning algorithm generation. The algorithm is briefly outlined below.

The nature chooses a random training exemplar in each round. Each protagonist $w \in V$ selects his actions $\theta_w \in \Theta_w$ deterministically. Considering the best responses to θ_w and to each other, the antagonist and zannis choose their actions. Then the protagonist utilities W_w^{prt} are calculated. The W_w^{prt} is affine in protagonist's action when the zanni and antagonist choices are considered. Corresponding to this each protagonist $w \in V$ update their strategies based on their utilities. It is well known that Nash equilibrium corresponds to critical point in training problem. This is generally considered as a local minimum but not a saddle point [35]. The backpropagation of the sampled sub-gradients computes best response actions from where protagonist affine utility is obtained. This discussion is further highlighted through experimental evaluation on the synthetic and real datasets in Sect. 3.

3 Experimental Evaluations

In this section the experimental evaluations of the methods discussed in Sects. 2.3 and 2.4 are highlighted. The experiments are conducted on the three synthetic datasets [18–20] mentioned in Sect. 2.2 and on real MNIST dataset [21].

We first present the experimental evaluations for the methods discussed in Sect. 2.3. It is to be noted that projected stochastic gradient descent and exponentiated weight algorithm have performance affected through step size. Optimal regret bounds are obtained through simple and logarithmic step sizes [30]. The step size parameters are regularly tuned to have the best results in place. The experiments are executed after convergence speeds towards lowest global values. Initial experiments considered the three synthetic datasets. The data dimension considered was 20 and 1000 training points are taken. The univariate prediction produced random hyperplane with labelled linearly separable data. All the results are shown in Fig. 1.

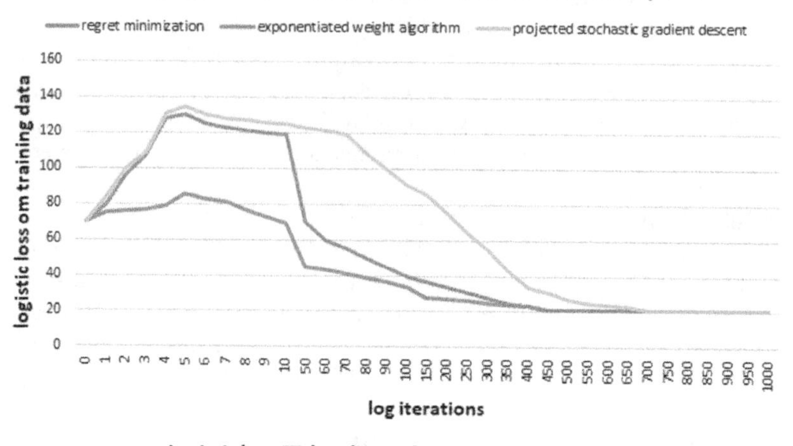

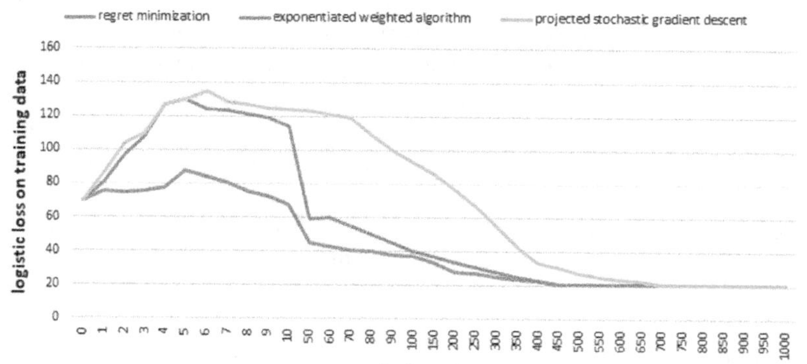

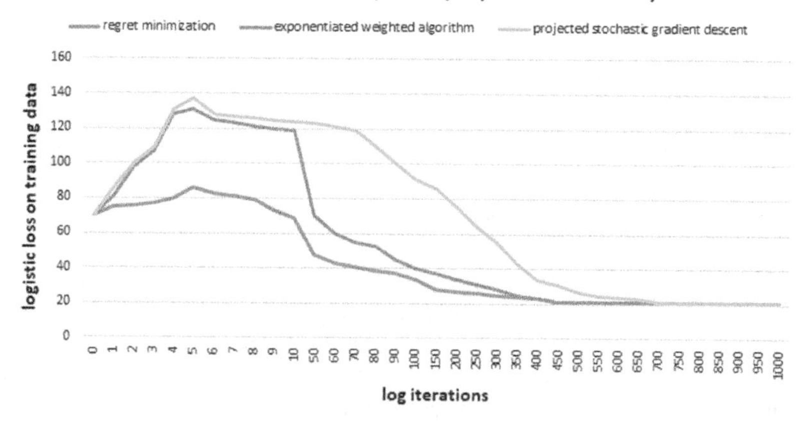

Fig. 1 No-regret algorithms training loss for RPS, CT and poker datasets

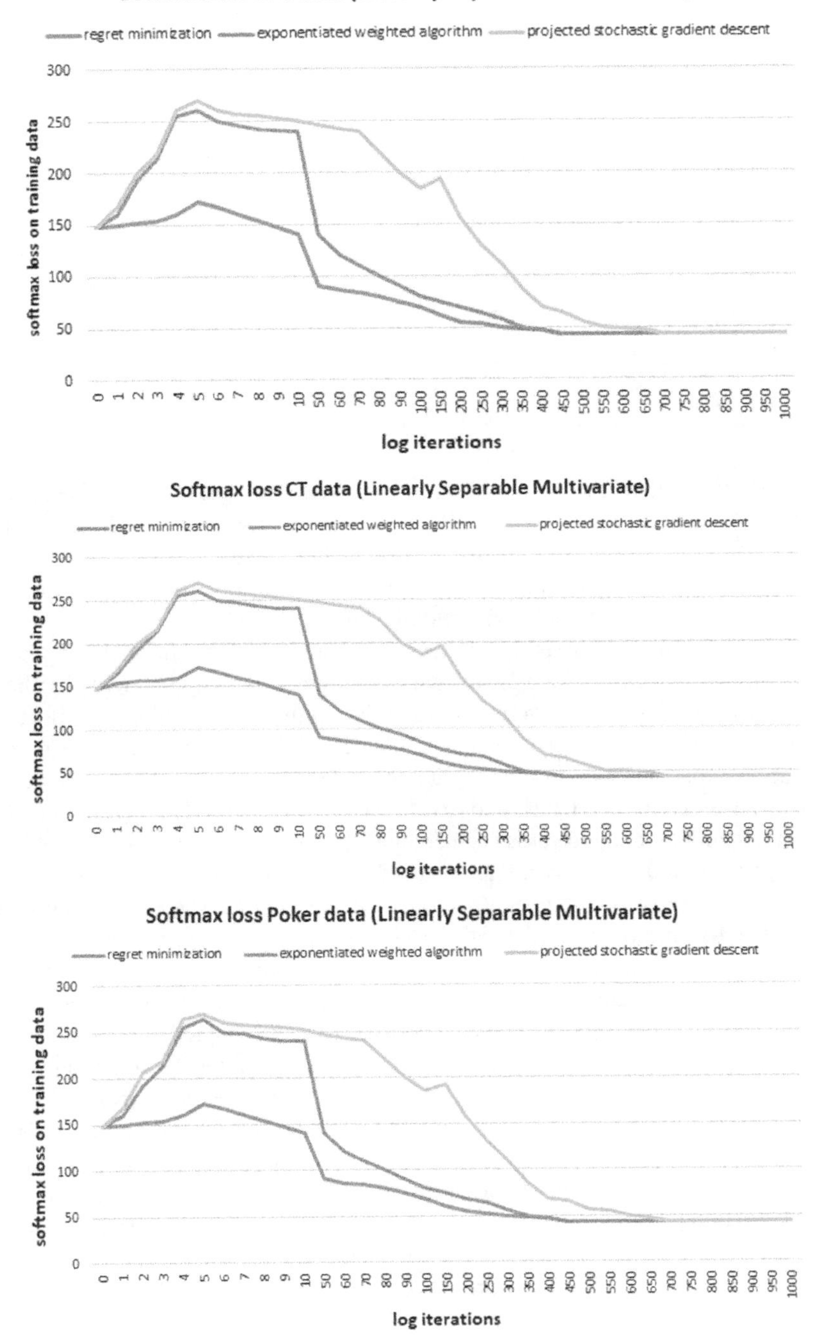

Fig. 1 (continued)

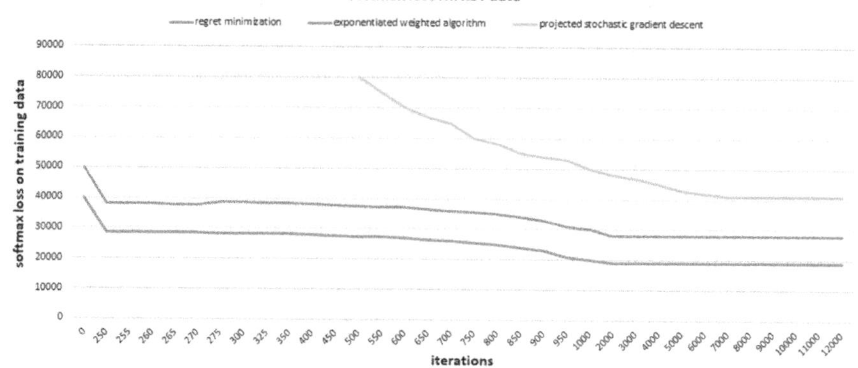

Fig. 2 No-regret algorithms training loss for MNIST dataset

For multivariate prediction all the results are shown in the figures below. The results for MNIST dataset (with mini-batch size = 200) are shown in Fig. 2. The regret minimization has proved to be the most effective here. Even after parameter tuning for projected stochastic gradient descent and exponentiated weight algorithm did not produce better results than regret minimization.

Now we present the experimental evaluations for the methods discussed in Sect. 2.4. The experiments are conducted to figure out the scope of applying these algorithms considering each vertex in HFDLN. To make the results more illustrative HFDLN is trained through models available in [36–39]. These methods do not impose any constraints in developing the optimization problem solution. However, they are un-regularized in nature and have weak generalization. The step size parameters are regularly tuned towards each comparison method involving each problem. The regret minimization algorithm produced better results than projected stochastic gradient descent and exponentiated weight algorithms. The regret minimization worked well through initialization of the cumulative regret vectors. To make the results more convincing experiments are first conduced on artificial parity and folded parity combinatorial problems [40]. The parity is approximated through single hidden layer with linearity in threshold gates [41]. The folded parity required at least two hidden layers [40]. The parity training is performed on fully connected $(m - 8m - 1); m = 10$ architecture. The folded parity training is conducted on $(m - 8m - 8m - 8m - 1); m = 10$ architecture. The L_1 constraint is bounded to 20 and the initialization scale was set as 200. The nonlinear activation function is used through a smooth approximation of the standard ReLU gate. The results in Fig. 3 show that regret minimization produces better competitive results for both parity and folded parity with sparsity models top to bottom.

Next the experiments are performed using MNIST dataset. The initial experiment considered a fully connected $784 - 2048 - 2048 - 10$ architecture with regret minimization execution bounded by 40 with scales of initialization as (60, 240, 60).

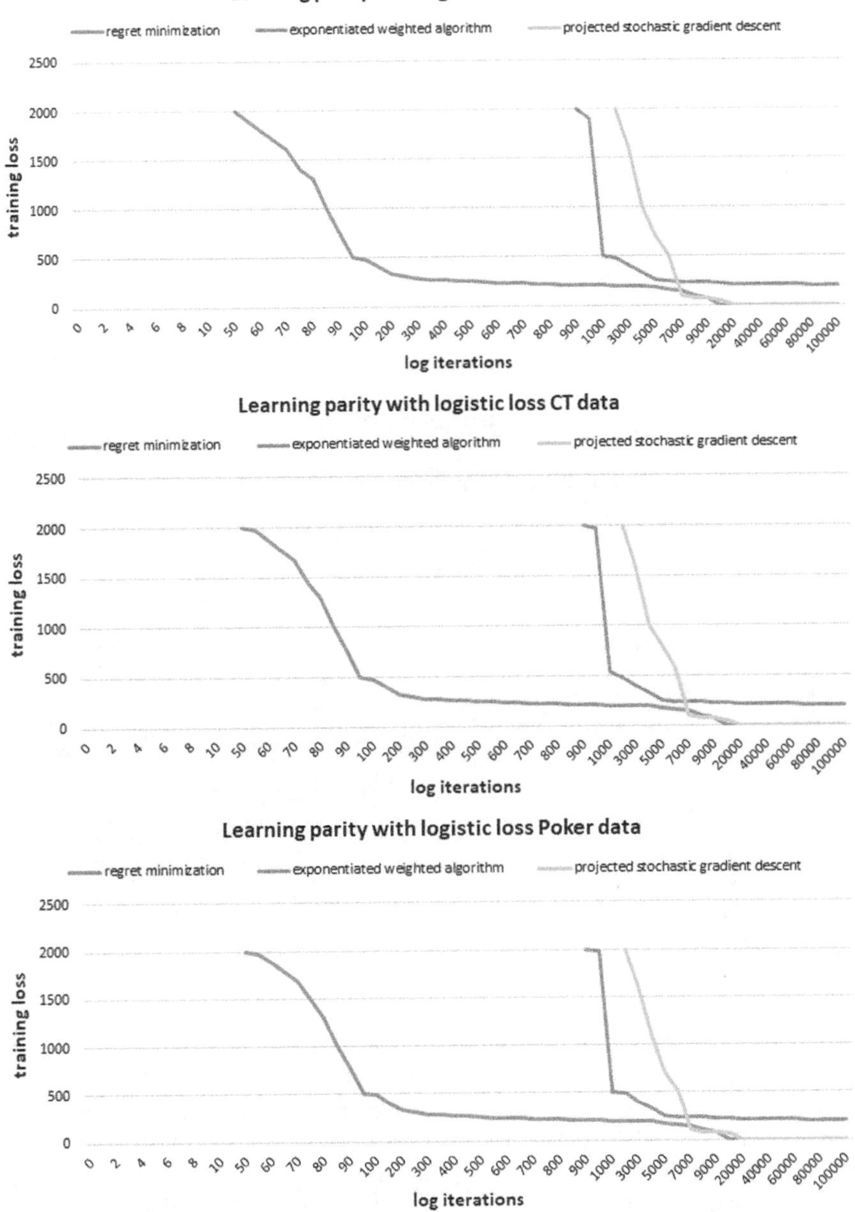

Fig. 3 Parity results for RPS, CT and poker datasets; training loss and testing error for MNIST dataset

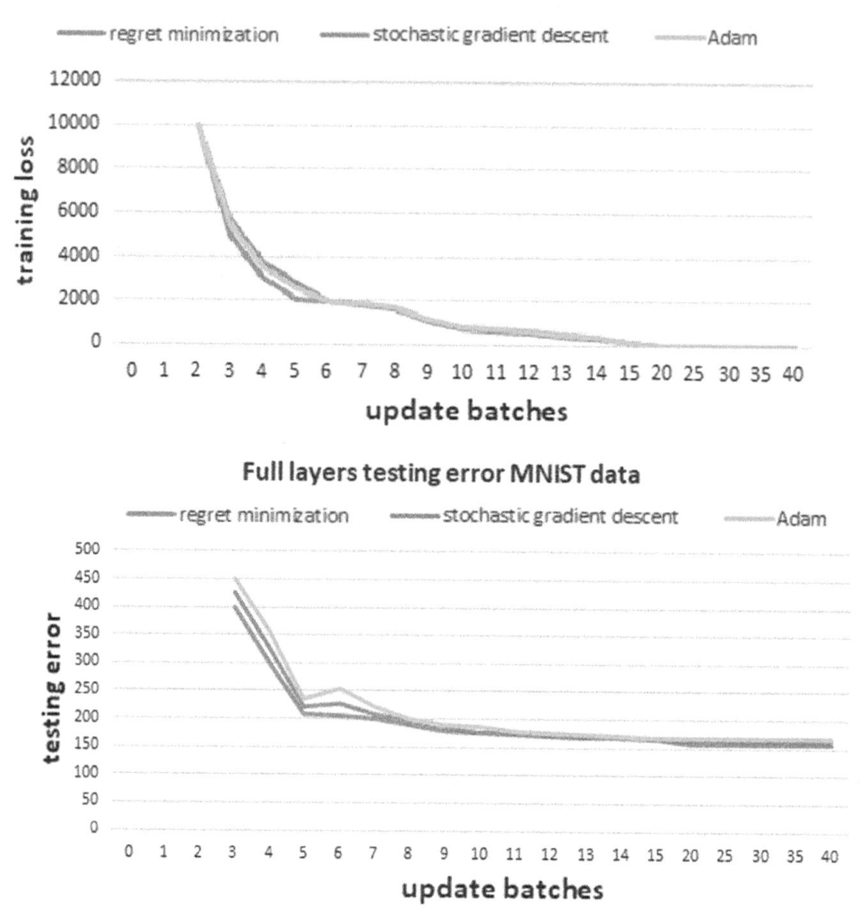

Fig. 3 (continued)

The second experiment has convolutional architecture $32 \times 32 - c(6 \times 6, 81) - -c(6 \times 6, 81) - c(6 \times 6, 81) - c(6 \times 6, 81) - 10$ (convolution window 6×6 and depth 81) with regret minimization run bounded by (40, 40, 40, 10) and initialization scale 600. The mini-batch size considered is 200 with results obtained after 800 mini-batches updates. Figure 4 show the training and test loss. These results highlight the evolution of training loss and test misclassification errors. The regret minimization for both fully connected and convolutional case superseded the results obtained through other methods. The regret minimization convergence rates are competitive as compared to other methods. The regret minimization results were sparse in nature and achieved lower test misclassification errors compared to other deep learning networks.

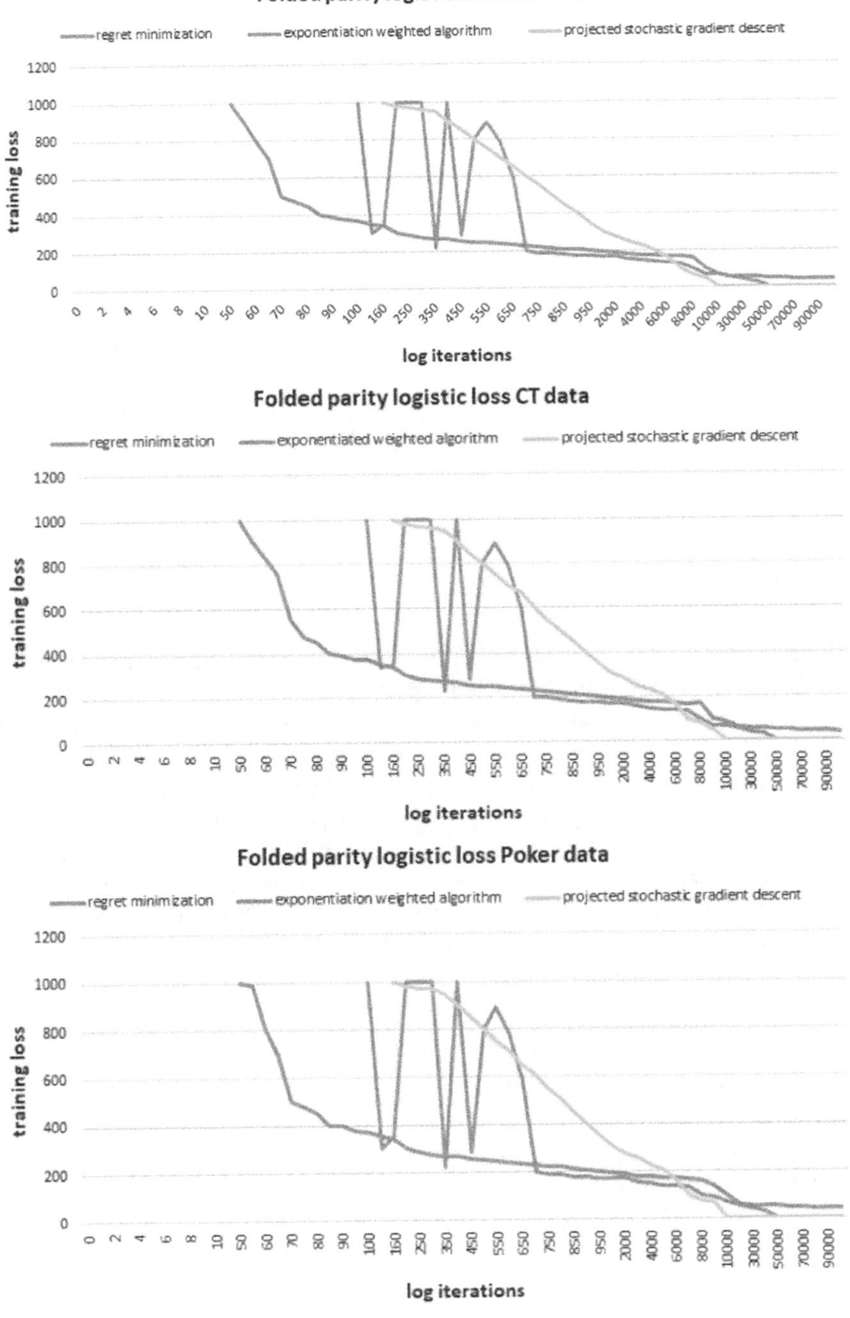

Fig. 4 Folded parity results for RPS, CT and poker datasets; training loss and testing error for MNIST dataset

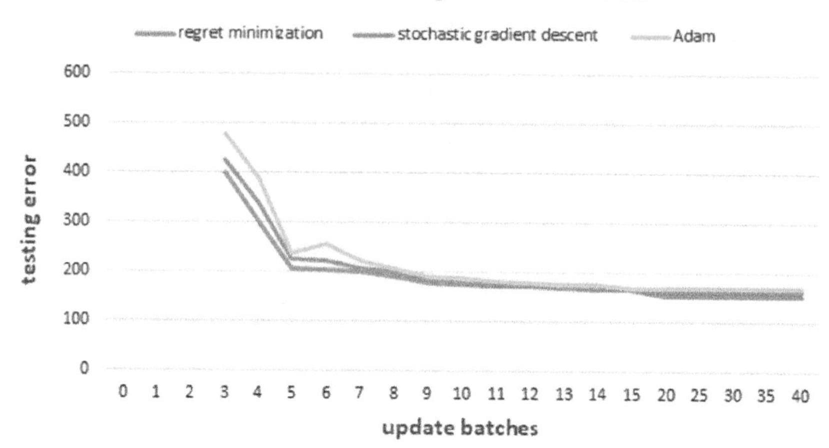

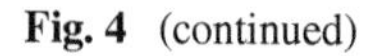
Fig. 4 (continued)

4 Conclusion

We have addressed here the reduction of HFDLN for predicting human behavior in strategic setups as a game centric object. The reduction presented in this chapter is the outcome of the research work that reveals the bridge between hierarchical deep learning networks and game theory. This has paved the way for the correspondence between critical points and Nash equilibria. Regret matching provides a better option in terms of achieving sparsity giving due consideration for speed or accuracy. This mathematical framework achieves new performance benchmark. This indicates higher performance of the hierarchical deep learning networks. The experimental evaluations highlight superiority of proposed framework. It remains an interesting work towards investigating alternative frameworks for deep learning games. However, the question which surfaces is that whether similar successes can be achieved.

References

1. A. Chaudhuri, S.K. Ghosh, Hierarchical fuzzy deep learning networks for predicting human behavior in strategic setups, in *Advances in Intelligent Systems and Computing*, ed. by R. Silhavy, et al., vol. 764 (Springer, Switzerland, 2019), pp. 111–121
2. Y. Freund, R. Schapire, Adaptive game playing using multiplicative weights. Games Econ. Behav. **29**(1–2), 79–103 (1999)
3. S. Shalev-Shwartz, Y. Singer, Convex repeated games and Fenchel duality, in *Advances in Neural Information Processing Systems*, vol. 19 (2006)
4. S. Shalev-Shwartz, Y. Singer, A primal-dual perspective of online learning algorithms. Mach. Learn. **69**(2–3), 115–142 (2007)
5. O. Tammelin, N. Burch, M. Johanson, M. Bowling, Solving heads-up limit Texas hold'em, in *Proceedings of International Joint Conference on Artificial Intelligence*, vol. 645–652 (2015)

6. T.M. Mitchell, *Machine Learning* (McGraw Hill, 2008)
7. S. Hart, A. Mas-Colell, A simple adaptive procedure leading to correlated equilibrium. Econometrica **68**(5), 1127–1150 (2000)
8. N. Cesa-Bianchi, A. Conconi, C. Gentile, On the generalization ability of on-line learning algorithms. IEEE Trans. Inf. Theory **50**(9), 2050–2057 (2004)
9. N. Ratliff, D. Bagnell, M. Zinkevich, Sub-gradient methods for structured prediction, in *Proceedings of 11th International Conference on Artificial Intelligence and Statistics*, vol. 2, pp. 380–387 (2007)
10. N. Ratliff, J.A. Bagnell, M. Zinkevich, Maximum margin planning, in *Proceedings of 22nd International Conference on Machine Learning*, pp. 729–736 (2006)
11. S. Shalev-Shwartz, Y. Singer, N. Srebro, A. Cotter, Pegasos: primal estimated sub-gradient solver for SVM. Math. Program. **127**(1), 3–30 (2011)
12. M. Zinkevich, Online convex programming and generalized infinitesimal gradient ascent, in *Proceedings of 20th International Conference on Machine Learning*, pp. 928–935 (2003)
13. M. Kearns, S. Suri, N. Montfort, An experimental study of the coloring problem on human subject networks. Science **313**, 824–827 (2006)
14. M. Zinkevich, M. Bowling, M. Johanson, C. Piccione, Regret minimization in games with incomplete information, in *Advances in Neural Information Processing Systems*, vol. 20 (2007)
15. D. Balduzzi, Cortical prediction markets, in *Proceedings of International Conference on Autonomous Agents and Multi-agent Systems*, pp. 1265–1272 (2014)
16. D. Balduzzi, Deep online convex optimization using gated games (2016). http://arxiv.org/abs/1604.01952
17. K. Hoeffgen, H. Simon, K. Van Horn, Robust trainability of single neurons. J. Comput. Syst. Sci. **52**(2), 114–125 (1995)
18. R. Tagiew, Hypotheses about typical general human strategic behavior in a concrete case, in *Emergent Perspectives in Artificial Intelligence. AI*IA*, ed. by R. Serra, R. Cucchiara, LNCS, vol. 5883 (Springer, Heidelberg, 2009), pp. 476–485
19. Y. Gal, A. Pfeffer, Modeling reciprocal behavior in human bilateral negotiation, in *Proceedings of the 22nd National Conference on Artificial Intelligence*, vol. 1, pp. 815–820 (2007)
20. M. Zinkevich, M. Littman, The AAAI computer poker competition. J. Int. Comput. Games Assoc. **29** (2006)
21. Y. LeCun, L. Bottou, Y. Bengio, P. Haffner, Gradient-based learning applied to document recognition. Proc. IEEE **86**(11), 2278–2324 (1998)
22. N. Cesa-Bianchi, G. Lugosi, *Prediction, Learning and Games* (Cambridge University Press, 2006)
23. A. Rakhlin, K. Sridharan, Optimization, learning, and games with predictable sequences, in *Advances in Neural Information Processing Systems*, vol. 26 (2013)
24. V. Syrgkanis, A. Agarwal, H. Luo, R. Schapire, Fast convergence of regularized learning in games, in *Advances in Neural Information Processing Systems*, vol. 28 (2015)
25. N. Littlestone, M. Warmuth, The weighted majority algorithm. Inf. Comput. **108**(2), 212–261 (1994)
26. N. Srinivasan, V. Ravichandran, K. Chan, J. Vidhya, S. Ramakrishnan, S. Krishnan, Exponentiated backpropagation algorithm for multilayer feedforward neural networks, in *Proceedings of International Conference on Neural Information Processing*, vol. 1 (2002)
27. G. Gordon, No-regret algorithms for structured prediction problems. Technical Report CMU-CALD-05112, Carnegie Mellon University (2005)
28. G. Gordon, No-regret algorithms for online convex programs, in *Advances in Neural Information Processing Systems*, vol. 19 (2006)
29. J. Duchi, S. Shalev-Shwartz, Y. Singer, T. Chandra, Efficient projections onto the l1-ball for learning in high dimensions, in *Proceedings of International Conference on Machine Learning*, pp. 272–279 (2008)
30. S. Shalev-Shwartz, Online learning and online convex optimization. Found. Trends Mach. Learn. **4**(2), 107–194 (2012)

31. D. Schuurmans, M. Zinkevich, Deep learning games, in *Advances in Neural Information Processing Systems*, vol. 29 (2016)
32. W. Karush, Minima of functions of several variables with inequalities as side constraints. Master's Thesis, University of Chicago, Chicago, Illinois (1939)
33. H. Kuhn, A. Tucker, Nonlinear programming, in *Proceedings of 2nd Berkeley Symposium*, University of California Press, pp. 481–492 (1951)
34. M. Johanson, N. Bard, N. Burch, M. Bowling, Finding optimal abstract strategies in extensive form games, in *Proceedings of AAAI Conference on Artificial Intelligence*, pp. 1371–1379 (2012)
35. J. Lee, M. Simchowitz, M. Jordan, B. Recht, Gradient descent only converges to minimizers, in *Proceedings of 29th Annual Conference on Learning Theory*, vol. 49, pp. 1246–1257 (2016)
36. L. Bottou, Stochastic gradient descent tricks, *Neural Networks: Tricks of the Trade,* 2nd edn, pp. 421–436 (2012)
37. J. Duchi, E. Hazan, Y. Singer, Adaptive sub-gradient methods for online learning and stochastic optimization. J. Mach. Learn. Res. **12**, 2121–2159 (2011)
38. D. Kingma, J.B. Adam, A method for stochastic optimization (2014). http://arxiv.org/abs/1412.6980
39. I. Sutskever, J. Martens, G. Dahl, G. Hinton, On the importance of initialization and momentum in deep learning, in *Proceedings of 29th International Conference on Machine Learning*, pp. 1139–1147 (2013)
40. A. Razborov, On small depth threshold circuits, in *Proceedings of Scandinavian Workshop on Algorithm Theory*, pp. 42–52 (1992)
41. A. Hajnal, Threshold circuits of bounded depth. J. Comput. Syst. Sci. **46**(2), 129–154 (1993)

Deep Learning in Gene Expression Modeling

Dinesh Kumar and Dharmendra Sharma

Abstract Developing computational intelligence algorithms for learning insights from data has been a growing intellectual challenge. Much advances have already been made through data mining but there is an increasing research focus on deep learning to exploit the massive improvement in computational power. This chapter presents recent advancements in deep learning research and identifies some remaining challenges as drawn from using deep learning in the application area of gene expression modelling. It highlights deep learning (DL) as a branch of Machine Learning (ML), the various models and theoretical foundations, its motivations as to why we need deep learning in the context of evolving Big Data, particularly in the area of gene expression level classification. We present a review, and strengths and weaknesses of various DL models and their computational power to specific to gene expression modeling. Deep learning models are efficient feature selectors and therefore work best in high dimension datasets. We present major research challenges in feature extraction and selection using different deep models. Our case studies are drawn from gene expression datasets. Hence we report some of the key formats of gene expression datasets used for deep learning. As ongoing research we will discuss the future prospects of deep learning for gene expression modelling.

1 Introduction and Motivations

The world is experiencing enormous growth of data at an unprecedented rate in all domains of life, industry, education, weather, health and scientific research to name a few. Extracting patterns and learning insights from this abundance of data is continuing and a developing research frontier. Fascinated by the prospects of patterns and knowledge hidden in data has motivated practitioners in the area of machine learn-

D. Kumar (✉) · D. Sharma
University of Canberra, Canberra, Australia
e-mail: DineshKumar@canberra.edu.au

D. Sharma
e-mail: Dharmendra.Sharma@canberra.edu.au

V. E. Balas et al. (eds.), *Handbook of Deep Learning Applications*,
Smart Innovation, Systems and Technologies 136,
https://doi.org/10.1007/978-3-030-11479-4_17

ing in developing several ML algorithms to help them extract relevant knowledge and insights. ML has matured over several decades of research and has churned out efficient data mining algorithms such as support vector machines, Bayesian models, decision tress and ANNs; and have been used to solve many interesting real world problems in various domains [1–3]. These include algorithms such as decision trees, Naïve Bayes (NB), Apriori rules, fuzzy logic, support vector machines (SVMs) and artificial neural networks (ANNs). The above algorithms described as conventional machine learning algorithms were limited in their ability to process data in its natural raw form [4]. Before using these algorithms serious data pre-processing is required. Data has to be curated and then transformed into a format suitable for the algorithm to work on. This also included manual feature extraction in the form of feature vectors as well. The learning algorithms could then only use these data to detect or classify patterns. Often this required researchers to have considerable domain knowledge. As data collection methods improves with the expansion of data storage capacity has enabled growth of complex high-dimensional datasets, both structured and unstructured. For existing machine learning systems to function properly data will need to be formatted to suit as valid input. This comes as a cost of losing considerable information which may have been embedded in the original structure of data. Hence to mine this huge datasets in their raw form has spawned the birth of new sets of models and algorithms referred to as Deep Learning (DL). Built on the computational paradigm of Artificial Neural Network, DL implements multiple hidden nodes and layers in their architecture. They are also referred to as deep structured learning or hierarchical learning models.

The need for DL architectures is further motivated by several factors [5] including shallow nets such as SVMs, NB, K-Nearest Neighbours (KNN) do not provide sufficient depth in their architectures to learn from high dimension data. Most of these algorithms perform well on labelled data and user-defined features a priori. When applied to high dimension datasets the number of parameters for the learning function as well as the computation time may grow very large making these algorithms inefficient for handling such datasets. Secondly AI intends to mimic the functions of the brain in terms of its reasoning capability. Just as the brain learns through examples, most machine learning algorithm are 'taught' using training data for classification and clustering purposes. The closest ML algorithm that represents how the neurons in the brain function is the ANNs. The brain however has deep architecture with several million interconnected neurons to process the input signals (features). Shallow ANNs with few hidden layers do not approximate the brain structure hence the need for deep learning architectures. Thirdly, the human visual system contains various levels of cell structures for information processing. Information filters from one cell layer to the next and in doing so it is able determine areas of importance in the feature maps that has a higher representational value. This means relevant neurons in the brain will be active than others for particular features. This simply illustrates that cognitive processes in the brain are deep. Ideas and concepts are organised hierarchically, simpler concepts are learnt first then composing them into more abstract ones; such as lines, circle, colors and curves on an image is detected first before combining them to learn the higher abstract level feature, the complete face. The solution to the problem is broken up into multiple levels of abstraction and processing. DL models are capable of identifying layers of abstractions or compositions from feature sets due to their complex deep neuron structures, hence are known be to excellent feature extractors from feature maps. They are proven to be able to pick features that lead to improved performance of the deep classifiers [1].

This chapter summarises recent advances on DL, the key models developed for DL and how they perform feature abstractions in supervised and unsupervised learning environments. Applications are drawn from the complex area of gene expression modeling. We first give a formal definition of DL and its origins. Next we discuss various key deep learning models and their applications. We will state the relevant theoretical foundations underpinning these models followed by their strengths and inadequacies. Further, we introduce the gene expression modeling problem (GEMP)—the expression datasets, kinds of questions that researches are seeking answers for from such data and how DL helps find solutions to these questions. Here we will summarise recent research work where DL is applied on gene expression data. Finally we will provide overall strengths of deep learning followed by conclusion and explore possible future investigations.

2 Deep Learning

DL algorithms accept raw data and are automatically able to discover patterns in them. They are able to do this by creating feature maps from which representations are discovered for detection of patterns or classification [4]. This leads deep learning methods to be able to detect higher level features defined in terms of lower level features. DL utilizes similar concepts from ANNs in machine learning. Since most ANNs in conventional machine learning are 'shallow' implementations (one or two hidden layers), DL implements 'deep' neural architectures (several hidden layers) which seemingly has proven apt architectures to solve problems in image and video processing, text analysis and object recognition. Deep learning is also sometimes referred to as multilayer perceptrons (MLPs) or deep neural networks (DNN). The implementations of these deep architectures however vary to suit different problems and goals. The popularity of deep learning has increased so much that many large companies such as Google, NVidia and Toyota have taken much interest in and have invested in research and development on deep leaning AI [4]. Figure 1 is a graph from Google Trends (https://trends.google.com/trends/) that shows the growth of interest in DL as a field of study in recent times. A rising trend illustrates the popularity of the term entered as a search-phrase by people d from around the world and in various languages as well.

Recent breakthroughs in DL research have led to solutions for many problems for which earlier techniques in machine learning were not quite able to solve or provide optimal solutions [4]. Many scientific hurdles faced in processing large datasets to extract insights using conventional shallow machine learning methods have been greatly reduced. The ability to drive programs to analyze large datasets to extract

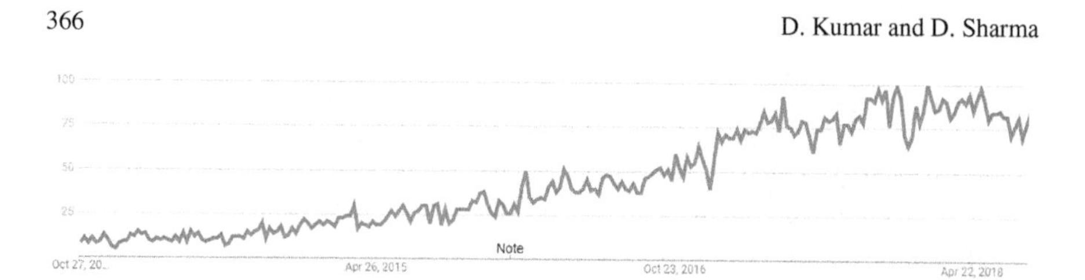

Fig. 1 Google trend chart showing interest in DL as field of study

meaningful patterns and inferences has always been a goal for researchers in the field of AI. DL now allows this to happen with higher degree of accuracy in various fields such as in image and video processing, text processing for sentiment analysis, object recognition, speech recognition and time series analysis. It is a rich paradigm to extract insights from experimental data to model and derive solutions through predictive analytics.

Research interest in DL has been increasing rapidly since a seminal paper on it in 1998 by LeCun et al. [6]. Significant research has been reported since resulting in some significant breakthroughs such as those reported in [4, 7, 8]. As explained by Benjio in [9], DL algorithms scan the input distribution to seek and exploit the unknown structure in an effort to discover good abstract representations. They aim to learn lower-level features and then use this as a basis to learn higher-level features. This allows DL to be used for unsupervised learning where data is unlabeled. Hence the key advantage of DL is its ability to learn good features automatically without considerable amount of reengineering or domain expert. Figure 2 extracted from Yoshua Bengio's lecture notes [10] on DL shows how different parts of an AI system relate to each other within different AI disciplines. It demonstrates DL learns features at multiple levels. At each level abstract features are discovered from features in the previous level. As features are discovered they are composed together at various levels to produce the output. The following are the three key DL architectures that have attracted considerable attention by machine learning enthusiasts;

- Feed Forward Networks (FFNs),
- Convolution Neural Networks (CNNs), and
- Recurrent Neural Networks (RNNs) [11].

In the following sections we describe properties of each of these three types of architectures, their advantages and disadvantages.

3 Deep Learning Models and Applications

Numerous DL models have been developed motivated by the application problems. In this section, we summarise the main models namely FFNs, CNNs and RNNs and their application to various problem domains.

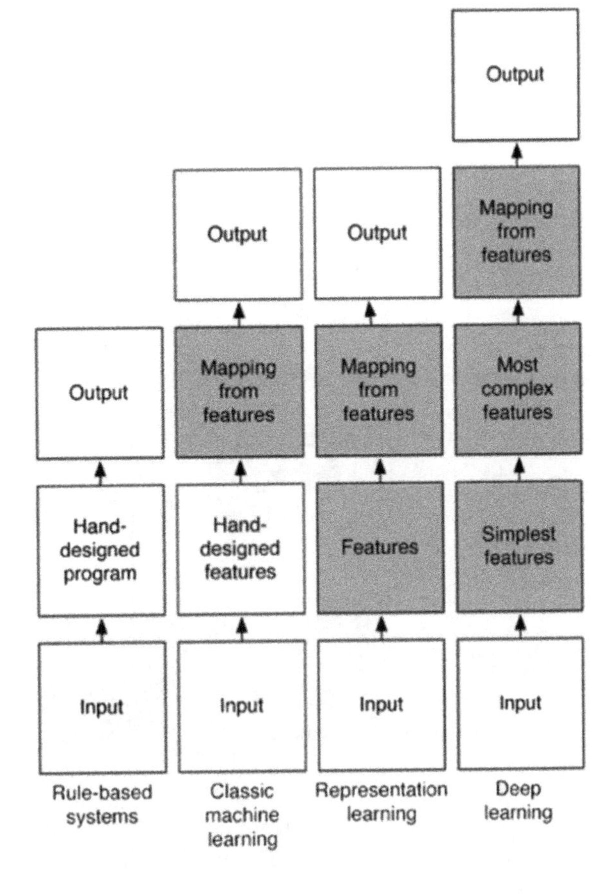

Fig. 2 Feature learning in various AI systems [10]

3.1 Feed Forward Networks (FFNs)

A feed forward network (FFN) shares the same architecture as a standard neural network however having several hidden layers. They are often referred to as MLPs and use the backpropagation procedure as the learning engine. Their operation is the same as a standard (ANN) model. They may have several hidden layers with several neurons in each layer. Each neuron is connected to other neurons in adjacent layers [12]. The neurons are grouped into 3 distinct layers namely, input, hidden and output layers. By definition all layers within the input and output layer are called hidden layers and there are several hidden layers in a deep net. Each input layer neuron is connected to the each hidden layer in the next level. Similarly hidden layer neurons are connected to each neuron in the next level hidden layer and so forth till the output layer neuron, which is usually limited to the number of output variables the learning algorithm is trying to predict. The inputs neurons get activated by sensors perceiving the environment. These usually are features discovered at various levels from the raw

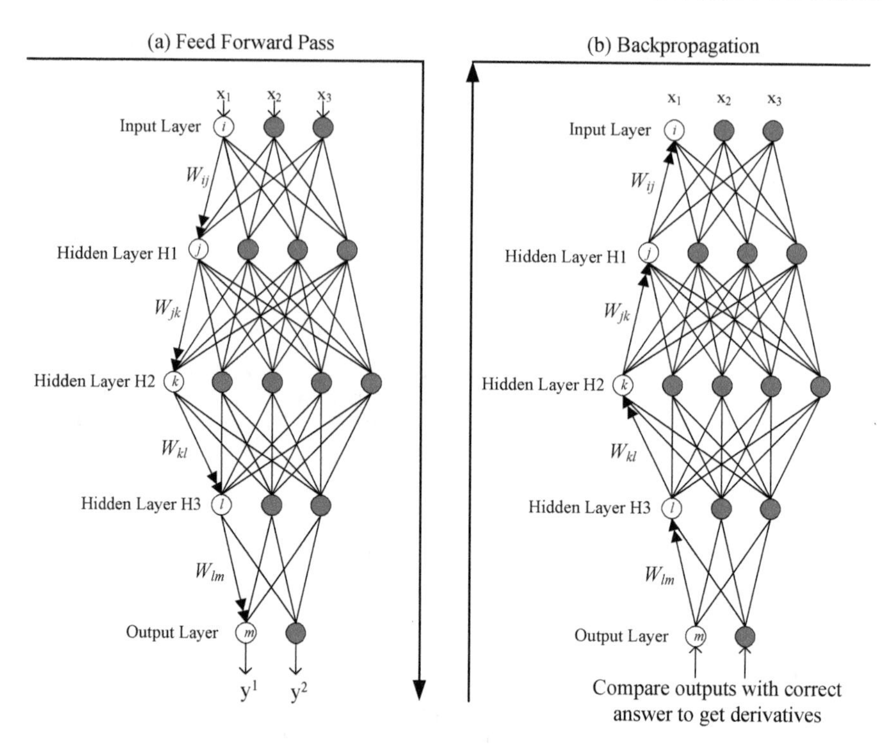

Fig. 3 Multi-layer perceptron model

feature map or dataset. These features filter though the network via weight activations from one neuron to the next. Finally the network learns by adjusting the weights via backpropagation rule over several epochs until the network exhibits desired behavior, such as achieving desired minimum training error for classification tasks.

Figure 3 shows an illustrative example of a MLP model with different types of layers. There is one input and output with three hidden layers in between. Training a MLP involves repetitive invocation of feed forward and backpropagation procedures. As illustrated in Fig. 3a during feed forward propagation through a network, a dot product is performed with the inputs and the associated weights connected to each neuron plus a bias factor and summed across all connections for each neuron. The output then passes through an activation function. The result is then used as an input for the next layer neurons. The output (activation) of a given node therefore is simply a function of its inputs. The summation at each layer is described using Eq. (1).

$$S_j = \sum_i W_{ij} a_i + b_{ij} \tag{1}$$

The activation function is defined in Eq. (2) as:

$$a_j = f(S_j) \tag{2}$$

where S_j is the sum of dot products of weights and outputs from the previous layer u, W_{ij} are represents the weights connecting layer i with layer j, a_i represents the activations of the nodes in the previous layer i, a_j is the activation of the node at hand, b_{ij} is the bias value between layer i and j; and f is a non-linear activation function. The purpose of the activation function is to introduce non-linearity into the function. In simple terms, activation functions normalize the output of a function to be within a given range. Several activation functions exist. However popular activation functions used with neural networks are sigmoid, softmax and ReLU [4]. One may find the ReLU activation function used for hidden layers in many neural network architectures. An output of a neuron processed with ReLU activation will return the same output or zero if input value less than zero; that is $f(S) = \max(0, S)$.

During the backward phase as shown in Fig. 3b, the weight formula is applied in the reverse direction. The weights at level $k + 1$ are updated before the weights at level k are updated. This approach allows the use of neurons at layer $k+1$ to estimate the errors for neurons at layer k [13]. A cost function (sometimes also referred to as loss function) measures the magnitude of error; and is given by Eq. (3) with the following definitions:

- E_p is total error over the training pattern,
- 1/2 is a value applied to simplify the function's derivative,
- n is the number of classes or output labels,
- t_{j_n} is the ground truth value for node n in output layer j, and
- a_{j_n} is the predicted output value node n in output layer j.

$$E_p = \frac{1}{2}\sum_{n}(t_{j_n} - a_{j_n})^2 \tag{3}$$

Backpropagation technique is normally used to train the network. Hence forward propagation and backpropagation is repeated over several epochs until the network reaches a minimum threshold error limit or the number of epochs has reached.

A FFN is usually attached at end of the representation model such as the convolution neural network CNN. This makes a MLP a directed graph structure where all neurons are liked to each other in-between layers. Feed forward neural network architectures are used in many applications of DL [4]. In 2006 the Canadian Institute for Advanced Research revived interest in deep feed forward networks [4]. They used the network in unsupervised learning to create layers of feature detectors without requiring labelled data. This allowed them to use the network for reducing the dimensionality of data which could then be fine-tuned using backpropagation [14, 15].

The most notable application of deep FFNs has been in the field of computer vision and speech recognition. Zeiler et al. in [16] used DNNs for acoustic modeling in speech recognition systems. They were able to show improved generalization of the model and claimed that the training of deep networks was much faster and using ReLU. The success of using DNNs in acoustic modeling for speech recognition motivated Lopez-Moreno et al. [17] who in their work used DNNs for language identification tasks. Using short-term acoustic features as input they were able to identify the language information. They showed with a great deal of success the capability of the network to identify from speech signals the discriminative language information.

3.2 *Convolutional Neural Networks*

CNNs or ConvNets is a DL model that uses convolutional layers for automatic feature extraction from input data [18]. The architecture of CNN utilizes a set of filters (called convolution layers) that preprocesses the data identifying key features in the data. Then these higher level features are fed into fully connected deep nets for training and classification. The aim of these convolution layers is to reduce the dimensionality of the data hence allowing fewer and faster calculations when these inputs make their way into the deep net. Figure 4 shows the architecture of a CNN. A CNN works well with image and video data. Other uses of CNNs are in the areas of sentence processing to identify sequence of words or characters for sentiment analysis (in web mining for example) and DNA sequences. The latter is one of the motivations of this chapter to use DL for gene expression modeling.

Inspired by how the visual cortex of mammals detects features through studies by Hubel and Wiesel in 1988 [19], algorithms such as the Convolutional Neural Network (CNN) have achieved great success in various computer vision tasks such as image classification, object detection, visual concept discovery, semantic segmentation and boundary detection. They were able to describe that the visual information processing system of mammals used layers of specialist cells organised in a hierarchical order with each layer capable of detecting particular stimulus from the retinal image. They found that cells in animal visual cortex are responsible for detecting light in receptive fields; and proved that in image detection the receptive fields detect features in localized areas of the image picking out patterns such as lines and curves (lower-level features), followed by shapes and colors (high-level features) leading up to detecting the complete image. This showed that from low-level features higher-level features are extracted. Using these results LeCun et al. [20] formalized the modern framework of CNNs. They applied the convolution technique to handwritten digits

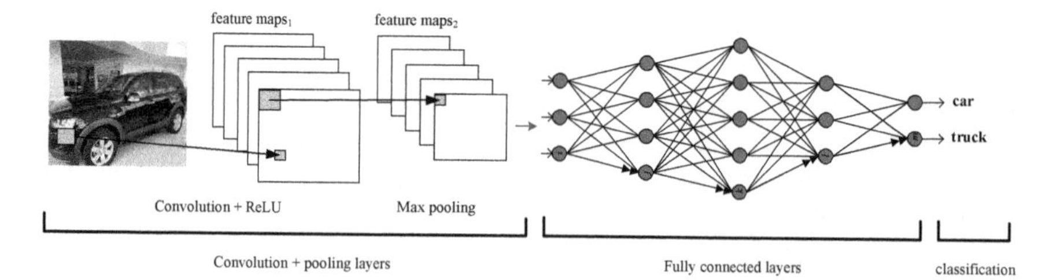

Fig. 4 Architecture of a convolution neural net

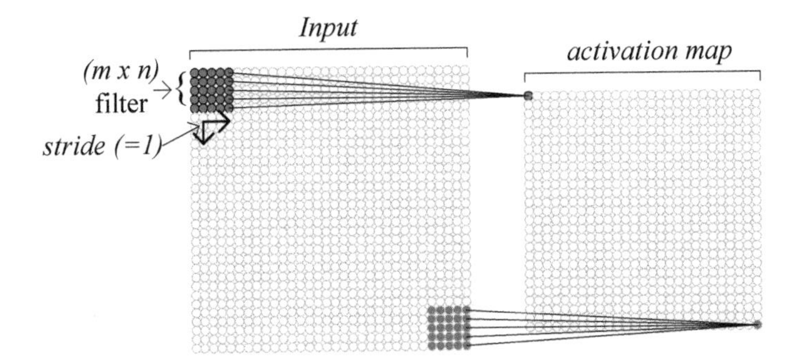

Fig. 5 The convolution process. A 5 × 5 filter convolving around the input space of size 32 × 32, using stride of 1, producing an activation map of size 27 × 27

recognition of postal zip codes. They later improved the algorithm in [6] and named it LeNet-5.

The basic underlying structure of a CNN can be divided into two distinct parts—convolution and fully connected layers (as illustrated in Fig. 4). The first stage comprised of the convolution layer following by the pooling layer in the next stage. In the convolution layer a feature map is formed by convoluting the input image with a filter of size $m \times n$. This filter is simply a series of weights. A filter is designed to detect a particular feature from the input space. As the filter strides over the input space dot operation is applied on the input values and the weights. The output of this local weighted sum is then processed through a non-linear function such as the ReLU. The purpose of this activation function is to introduce nonlinearity to the model in order to detect non-linear features and also to avoid overfitting. The result is added to the feature map. Successive strides from left to right top to bottom forms a complete feature map using a single filter. Figure 5 explains this process. A feature map is sometimes also referred as an activation map. Several filters can be applied on the input space to detect different features (as each filter has different weights) resulting in multiple feature or activation maps.

The number of filters (also called kernels in some literature) applied in a convolutional layer results in the same number of feature maps produced as output. This results in a lot of parameters and increases the computation time for the network. Pooling is therefore applied to down sample the activation map and also avoids overfitting. It makes the representation smaller and more manageable and is applied to each activation map independently. One common pooling method is *Max Pooling* which takes the activation map as input and computes the maximum of a local patch or region of units (as illustrated in Fig. 6). The process of convolution and pooling can be iterated several times before feeding the final activation map through the fully connected later.

CNNs are best used on data that are in two-dimensional form; for example 2D images with one color (grey scale) or three color channels and audio spectrograms [4]. Hence CNNs have been applied in image classification problems with a great

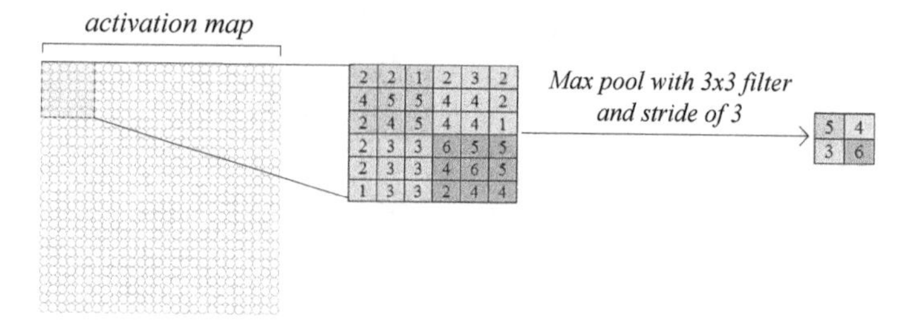

Fig. 6 Max pooling with 3 × 3 filter and stride of 1 is applied on activation map of size 27 × 27 down sampling it to size 9 × 9

success [21, 22]. The AlexNet developed by Krizhevsky et al. [21] achieved the best performance in the ILSVRC challenge in 2012 [23]. Following suit other deep CNNs developed over the years were ZFNet (2013) [24], GoogleNet (2014) [25], VGGNet (2014) [26] and ResNet (2015) designed by Microsoft Research Team [27].

CNNs have also been used for one dimensional datasets such as signals and sequences such as in area of text detection [28, 29] and recognition [30]. In three dimensional data spectrum CNNs have been mainly used for video and volumetric images such as recognizing actions in video sequences. Since CNNs are designed to work efficiently on two dimensional data; applying the model on such data has been a challenge to researches in this field. This is essentially because the video sequences are temporal. One solution to this problem was to treat the video composed of images of consecutive frames and apply two dimensional convolution techniques on it as proposed in [31]. Finally Gu et al. in [23] provide a good review of the recent advances in CNNs and several of its application areas.

3.3 Recurrent Neural Networks

Recurrent neural nets (RNNs) employs a similar structure as FFN but their representation advantages power comes from feedback loops within the hidden neurons. This allows RNN to work best for time series data as the network remembers the learnt sequence. Hence these types of nets are well suited for forecasting problems such as forecasting demand and supply chain planning. Other uses of RNNs are for clustering, image captioning, sentence classification and embedding, neural-machine translation and sequence classification. RNNs however are difficult to implement and suffer from vanishing gradient or exploding gradient.

RNNs allow information to pass from a previous time step to the next, hence provides for loops within the network. Figure 7 shows the architecture of RNNs. It shows the hidden neurons grouped under node s with values s_t at time t and W is the shared weights.

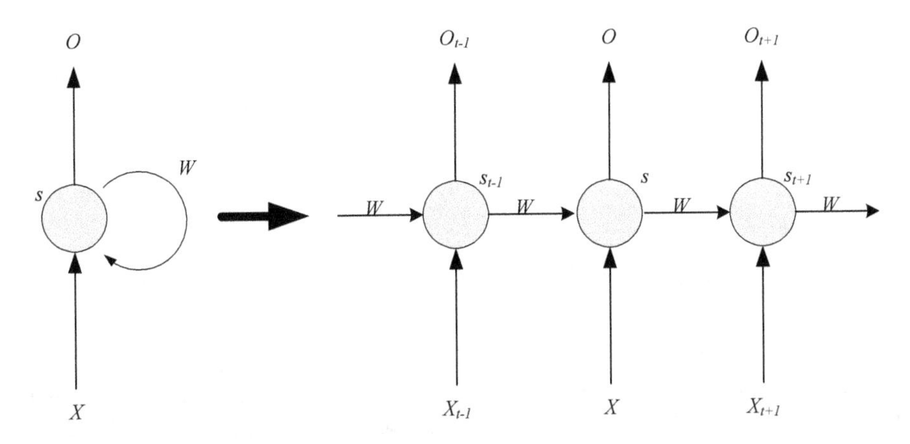

Fig. 7 A recurrent neural network with an illustration of present loops in its forward computation [4]

Natural language processing is an area where RNNs have gained considerable success [32]. Given this success, they are now being applied to other areas such as biomedical text analysis [33].

4 Deep Learning Challenges

The research community widely agrees DL works best if the problem satisfies certain specified constraints. Firstly DL requires high volumes of data and this includes high dimensionality in features. This avoids the model from overfitting. However acquiring high volumetric data is usually a challenge to acquire or curate. Secondly, data needed for training deep nets need to be strong and clean. Thirdly deep nets in the current time work well for only well-defined data structures such as image, speech, natural language processing and video.

Research in the area of DL attempts to create models to learn features from large-scale, unlabeled data automatically. DL attempts to model similar learning mechanism of out biological nervous system—the brain learning from examples (training data). This leads to the main criticism of DL in that it does not truly capture the biological mechanisms of the brain. Instead DL leans on the concept of Computational Equivalence and can possibly have equivalent capabilities as a biological brain albeit using different computational mechanisms. Further a learning system is only able to derive the *Cause* while observing the *Effect* [34] and that it is unable to predict the *Effect* from *Cause*. As such DL suffers from this limitation as well.

Current trends in the areas for image classification and generation are showing promise by combining CNNs with FFNs and for video modeling by using CNNs and RNNs. DL however do face some issues such as slow learning and local traps. Processing time of deep nets have however reduced drastically given they can now be offloaded on high performance accelerated graphics processing unit (GPU) cards. To avoid local traps have been solved to a large extent by using ReLU activation function. In summary, despite the advancements and the robustness of the model, deep nets like other machine learning models do suffer from data model uncertainty and overfitting.

5 Gene Expression Modeling Problem

Since one of the requirements for DL to work is to have a substantial size of dataset, we are motivated to show results of these techniques and models in one key and complex area of bioinformatics—gene expression analysis [35] as part of the Gene Expression Modeling Problem (GEMP). The advancement of Microarray technology [36] presents an opportunity to identify traits of diseases through analysis of gene expressions. This has led to the rise of the field of Genomics.

5.1 *What Is Gene Expression Analysis?*

Genomics is a field of study in statistical data mining. What makes this area interesting to explore is the availability of huge amounts of gene expression data for mining and the need to get insights and patterns from these data to solve problems such as identification of cancerous genes, effects of drugs and treatment of an individual at gene level. A living organism contains millions of genes. Each gene can be thought of as containing instruction code for the functionality and make up of various organs and parts of the body. The translation of information encoded in a gene into protein produces gene expression data [37]. It provides information regarding the complex activities within corresponding cells, measured by messenger ribonucleic acid (mRNA) produced during transcription [38]. We use the term gene expression to describe the process of transcribing a gene's DNA sequence into RNA.

5.2 *Uses of Gene Expressions Data and Limitations*

A gene is said to be expressed when the gene's RNA is used to produce proteins [39]. Researches have concentrated on studying the change and patterns in the expression values of genes as an indicator of certain diseases such as various forms of cancer. Normal cells transform into cancerous cells due to mutations driven by gene expression level changes. Analysing gene expression data therefore helps in identifying different gene functions and diagnosis of disease such as cancer.

One of the recent technologies to produce gene expression data is by Microarrays, or DNA chips which allows recording of gene expression. Since thousands of genes

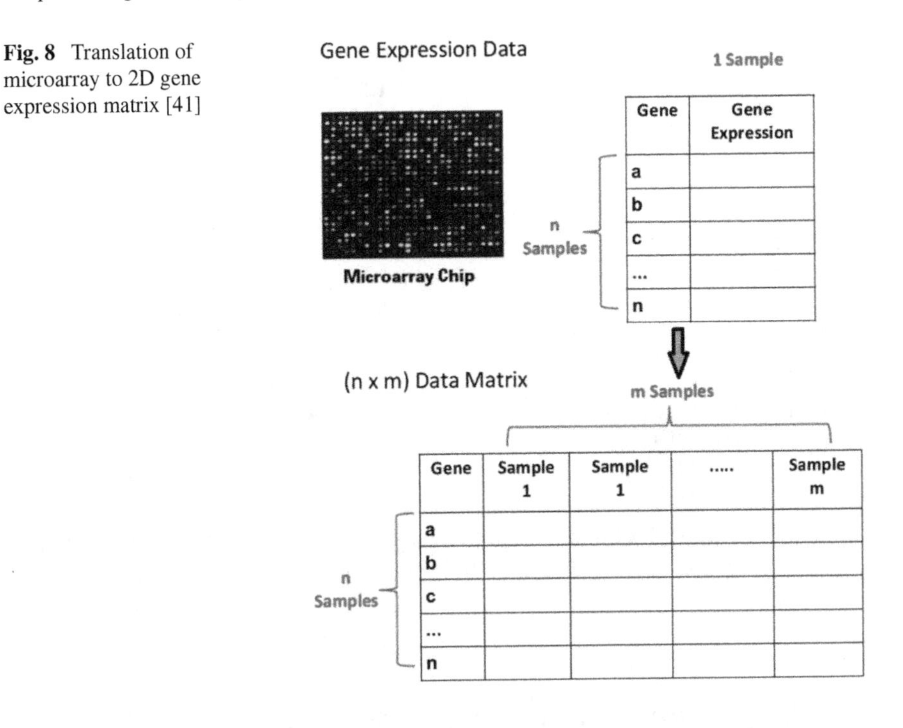

Fig. 8 Translation of microarray to 2D gene expression matrix [41]

are analysed in parallel results in huge amounts of data. Typically, a gene expression matrix (GEM) is extracted from microarray image analysis. In this matrix, genes and samples (tissues or experiments) are represented by rows and columns respectively. The value in each cell of the GEM denotes the expression level of that particular gene in the particular sample [40]. Figure 8 explains the translation of Microarray image to a $n \times m$ dimensional gene expression data. The challenge is the development of analytical techniques to make sense out of these large amounts of biological data [41]. By their nature these datasets are quite different from conventional datasets used for classification algorithms to work on. Amidst this, gene expression data does have some properties as pointed by Lu and Han in [39] which limit the behavior of classifiers, namely they have very high dimensionality usually containing tens of thousands of genes but due to small number of samples the data size is very small. Further most genes may not be relevant to any particular disease which opens up the problem of finding the right gene which has malfunctioned.

5.3 Publically Available Data Repositories for Gene Expressions

The Microarray Gene Expression Data society, founded in 1999 by microarray users and producers—Affymetrix, Stanford and EBI, established standards for data quality,

Table 1 Official gene expression data repositories

Gene expression repositories	URL
• CGAP and SAGEmap	https://cgap.nci.nih.gov/
• ExpressDB	http://arep.med.harvard.edu/ExpressDB/
• GEO at the NCBI	https://www.ncbi.nlm.nih.gov/geo/
• ArrayExpress at the EBI	https://www.ebi.ac.uk/arrayexpress/
• CIBEX	http://cibex.nig.ac.jp/data/index.html
• Expression Atlas	https://www.ebi.ac.uk/gxa/home
• The Cancer Genome Atlas (TCGA)	https://cancergenome.nih.gov/
• Ensembl	http://www.ensembl.org/index.html
• NIH Roadmap Epigenomics Mapping Consortium (REMC)	http://www.roadmapepigenomics.org/

storage, management, annotation and exchange at the genomics, transcriptomics, and proteomics levels. Table 1 shows the links to some current official gene expression repositories.

5.4 Deep Learning on Gene Expressions

Several methods have been developed which analyzes gene expression data for cancer diagnosis and classification, clustering of cancerous cells from normal cells using standard machine learning methods such as in [39, 42–44]. Siang et al. in [45] in her review paper provides a detailed list of data mining software used for analysis on gene expression data based on standard machine learning methods. More recently, researchers have started to apply DL techniques on the same. Min et al. [35] in their research analysed the application of DL in three key fields of bioinformatics, namely omics, biomedical imaging and biomedical signal processing. Then they identified applications from literature within the above fields and categorized them according to the type of DL architecture that was used in those researches. Namely they placed the applications into categories DNNs, CNNs, RNNs and emergent architectures. Table 2 shows their analysis. They encompassed DNNs to not only refer to deep multilayer perceptrons but also to other variants of DNNs such as stacked auto-encoder, deep belief networks, autoencoders and Restricted Boltzmann Machines. For the purpose of this chapter we will describe some recent applications of DL on gene expression data and analysis. Several other papers such as in [46–48] show the emerging use of DL techniques applied to gene expression data. Angermueller et al. in [49] provides a comprehensive review of DL in the area of computational biology.

Table 2 DL applied research in bioinformatics [35]

	Omics	Biomedical imaging	Biomedical signal processing
	Research topics		
Deep neural networks	Protein structure	Anomaly classification	Brain decoding
	Gene expression regulation	Segmentation	Anomaly classification
	Protein classification	Recognition	
	Anomaly classification	Brain decoding	
Convolutional neural networks	Gene expression regulation	Anomaly classification	Brain decoding
		Segmentation	Anomaly classification
		Recognition	
Recurrent neural networks	Protein structure		Brain decoding
	Gene expression regulation		Anomaly classification
	Protein classification		
Emergent architectures	Protein structure		Brain decoding

5.5 *Deep Neural Nets*

In terms of gene expression analysis, several variants of deep neural nets have been used such as deep belief nets, autoencoders and restricted Boltzman machines. Some have been applied as classifiers to solve prediction problems while others have been used as feature selectors and extractors during *pre-processing* step. For example, to cluster gene expression data of yeasts, Gupta et al. [46] in their research used deep networks as a pre-processing step. In particular they used deep networks in the form of denoising autoencoders as a pre-processing step. Their autoencoders were used in an unsupervised manner to learn a compact representation of input from which features were extracted for further supervised learning. Their aim was to identify characteristics of the underlying distribution of gene expression vectors and use it to explain for example interactions and correlations among two genes. In a similar manner a fair amount of research has been done on using deep architectures as feature selectors as a pre-processing step with promising results [48, 50].

In another research D-GEX, a deep multi-task multi-layer feed forward network was designed to use expression of landmark genes to deduce expressions of target genes. In biology landmarks genes are a set of genes (approximately 1000) than can represent approximately 80% of the information in the CMap data (http://www.lincscloud.org/). This was discovered by LINCS program researchers (http://www.lincsproject.org) while analyzing the gene expression profiles from the CMap data using principal component analysis (PCA). Their experiments were conducted on the GEO microarray expression data from the GEO database which consisted of 129,158 gene expression profiles. Each profile contained 22,268 probes corresponding to 978 genes landmark genes and 21,290 target genes. Using microarray and RNA-seq expression data from the GEO database as input into a MLP model, LINCS researchers were able to infer expression of up to 21,290 target genes. Challenged by the high-dimensionality of gene expression datasets, Fakoor et al. [38] in their research applied principal component analysis (PCA) on their gene expression microarray data. The resultant low-dimension dataset was then used to train a sparse autoencoder classifier. Later they were able to use their model to classify various types of cancers such as acute myeloid leukemia, breast cancer, ovarian cancer and colon cancer. On one hand where gene expression data is high-dimensional, they are usually limited by the small sample size present in the dataset. However Fakoor et al. were able to show that cancer data derived from the same platform with the same genes can be used as unlabeled data for training classifiers. Such a method allowed them to generalize the feature sets across different cancers. This means for example using data from other cancers such as prostrate, colon, lung, they were able to solve breast cancer detection and classification problems.

5.6 Convolutional Neural Networks

Convolutional neural networks have been designed to work best when data is in the form of images or in two-dimensional form. As such there have been quite a number of studies that have used images such as images of tumor signatures for classification [51]. However few studies have been conducted that directly utilizes convolutional nets on gene expression or microarray data. This is mostly due to data available only in biological sequence formats to solve gene expression regulation problems [52–54]. For example in a recent publication by Nguyen et al. [18] the authors used CNNs to classify DNA sequences by treating DNA sequences as text data. Most of their DNA sequencing data was sourced from the research of Pokholok et al. [55]. They translated the DNA sequence into sequence of words using the sliding window technique (window size = 3, stride = 1). Using these parameters for convolution a dictionary of 64 words was formed. After not encoding each word resulted in a two dimensional structure which was then used an input for their CNN model for classification. Figure 9 shows their DNA sequence converted as one-hot vector representation.

There have been some researchers that have used other forms of biological data to predict gene expressions levels instead. Singh et al. [56] proposed a CNN framework to classify expression levels of genes using histone modification data as input. They emphasized histone modification signal data contained important information about expression levels of genes. To be able to predict these expression levels using DL therefore would prove useful in understanding their effects in gene regulation.

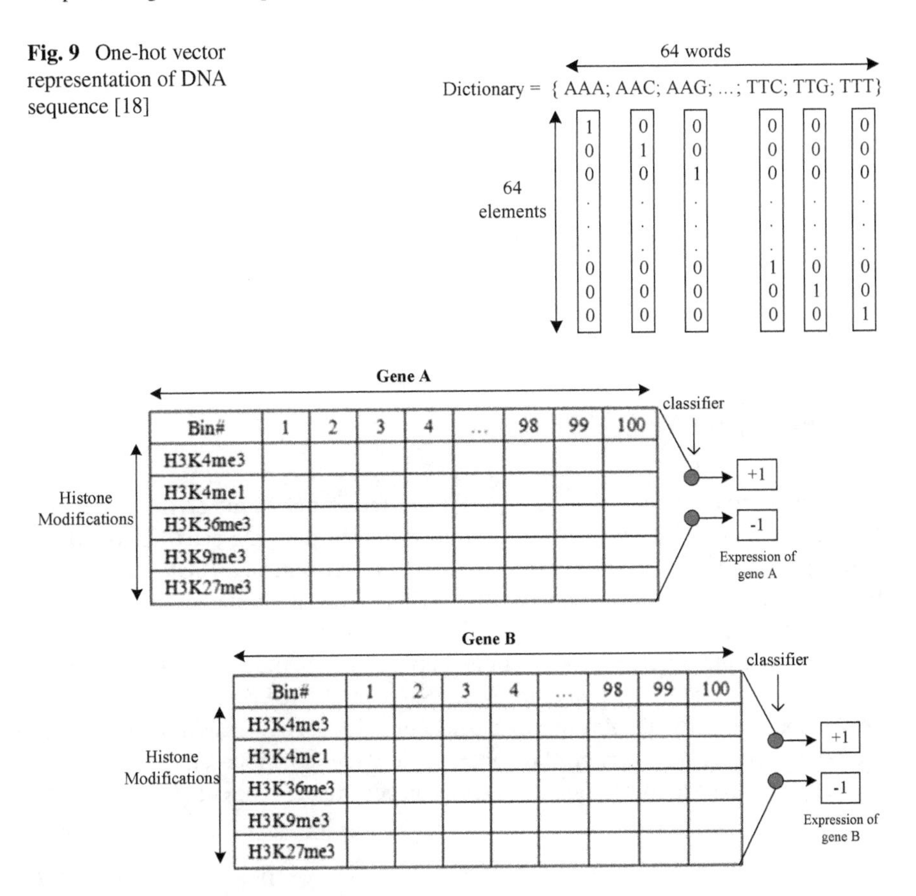

Fig. 9 One-hot vector representation of DNA sequence [18]

Fig. 10 Gene histone modification marks in bin format [56]

Such knowledge can help in developing 'epigenetic drugs' for diseases like cancer. In their research they first extracted 5 core histone modification marks from the REMC database (http://www.roadmapepigenomics.org/). For each histone, they selected bins of 100 base-pairs (bp) extracted from regions flanking the transcription start site (TSS) of each gene. The result was a 5 $\times$ 100 matrix for each gene where rows represent the histone modification marks and columns represent different bins. The expression level of a gene whether it is highly expressed (+) or lowly expressed (−) could therefore be determined from the signal values of the 5 histone modifications in bin form. Figure 10 shows the 2D structure of histone modification marks in bins for single genes and Fig. 11 shows their CNN model for gene expression level prediction.

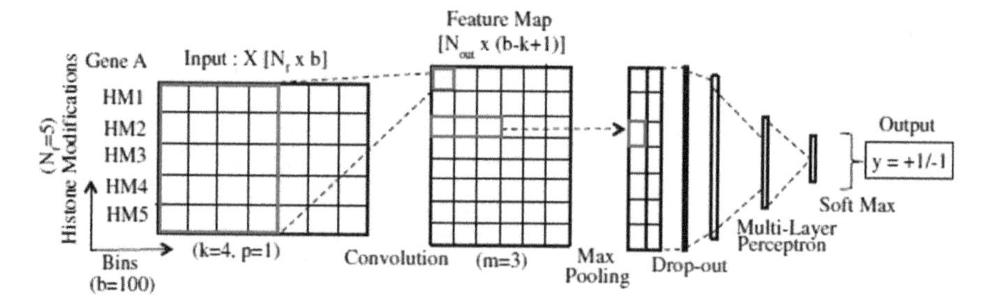

Fig. 11 DeepChrome convolutional neural network model [56]

5.7 *Recurrent Neural Networks*

Gene expression data can also be expressed in the form of time-series expression data [46]. Xu et al. [57] in their research investigated the gene activities of whole genomes using time-series gene expression data to infer genetic regulatory networks. To determine the network parameters they trained their RNN model using particle swarm optimisation (PSO) technique. Their hybrid RNN/PSO model helped in understanding the regulatory interactions between genes. In another work Noman et al. [58] proposed using a variant of the RNN model referred to as "*decoupled RNN*" model. Their main aim was to use decoupled RNN to successfully infer gene expressions from expression profiles. Several other literature report the use of RNNs (some hybridised with other models) applied on gene related data such as in [59, 60].

6 Conclusion

Much progress has been made through research in learning algorithms. The next frontier is to investigate and further develop deep learning as it shows much promise especially with the rising computational power of systems. The chapter has summarized the motivations and theoretical foundations of the significant deep learning models. Their applications to difficult real world problems are reviewed. Some promising results from the complex problem of gene expression modeling are presented. Many future challenges remain and are identified for investigation as future work. They include investigating algorithms and deep learning structures for incremental learning, enhancement of intelligent algorithms for abstracting ontologies, intelligent pre-processing of data ensuring minimal loss of information, extracting information from outliers and possible models for a multi-agent based pattern extraction through deep learning.

Together with wet-lab techniques, Danaee et al. [61] emphasized the importance of computer analysis of gene expression data for better understanding of the human genome. However this requires shifting our attention to finding and recognising those subtle patterns in data which may hold the key for the discovery of disease diagnosis and drug discovery. Although the research community has embraced applying data mining methods on gene expression data to find causes of diseases such as cancer there is still a critical need to identify genes that cause mutations in cells leading to cancer; and ultimately find cures for these diseases.

References

1. S. Agarwal, G. Pandey, M.D. Tiwari, Data mining in education: data classification and decision tree approach. Int. J. e-Educ. e-Bus. e-Manag. e-Learn. **2**(12) (2012)
2. Y. Li, W. Ma, Applications of artificial neural networks in financial economics: a survey, in *International Symposium on Computational Intelligence and Design (ISCID)* (2010)
3. H. Xie, J. Shi, W. Lu, W. Cu, Dynamic Bayesian networks in electronic equipment health diagnosis, in *IEEE Prognostics and System Health Management Conference (PHM-Chengdu)* (2016)
4. Y. LeCun, Y. Benjio, G. Hinton, Deep learning. Nature **521**(7553), 436–444 (2015)
5. L. Weifeng, V. Benjamin, X. Liu, Hsinchu, Deep learning: an overview—lecture notes, University of Arizona, Arizona, Apr 2015
6. Y. LeCun, L. Bottou, Y. Benjio, P. Haffner, Gradient-based learning applied to document recognition. Proc. IEEE **86**(11), 2278–2324 (1998)
7. T. Raiko, H. Valpola, Y. LeCun, Deep learning made easier by linear transformations in perceptrons, in *Proceedings of the 15th International Conference on Artificial Intelligence and Statistics (AISTATS)*, La Palma, Canary Islands (2012)
8. Y. Benjio, Scaling up deep learning, in *Proceedings of the 20th ACM SIGKDD International Conference on Knowledge Discovery and Data Mining*, New York, USA (2014)
9. Y. Bengio, Deep learning of representations for unsupervised and transfer learning, in *JMLR: Workshop and Conference Proceedings*, vol. 27, pp. 17–37 (2012)
10. Y. Benjio, Deep learning: theoretical motivations, 3 Aug 2015, http://videolectures.net/deeplearning2015_bengio_theoretical_motivations/. Accessed 26 June 2017
11. Y. Benjio, Deep learning architectures for AI. Found. Trends Mach. Learn. **2**(1), 1–127 (2009)
12. J. Schmidhuber, Deep learning in neural networks: an overview. Neural Netw. **61**, 85–117 (2015)
13. P.-N. Tan, M. Steinbach, V. Kumar, *Introduction to Data Mining* (Pearson Education Inc., Boston, 2006)
14. G. Hinton, R. Salakhutdinov, Reducing the dimensionality of data with neural networks. Science **313**, 504–507 (2006)
15. M. Ranzato, C. Poultney, S. Chopra, Y. LeCun, Efficient learning of sparse representations with an energy-based model. Adv. Neural. Inf. Process. Syst. **13**, 1137–1144 (2006)
16. M.D. Zeiler, M. Ranzato, R. Monga, M. Mao, K. Yang, Q.V. Le, P. Nguyen, A. Senior, V. Vanhoucke, J. Dean, G.E. Hinton, On rectified linear units for speech procesing, in *IEEE International Conference on Acoustics, Speech and Signal Processing (ICASSP)*, pp. 3517–3521 (2013)
17. I. Lopez-Moreno, J. Gonzalez-Dominguez, D. Martinez, O. Plchot, J. Gonzalez-Rodriguez, P.J. Moreno, On the use of deep feedforward neural networks for automatic language identification. Comput. Speech Lang. **40**, 46–59 (2016)
18. N.G. Nguyen, V.A. Tran, D.L. Ngo, D. Phan, F.R. Lumbanraja, M.R. Faisal, B. Abapihi, M. Kubo, K. Satou, DNA sequence classification by convolutional neural network. J. Biomed. Sci. Eng. **9**, 280–286 (2016)
19. D.H. Hubel, T.N. Wiesel, Receptive fields and functional architecture of monkey striate cortex. J. Physiol. (1968)

20. Y. LeCun, J.S. Denker, D. Henderson, R.E. Howard, W. Hubbard, L.D. Jackel, Handwritten digit recognition with a back-propagation algorithm, in *Proceedings of NIPS* (1990)
21. A. Krizhevsky, I. Sutskever, G.E. Hinton, Imagenet classification with deep convolutional neural networks, in *Proceedings of NIPS* (2012)
22. D.C. Ciresan, U. Meier, J. Masci, L.M. Gambardella, J. Schmidhuber, Flexible, high performance convolutional neural networks for image classification, in *Proceedings of IJCAI* (2011)
23. J. Gu, Z. Wang, J. Kuen, L. Ma, A. Sharoudy, B. Shuai, T. Liu, X. Wang, G. Wang, Recent advances in convolutional neural networks (2017). arXiv:1512.07108
24. M.D. Zeiler, R. Fegus, Visualizing and understanding convolutional networks (2013). arXiv: 1311.2901
25. C. Szegedy, W. Liu, Y. Jia, P. Sermanet, S. Reed, D. Anguelov, D. Erhan, V. Vanhoucke, A. Rabinovich, Going deeper with convolutions, in *Proceedings of the IEEE Conference on Computer Vision and Pattern Recognition* (2015)
26. K. Simonyan, A. Zisserman, Very deep convolutional networks for large-scale image recognition (2014). arXiv:1409.1556
27. K. He, X. Zhang, S. Ren, J. Sun, Deep residual learning for image recognition, in *Proceedings of the IEEE Conference on Computer Vision and Pattern Recognition* (2016)
28. M. Delakis, C. Garcia, Text detection with convolutional neural networks, in *Proceedings of VISAPP* (2008)
29. H. Xu, F. Su, Robust seed localization and growing with deep convolutional features for scene text detection, in *Proceedings of ICMR* (2015)
30. M. Jaderberg, K. Simonyan, A. Vedaldi, A. Zisserman, Deep structured output learning for unconstrained text recognition, in *Proceedings of ICLR* (2015)
31. A. Karpathy, G. Toderici, S. Shetty, T. Leung, R. Sukthankar, L. Fei-Fei, Large-scale video classification with convolutional, in *Proceedings of CVPR* (2014)
32. W. Yin, K. Kann, M. Yu, H. Schutze, Comparative study of CNN and RNN for natural language processing (2017). arXiv:1702.01923 [cs.CL]
33. L. Li, L. Jin, D. Huang, Exploring recurrent neural networks to detect named entities from biomedical text, in *Chinese Computational Linguistics and Natural Language Processing Based on Naturally Annotated Big Data,* ed. by M. Sun, Z. Liu, M. Zhang, Y. Liu, Lecture Notes in Computer Science, vol. 9427 (Springer, Cham, 2015)
34. D. Lopez-Paz, K. Muandet, B. Schölkopf, I. Tolstikhin, Towards a learning theory of cause-effect inference, in *International Conference on Machine Learning*, pp. 1452–1461 (2015)
35. S. Min, B. Lee, S. Yoon, Deep learning in bioinformatics. Brief. Bioinform. (2016)
36. V. Trevino, F. Falciani, H.A. Barrera-Saldaña, DNA microarrays: a powerful genomic tool for biomedical and clinical research. Mol. Med. **13**(9–10), 527–541 (2007)
37. D.R. Edla, P.K. Jana, S. Machavarapu, KD-tree based clustering for gene expression data, *Encyclopedia of Business Analytics and Optimization*, p. 15 (2014)
38. R. Fakoor, F. Ladhak, A. Nazi, M. Huber, Using deep learning to enhance cancer diagnosis and classification, in *Proceedings of the 30th International Conference on Machine Learning*, Atlanta, USA (2013)
39. Y. Lu, J. Han, Cancer classification using gene expression data. Inf. Syst. Data Manag. Bioinform. **28**(4), 243–268 (2003). Elsevier
40. A. Brazma, J. Vilo, Gene expression data analysis. FEBS Lett. **480**(1), 17–24 (2000)
41. J. Clemente, gene expression data analysis, LinkedIn, 28 Jan 2013, http://www.slideshare.net/JhoireneClemente/gene-expression-data-analysis. Accessed 8 Jan 2017
42. J. Liu, W. Cai, X. Shao, Cancer classification based on microarray gene expression data using a principal component accumulation method. Sci. China Chem. **54**(5), 802–811 (2011)
43. H. Zhang, C.-Y. Yu, B. Singer, Cell and tumor classification using gene expression data: construction of forests. PNAS **100**(7), 4168–4172 (2003)
44. O. Dagliyan, F. Uney-Yuksektepe, I.H. Kavakli, M. Turkay, Optimization based tumor classification from microarray gene expression data. PLOS **6**(2) (2011)
45. T.C. Siang, W.T. Soon, S. Kasim, M.S. Mohamad, C.W. Howe, S. Deris, Z. Zakaria, Z.A. Shah, Z. Ibrahim, A review of cancer classification software for gene expression data. Int. J. Bio-Sci. Bio-Technol. **7**(4), 89–108 (2015)

46. A. Gupta, H. Wang, M. Ganapathiraju, Learning structure in gene expression data using deep architectures, with an application to gene clustering, in *IEEE Workshop on Biomedical Visual Search and Deep Learning*, Washington D.C (2015)
47. Y. Chen, R. Narayan, A. Subramanian, X. Xie, Gene expression inference with deep learning. Bioinformatics (2015)
48. J. Tan, J.H. Hammond, D.A. Hogan, C.S. Greene, ADAGE-based integration of publicly available pseudomonas aeruginosa gene expression data with denoising autoencoders illuminates microbe-host interactions. mSystems **1**(1) (2016)
49. C. Angermueller, T. Parnamaa, L. Parts, O. Stegle, Deep learning for computational biology. Mol. Syst. Biol. **12**(878) (2016)
50. J. Tan, M. Ung, C. Cheng, C.S. Greene, Unsupervised feature construction and knowledge extraction from genome-wide assays of breast cancer with denoising autoencoders, in *Pacific Symposium on Biocomputing* (2015)
51. Q.V. Lee, J. Han, J.W. Gray, P.T. Spellman, A. Borowsky, B. Parvin, Learning invariant features of tumor signatures, in *Proceedings of ISBI*, pp. 302–305 (2012)
52. B. Alipanahi, A. Delong, M. Weirauch, B.J. Frey, DeepBind: predicting the sequence specificities of DNA-and RNA-binding proteins by deep learning. Nat. Biotechnol. (2015)
53. H. Zeng, M. Edwards, G. Liu, D.K. Gifford, Convolutional neural network architectures for predicting DNA–protein binding. Bioinformatics **32**(12) (2016)
54. D. Kelly, J. Snoek, J.B. Rinn, Learning the regulatory code of the accessible genome with deep convolutional neural networks. bioRxiv (2015)
55. D. Pokholok, C. Harbison, S. Levine, M. Cole, N. Hannet, T. Lee, G. Bell, K. Walker, P. Rolfe, E. Herbolsheimer, J. Zeitlinger, F. Lewitter, D. Gifford, R. Young, Genome-wide map of nucleosome. Cell **122**, 517–527 (2005)
56. R. Singh, J. Lanchantin, G. Robins, Y. Qi, DeepChrome: deep learning for predicting gene expression from histone modfications. Bioinformatics **32**, i639–i648 (2016)
57. R. Xu, D.C. Wunsch II, R.L. Frank, Inference of genetic regulatory networks with recurrent neural network models using particle swarm optimisation. IEEE Trans. Comput. Biol. Bioinform. (2006)
58. N. Noman, L. Palafox, H. Iba, Reconstruction of gene regulatory networks from gene expression data using decoupled recurrent neural network model, in *Proceeding in Information and Communication Technology (PICT 6)*, pp. 93–103 (2013)
59. B. Lee, J. Baek, S. Park, S. Yoon, DeepTarget: end-to-end learning framework for microRNA target prediction using deep recurrent neural networks, in *Proceedings of BCB '16*. ACM (2016)
60. A. Khan, S. Mandal, R.K. Pal, G. Saha, Construction of gene regulatory networks using recurrent neural networks and swarm intelligence. Scientifica (2016)
61. P. Danaee, R. Ghaeini, D.A. Hendrix, A deep learning method for cancer detection and relevant gene identification, in *Pacific Symposium on Biocomputing*, vol. 22, pp. 219–229 (2016)